MEDICAL CODING SPECIALIST'S EXAM REVIEW
HOSPITAL

To the good hearts who give us hope, especially my sister, Becky, and my aunts, Alice and Donna.

MEDICAL CODING SPECIALIST'S EXAM REVIEW HOSPITAL

•••

LYNETTE OLSEN

THOMSON
DELMAR LEARNING   Australia   Canada   Mexico   Singapore   Spain   United Kingdom   United States

Medical Coding Specialist's Exam Review—Hospital
by Lynette Olsen

Vice President,
Health Care Business Unit:
William Brottmiller

Editorial Director:
Matthew Kane

Acquisitions Editor:
Rhonda Dearborn

Developmental Editor:
Sherry Conners

Editorial Assistant:
Debra S. Gorgos

Marketing Director:
Jennifer McAvey

Marketing Channel Manager:
Tamara Caruso

Marketing Coordinator:
Kimberly Duffy

Production Director:
Carolyn Miller

Production Coordinator:
James Zayicek

Technology Project Manager:
Mary Colleen Liburdi

COPYRIGHT © 2006
Thomson Delmar Learning, a part of the Thomson Corporation. Thomson, the Star logo, and Delmar Learning are trademarks used herein under license.

Printed in United States of America
1 2 3 4 5 6 7 8 XXX 09 08 07 06 05

For more information, contact Thomson Delmar Learning, 5 Maxwell Drive, Clifton Park, NY 12065

Or find us on the World Wide Web at
http://www.delmarlearning.com

ALL RIGHTS RESERVED. No part of this work covered by the copyright hereon may be reproduced or used in any form or by any means—graphic, electronic, or mechanical, including photocopying, recording, taping, Web distribution or information storage and retrieval systems—without the written permission of the publisher.

For permission to use material from this text or product, contact us by
Tel (800) 730-2214
Fax (800) 730-2215
www.thomsonrights.com

Library of Congress Cataloging-in-Publication Data

Olsen, Lynette.
  Medical coding specialist's exam review—hospital / Lynette Olsen.
    p. ; cm.
  Includes bibliographical references and index.
  ISBN 1-4018-3750-6 (alk. paper)
  1. Nosology—Code numbers—Examinations, questions, etc.
  2. Clinical medicine—Code numbers—Examinations, questions, etc.
  I. Title.
  [DNLM: 1. Forms and Records Control—methods—United States—Examination Questions.
  W 18.2 O52m 2006]
  RB115.O476 2006
  616'.0076—dc22

2005018827

Notice to the Reader

Publisher does not warrant or guarantee any of the products described herein or perform any independent analysis in connection with any of the product information contained herein. Publisher does not assume, and expressly disclaims, any obligation to obtain and include information other than that provided to it by the manufacturer.

  The reader is expressly warned to consider and adopt all safety precautions that might be indicated by the activities described herein and to avoid all potential hazards. By following the instructions contained herein, the reader willingly assumes all risks in connection with such instructions.

  The publisher makes no representations or warranties of any kind, including but not limited to, the warranties of fitness for particular purpose or merchantability, nor are any such representations implied with respect to the material set forth herein, and the publisher takes no responsibility with respect to such material. The publisher shall not be liable for any special, consequential, or exemplary damages resulting, in whole or part, from the reader's use of, or reliance upon, this material.

CONTENTS

Preface		vii
Introduction		vii
Content		vii
How to Use This Book Effectively		viii
Acknowledgements		viii
About the Author		viii
Reviewers		viii
Chapter 1	Introduction	1
Chapter 2	Importance of Being a Coder	6
Chapter 3	Preparing for the Exam	14
Chapter 4	Taking the Exam	19
Chapter 5	After the Exam	26
Chapter 6	ICD-9 Coding	30
Chapter 7	Current Procedural Terminology (CPT)	92
Chapter 8	Evaluation and Management Codes (E/M)	136
Chapter 9	Health Care Procedure Coding System (HCPCS)	144
Chapter 10	ICD-9-CM Procedural Coding	147
Chapter 11	Abstracting	155
Chapter 12	Reimbursement	170
Appendix A	CPC-H Mock Exam 1	189
Appendix B	CCS Mock Exam 1	223
Index		257

PREFACE

INTRODUCTION

One thing is certain in life—change—and nowhere is this more true than in the medical field. Just as procedures and knowledge change, so do the codes used to classify them. And so, it makes it critical in the medical field that qualified coders and billers are highly knowledgeable, since it is from them that the money flows and that the existence of a medical practice rests. As misinformed as people are about insurance, taxes, and other matters in life, they are even more misinformed about medical practice processes that sustain life and ensure the future of medical care in this country. Whether it is a good system or not is not debatable at this point, since it exists already and most medical care is not provided without proper billing and coding. So be forewarned, physicians and patients! Billers and coders can be your best friends and deserve your respect.

With this warning to others, there also is an admonishment to billers and coders—you need to maintain your knowledge and skills. Otherwise, because changes are occurring so quickly, you can be outdated rapidly, which could be disastrous not only to your career but also to the physicians and patients who depend on you. Certification by a well-known national organization is the means by which standards of knowledge and skill are established and maintained.

Just as coding knowledge and expertise is important to physicians and patients, it is just as important to coders, as it provides a wide variety of excellent career opportunities. In the progression of careers in medical coding, a coding position in a hospital usually follows experience as a coder in a physician's office. It is similar with national certifications. Usually a person is certified as a physician coder and may later earn certification for hospital coding.

Hospital coding generally involves advanced skills and a greater knowledge of the medical field than is required for physician coding, although this certainly does not diminish the expertise of coders in the physician offices. Hospital coding certification is designated as Certified Professional Coder for the Hospital (CPC-H) from the American Association of Professional Coders (AAPC) or Certified Coding Specialist (CCS) from the American Health Information Management Association (AHIMA).

This book focuses on knowledge and skills for coders in hospitals, whether they code for outpatient or inpatient services. There are important distinctions between the two, just as there are distinctions between hospital and physician coding. The AAPC exam focuses on outpatient hospital coding and the AHIMA exam focuses on outpatient and inpatient hospital coding. While the certification exams for physician coders is very difficult, the hospital exams take coding a step beyond, so be prepared if you plan on taking the hospital coding exam!

Just as the physician's certification exam book was not introductory, so this book is also not introductory. In fact, this book takes the physician's book one step further. If you are seriously considering taking the hospital exam, you also should take other preparatory courses that focus on hospital coding. Be careful—remember that there are no quick ways to prepare for coding exams, so invest enough time and use proper materials and courses to prepare. Courses on advanced billing and coding are helpful, especially those that focus on hospital issues.

CONTENT

This book includes sections similar to the physician's coding books, including ICD-9-CM, CPT-4, reimbursement, abstracting, and modifiers, but there are many changes and additions to the chapters. Changes include a chapter on ICD-9-CM procedure codes, which are contained in Volume 3; groupers; and different billing forms. Information on test-taking skills, national organizations, and types of exams are provided that are specific to the national coding exams. It is important to stress that the book focuses on the major areas of information and does not attempt to explain every difference between every payer throughout many geographical areas. The book follows the format of the national exams. In addition, there are mock exams for the two types of national exams. These mock exams follow the format of the national exams, although they are not mirror images of the exams with the same questions. It should be remembered that the two national agencies differ in how the exams are provided as well as their structure.

This book begins with descriptions of the national certifying agencies and then progresses into general test-taking strategies and suggestions for preparation before and on the day of the exam. Subsequent chapters provide detailed information about the different coding systems. Later chapters focus on areas of application, including abstracting and reimbursement.

HOW TO USE THIS BOOK EFFECTIVELY

This book is a supplement to other preparations and does not presume to be adequate if used alone to successfully pass the national exams. This book should be used with a combination of the following (if possible), which are particularly important for hospital coding: experience, courses, relevant publications, study groups, and other textbooks. Most importantly, a person preparing to take this exam should make the exam their top priority for at least six months before the exam. Study time should total at least several hundred hours.

ACKNOWLEDGEMENTS

As always my students and employers, many of whom I now call friends, are my constant source of inspiration and motivation, for if they had not believed in me, I may never have persevered. They include Cathy Estrada, Gloria Morelos, Anna Dicochea, Ave Campbell, George Kaminski, Joanne Aguilar, Nancy Haverlack, Lilia Leavitt, Sylvia Gonzalez, Rick Reyna, Jorge Reyna, Connie Peve, and Teresa Pickard.

ABOUT THE AUTHOR

The author has been an instructor in the medical office programs and program manager at Pima Community College in Tucson, Arizona, for the past ten years, which has resulted in great successes and continuing education for her students. She is a Certified Professional Coder with the American Association of Professional Coders and a Certified Coding Specialist for the Physicians with the American Health Information Management Association. She has a bachelor's degree in psychology from the University of Montana and a master's degree in public administration from Montana State University. She is ABD for her Ph.D. in higher education from the University of Arizona. Prior to her career with Pima Community College, she owned her own business providing medical office services and was a safety and health inspector for U.S.F.& G. Insurance.

HOW TO CONTACT THE AUTHOR

gtolsen@att.net

REVIEWERS

Marsha S. Diamond, CPC, CPC-H
Central Florida College
Orlando, FL

Karen R. Fisk CBCS, RHE
Indiana Business College
Terre Haute, IN

Norma Mercado, MAHS, RHIA
Austin Community College
Austin, TX

Stacey Mosay, RHIA, CCS-P, CPC-H
Trident Technical College
Charleston, SC

Chapter 1   INTRODUCTION

Objectives

(1) Evaluate your preparedness for the national hospital-based coding exam.

(2) Explore realism relative to your future encounter with the national exam.

(3) Understand the formats and usefulness of the various coding books available.

(4) Learn about the history, structure, role, and application process of national coding organizations.

Key Terms

AAPC—American Association of Professional Coders

AHIMA—American Health Information Management Association

CCS-P—Certified Coding Specialist–Physician issued through AHIMA

CPC—Certified Professional Coder through AAPC

CMS—Centers for Medicare and Medicaid Services

CPT-4—Current Procedural Terminology produced by American Medical Association

HCPCS—Healthcare Common Procedural Coding System

ICD-9-CM—International Classification of Diseases, Ninth Revision, Clinical Modification

A STEP BEYOND

If you are considering taking the hospital-based coding exam, you have probably gained experience and perhaps have some type of medical office certification. Most people begin with physician coding and/or other medical office experience, then progress to hospital-based coding. This is the most common process since hospital-based coding exams are more difficult than the physician coding exams and most employers hire and promote in this way.

This book is a supplement to prior certifications and experiences as supplemented by relevant courses and books. This book focuses on information and preparation for the national coding exams. However, the importance of prior inpatient hospital coding cannot be underestimated, particularly since educational resources for inpatient coding and reimbursement are not as extensive as for physician coding and reimbursement.

A first step in preparing for the hospital-based **Certified Coding Specialist (CCS)** or Certified Professional Coder Hospital (CPC-H) national coding exam usually is to obtain the physician-based coding (**CCS-P** or **CPC**) national certification. Most hospital-based employers prefer that coders have worked in the field for several years as physician-based certified coders before progressing to hospital-based coding. Even if you have earned a physician-based national certification (CCS-P or CPC) and want to earn a hospital-based coding certification, the hospital-based exam will be difficult.

Some hospital-based coding rules differ from physician-based coding. Many of these will be described in this book, but it is not possible for one book to capture the entire essence or content of all the coding requirements and changes since medical coding includes a vast field of knowledge and experience, which is flexible and constantly changing.

A much wider variety of medical information is needed for the hospital-based national coding exam than for the physician-based coding exam. In addition to information such as the **Internal Classification of Diseases, Ninth Revision, Clinical Modification**

(ICD-9-CM) Volumes 1 and 2, Current Procedural Terminology (CPT), **Healthcare Common Procedural Coding System (HCPCS)**, modifiers, medical terminology, etc., you must also know how to properly apply the codes from Volume 3 of the ICD-9-CM coding book, disease processes, pharmacology, laboratory, testing, diagnosis-related groupings (DRGs), Ambulatory Patient Classification (APCs), hospital billing forms, and hospital and regulatory requirements.

CODING BOOKS

Be careful and stay current! The national coding organizations may change the format and requirements for their exams, so be careful when reading the instructions as they are very specific. Misreading instructions can cost you valuable points! The information in this book is based on the information and instructions available as of 2004.

There are two coding books required for the hospital-based national coding exam for the CCS, the ICD-9-CM (Volumes 1, 2 and 3) and CPT. For the CPC-H, three books are required, ICD-9-CM (Volumes 1, 2 and 3), CPT, and HCPCS Level II. Be sure that for the hospital-based coding, you have an ICD-9-CM coding book that includes Volume 3 since Volume 3 lists the procedures that are used in hospital coding.

INTERNATIONAL CLASSIFICATION OF DISEASES, NINTH REVISION, CLINICAL MODIFICATION (ICD-9-CM-CM)

The International Classification of Diseases, Ninth Revision, (ICD-9) was created by the World Health Organization (WHO) in 1979. Diagnostic coding, however, dates back to the London Bills of Mortality which collected statistical information in the 17th century. These codes were originally intended to provide statistical information for morbidities and mortalities; however, they have been revised within the United States to be used for billing purposes as well. Today these codes are used to identify diagnoses (symptoms, conditions, or other reasons), justify medical necessity, and transform words in a patient's chart into the appropriate numbers for reimbursement.

In 1983, Medicare adopted the ICD-9-CM as the basis for determining reimbursement for Medicare patients. The United States revision (as modified by the National Center for Health Statistics, NCHS) is known as ICD-9-CM which signifies Clinical Modification (CM). Although the ICD-9-CM book contains three-digit codes, ICD-9-CM contains codes with additional fourth and fifth digits that are required for coding diagnoses. A three-digit code cannot be used if a fourth or fifth digit exists. It is critical when coding that the highest degree of specificity (fourth/fifth digits) is used for coding all diagnoses. In the first and second volume of most versions of ICD-9-CM, there are notations indicating the need for a fourth or fifth digit. Be sure your book includes these notations as they are very helpful during the examination. The third volume of the ICD-9-CM does not include these notations but simply lists all related codes following the main code.

The ICD-9-CM is composed of three volumes. The first volume (tabular) contains the actual codes for disease and injuries and includes appendices and supplementary classifications such as V codes and E codes. V codes encompass a wide array of circumstances for patient visits other than disease or injury such as personal histories or vaccinations. E codes are for external causes of injuries or diseases and are not required on the national coding exams except for instances involving drug use/abuse. There is an alphabetic index for E codes and appendices, such as listings of morphology codes and classification of drugs. The second volume is an alphabetical index.

The third volume of ICD-9-CM lists the procedural codes that are used for inpatient hospital coding. The codes in Volume 3 differ significantly from the diagnostic codes since this volume is for procedural coding. Like Volumes 1 and 2, there is a tabular and alphabetic index for the procedures in the third volume. In contrast to the diagnostic codes which can be up to five digits in length, Volume 3 codes have two digits before the period and one or two digits after. There are no notations explaining how many digits are required; however, codes with additional digits are listed after the two and three digit codes, so they are easy to find. It still is critical that you code to the highest specificity by using the proper number of digits when coding. Remember, if one digit is omitted, the code is wrong.

Volume 3 is very detailed and includes references known as "excludes" and "code also". It is critical that you pay attention to the detail!! Read the notes!! The tabular list is divided primarily by organ systems. The first section listed, however, is for procedures and interventions not classified elsewhere. The last section is for miscellaneous procedures and there is a separate section for obstetrical procedures.

All providers, whether physician offices or hospitals, use Volumes 1 and 2 of the ICD-9-CM code book for diagnostic coding, which provides the justification for patient care. Only facilities such as hospitals and skilled nursing facilities use Volume 3 for coding services.

Outpatient facilities, such as ambulatory surgery centers (ASCs), use CPT codes level I.

HEALTHCARE COMMON PROCEDURAL CODING SYSTEM/CURRENT PROCEDURAL TERMINOLOGY

The Health Care Financing Administration, which changed its name to the **Centers for Medicare and Medicaid Services (CMS)** first created the procedural system for reporting medical services and procedures in 1983. This system is called the Healthcare Common Procedure Coding System (HCPCS). Originally, HCPCS had three levels: Level I is Current Procedural Terminology (CPT) as incorporated from the American Medical Association, Level II are the national codes for supplies and materials (HCPCS), and Level III are local codes established by Medicare. Level III codes are not included on the national coding exams. CPT codes, like ICD-9-CM codes, are used not only for billing but also for statistical and management purposes.

Level I HCPCS are CPT codes that are produced yearly by the AMA. CPT codes are five digit codes used for coding for physicians, outpatient hospital services, and ambulatory surgery centers. Two digit modifiers can be appended to the five-digit codes. For outpatient hospital services, CPTs are incorporated into the Ambulatory Payment Classification (APC) system for reimbursement. CPT codes are revised continuously with new procedures added, changes in some codes, and deletions. These changes are incorporated yearly in the annual CPT book which is available in October. The CPT book begins with Evaluation and Management codes and then continues with Anesthesia, Surgery, Radiology, Pathology/Laboratory and Medicine. The book includes an index and appendices which include a list of modifiers and yearly changes.

Level II HCPCS, known as the national codes, are produced by CMS. They are alphanumeric and are used to report in detail the use of supplies and materials, non-physician services, durable medical equipment, and medications. Temporary codes are available in sections G, K or Q. An alphabetical index is provided. Modifiers are listed which are alphanumeric or letters. These level II HCPCS codes are not used on the CCS exam.

THE NATIONAL ORGANIZATIONS

There are two national organizations that provide coding certification although there are some major differences in their exams and in their organizational structure.

AMERICAN HEALTH INFORMATION MANAGEMENT ASSOCIATION (AHIMA)

American Health Information Management Association (AHIMA) is a national organization that has certified health information specialists for over 75 years. AHIMA was formed in 1928 by the Association of Record Librarians of North America to standardize and professionalize the field of health information, which included medical records in hospitals and other medical institutions. Currently AHIMA includes over 40,000 registered members with headquarters are in Chicago. The organization provides information for the medical office community through audio seminars, web-based training, publications, local, state and national conventions and meetings, coding seminars and continuing education quizzes. AHIMA has served the medical community for many years and has been involved in the development of many governmental and industry standards for the medical office, including coding and reimbursement.

Certifications

Most importantly from the perspective of this book, AHIMA provides coding certification for both physician (CCS-P) and hospital (CCS). Hospital coding is considered an advanced level position within the medical community that requires greater experience and knowledge. There also is certification for entry-level coders (CCA; Certified Coding Associate). You must be a high school graduate or have an equivalent educational background to be eligible to take the coding exams. It is also recommended that you have several years experience in the coding and billing fields before taking the CCS-P and CCS exams.

AHIMA is composed of two elected bodies, the House of Delegates and the Council on Certification. The House of Delegates comprises elected representatives from each of AHIMA's 52 state associations. The Council on Certification is composed of eight members who are elected by the AHIMA membership.

AHIMA Services

AHIMA can be contacted at their website (www.ahima.org) where applications, forms and information are provided. Membership is not included with certification, although it provides additional benefits when combined with certification. Membership costs are $20 for students, $145 for active membership, or $135 for associate membership as of 2004. AHIMA's website also provides sample tests, descriptions of available references, a listing of AHIMA-certified educational resources, correspondence material and other coding-related information. Materials can be purchased for a home study course in coding, but not for Record Health Information Technician (RHIT) or Record Health Information Administrator (RHIA) courses. Each state association provides meetings, seminars, coding roundtables, and other forms of ongoing education to its members.

Testing Company

An outside company, Thomson Prometric, is contracted to administer and score the AHIMA exams, which are now computerized. You will receive an Authorization to Test (ATT) letter after you have submitted your application. You will then have six months to arrange a time to take the test with Thomson Prometric. Testing is available Monday through Friday with some locations also open on Saturdays. Testing centers are located throughout many cities, which can be found online at ahima.org. A tutorial is provided before the test begins which provides step-by-step instructions.

There are rules for testing, which are included in your application packet. Be sure to check either the application book or website for deadlines!

Application Information

The information contained in this book reflects current practices of AHIMA, but can be changed at any time by AHIMA; therefore, it is important to check their current information thoroughly.

Applications are available online or by ordering one from AHIMA. Your application must be mailed and received 60 days before the exam date. Costs for the exam in 2004 were $380 for nonmembers and $300 for members as of 2005. Membership fee is additional. Complete the application with a pen!! Your full name must be provided. The number for the testing center where you will be taking the exam must be provided and is found within the application materials. If you need any special accommodations due to disabilities, you must complete the form specifically designed for this. A professional letter testifying to the disability needs to accompany the application.

Don't Wait to the Last Minute

Do not wait until the last minute to mail the application. Be prepared. It is wise to mail the application certified so you will know when AHIMA receives it.

Withdrawal

You may withdraw from the exam but must meet deadlines to do so. A written request must be sent. A refund will be issued but a processing fee is charged.

AMERICAN ASSOCIATION OF PROFESSIONAL CODERS (AAPC)

The **American Association of Professional Coders (AAPC)** was formed in 1988 and provides national coding certification for both physician (CPC) and hospital outpatient (CPC-H) coding. AAPC is composed of a national advisory board with members selected from state organizations and is supported by a national physician advisory board. There are now over 35,000 members throughout 250 local chapters. There are state and local groups that provide educational opportunities and meetings for its members. AAPC offers continuing education (CE) through local chapters, workshops, monthly publications and their annual conference.

AAPC can be contacted at their website (www.aapc.com) where forms, dates, application, and information are provided. In addition, the website provides sample tests, description of available references, a listing of AAPC-certified educational resources, correspondence material and other coding-related information. Coursework can be purchased for a home study course. Dates for exams are available in AAPC's publication, "The Coding Edge" or by calling the national office. Locations and dates of exams are determined by local AAPC Chapters.

Application Information

The information in this book reflects current practices of AAPC, which can be changed at any time by AAPC; therefore, check their current information thoroughly.

You can obtain an application online or by ordering one from AAPC. When completed, the application must be mailed and received six weeks before the exam date. Applications received less than six weeks before the exam but more than 20 days will be accepted by special arrangement and an additional fee. The cost of the exam in 2004 is $285.

Remember, hospital coding requires a more advanced level of experience and knowledge than typically expected for physician coding. It is recommended that you have several years experience in the coding and billing fields before taking the exam. Two letters of recommendation must accompany the application that address your coding qualifications and professionalism and verify that you have at least two years of coding experience. One year may be waived if you have completed a coding course that is at least 80 hours in length. If you do not meet these qualifications but pass the exam, you would be designated as an apprentice. Appropriate people to write recommendation letters include supervisors, instructors, physicians, or other colleagues who are familiar with your level of experience and/or education.

Complete the application in pen and provide your full name!! The index number which signifies the particular exam and its location and date must be included on the application. These numbers are provided where you access the original information regarding test dates, times and locations. Use certified mail when sending the application so you will know when AAPC receives it.

Remember, contact AAPC or the proctor as early as possible before the exam if you have not received any information or confirmation.

SUMMARY

Evaluate your preparedness for taking the national coding exams as they are difficult, particularly the hospital-based exams. You should also assess if certification as a hospital coder is the career direction you want to take. Whereas coding certification for physicians opens up a new field of career possibilities, hospital coding is specialized and demanding. If you are more interested in medical office management and prefer variety in your career, then pursuing medical office management certification, such as RHIT or RHIA might be a more suitable next step for your career building. However, if your goal is to advance within the coding field, the hospital-based national coding certification is the next step on the career ladder after certification as CCS-P or CPC.

If you do not believe that you are fully prepared for the exam but still want to take it, there are ways to prepare for the exam. Being familiar with the national organizations that provide certification is an important beginning step in earning national coding certification, particularly regarding regulations and the application process. These organizations also offer continued participation as a certified member once you have passed your exam. Other resources, including government resources and medical office/coding courses, can help you achieve your goals.

Chapter 2: IMPORTANCE OF BEING A CODER

Objectives

(1) Understand the importance of the professional coder in the hospital environment.

(2) Understand the legal and ethical issues for a professional coder.

(3) Understand the importance of proper coding and its effect upon the operations of the hospital, particularly regarding reimbursement and regulation.

(4) Recognize the repercussions from failure to comply with legal and ethical responsibilities.

(5) Evaluate employment possibilities and career paths for certified coders.

(6) Be able to recognize what constitutes fraud and abuse and how to prevent them.

Key Terms

Abuse—Failure to perform fair and reasonable billing practices or providing other reasonable services.

Compliance—Acting in accordance with rules and regulations of billing agencies, such as government or private carriers.

Fraud—Purposeful intent to gain funds in an illegal manner.

HIPAA—Health Insurance Portability and Accountability Act of 1996 which established guidelines for privacy and coding/billing processes.

THE IMPORTANCE OF BEING A CODER

Because of financial, legal, and regulatory issues, making or breaking it in the medical field today depends on good coding skills combined with proper billing practices. A coder's ability to increase financial reimbursement by streamlining and optimizing reimbursement processes, reducing errors and enhancing the quality of overall hospital operations ensures the continuation of medical practices. Correct coding also helps ensure that hospitals can provide proper medical care to patients and payments will be prompt. Patients are less likely to be denied proper medical care if coding is correct nor will patients be harassed with burdensome charges they cannot afford. In addition, coders can expedite the reimbursement process with insurance and governmental agencies, of which most patients are not knowledgeable.

Coders can achieve optimal efficiency and effectiveness for the hospital by insuring that documentation is complete and ready for decision making by carefully analyzing and interpreting patient chart information. Proper execution of coding duties helps ensure patient privacy and security within the hospital and also assists with rules and **compliance**, particularly with auditing and accreditation processes.

One other contribution that coders can provide to the hospital is educating physicians and staff on proper coding methods. Education is necessary for everyone, particularly hospital personnel, regarding reimbursement, dictation of reports, selection of proper codes, data entry of billing information, and follow up for rejected claims.

ETHICAL/LEGAL RESPONSIBILITIES

National certification requires responsibility and integrity to maintain high standards of conduct. A prominent responsibility of a medical professional is observing strict privacy rules and regulations concerning respect for patients, as well as staff and other professionals within the medical field. You must act as a professional and maintain your level of expertise by continuing your education and promoting the well being of the coding field. Do not participate in illegal or unethical activities that threaten not only your professionalism, but the coding community as well. You represent yourself and your institution, other staff members, payers, physicians, national coding organizations, and the medical coding profession.

A nationally certified coder has legal protection, as well as responsibilities. It is assumed that if you are certified, you are professional, knowledgeable and able to perform your duties correctly. Certification is advantageous for coders as it promotes the respect from professionals, such as physicians, insurance carriers, and governmental agents. Likewise, it is assumed that you will act professionally and ethically in accordance with the code of conduct of the national certifying organization (AHIMA or AAPC). Continuing education is particularly important to ensure that you remember your responsibilities and are up-to-date with coding information.

As a certified coder, you must be able to respond professionally and ethically to the wide variety of situations. Be sure you follow proper procedures and can verify that you have followed the correct coding guidelines. Respond appropriately and professionally, whether or not your organization agrees with you regarding your recommended procedures. Maintaining copies of memos or other correspondence with physicians or hospital staff demonstrates that you have properly followed accepted coding guidelines. Always code with this perspective: based on the medical record documentation can you justify to an auditor why you selected the codes you did? Remember in billing, someone who is not at the office will be looking at the codes you submitted without any additional information, and must understand what you did.

At the forefront of legal responsibility is federal, state, and local legislation. The most important recent legislation is the **Health Insurance Portability and Accountability Act (HIPAA)** of 1996 which established guidelines for privacy, security, and standardization of coding and billing processes. According to this legislation, the procedures and practices regarding patient information is a top priority within the medical office.

As a coder with access to patients' files, you have a legal responsibility to maintain the patient's right to privacy and a duty to ensure that information in the file is correct and complete. You must not divulge any patient information. It is essential that you do not code incorrectly, diagnose, or misinterpret patient's diagnoses and care. You must help guarantee that charts and reports are completed properly. It is essential, for legal purposes, that patient files are correct and complete.

In addition to the content of the patient's files, the types of communication used are a critical legal issue. Use of electronic files and the internet, computer access, and interactions with staff, physicians, administrators, patients and personal contacts pose threats to the privacy and security of patient information. Coders often discuss coding situations among themselves or share information. Be very discrete because it is easy to cross over the line and divulge protected information to inappropriate people without even knowing that patient rights have been violated. Do not discuss personal patient information or provide identifiable information, which may even include general medical information. If a person has information about a patient and learns additional information from you unknowingly, then she or he may be able to piece together personal information about the patient that could be harmful. For instance, if it were known that a certain patient visits a physician who specializes in communicable diseases, it might be easy for an unauthorized person to conclude that the patient has a communicable disease, such as syphilis. HIPAA requires that only the minimum amount of information be provided to those who are authorized; as a coder you should not be providing any additional information to anyone, particularly if you are not designated as the person responsible for releasing information.

CODER'S CONTRIBUTIONS

Another important responsibility of the coder is assisting medical practices to establish a compliance program for the medical practice. Resources regarding compliance plans can be obtained through textbooks, other reference materials, and the internet. The Office of Inspector General (OIG) website provides guidance and sample compliance plans as templates for various settings.

Auditing

Auditing is a critical part of any good compliance program. Auditing allows a medical practice to locate and correct billing and coding errors. The compliance program must be written and must outline compliance policies that are adhered to by hospital and office staff. Communication and enforcement policies ensuring effective compliance should be described and outlined in writing. There should be a designated compliance officer who has the authority to enforce the compliance program. Physician and hospital staff education is also an important part of a good compliance program. This education extends to the development of office policies and insurance contracts regarding coding and billing issues. Finally, coders should be involved in follow-up claim procedures so that coding errors can be corrected and/or appeals made. A coder should be an advisor for the hospital staff, physician and the patient and should help obtain proper reimbursement as a way to ensure future services.

Legal and Economic Reasons

It is critical that codes are precise for legal and economic reasons. Failure to properly code and bill can result in loss of business or even failure. This can be caused by loss of clientele and reputation, as well as significant penalties, especially if **fraud** and **abuse** is suspected. The omission of a simple code or its improper use can cost the hospital thousands of dollars a month or more! Remember in coding—not documented, not done. To code when the documentation does not validate that code can be considered in some instances fraud or abuse. Illegible physician handwriting and unsigned reports may be considered the same as undocumented. Lack of proper hospital procedures, such as missing physician signatures or proper supervision of assistants (i.e., physician assistants) can be the basis for misconduct charges. The possible hospital costs resulting from poor coding practices can be. These can include fines, unpaid bills, and termination of contracts with third-party payers and government.

ACTIVITY 2.1

In the following coding scenarios, costly facility errors have been made, either as inpatient or outpatient. There may be several errors in one scenario. Explain how these errors should be corrected. Some scenarios may have several incorrect or missing codes. Remember that we are not coding for the physician, but for the facility.

1. A 32-year-old female was admitted to the hospital for an appendectomy. Services were coded as 44590 and the diagnosis was coded as 541.

2. This 45-year-old male came to the Emergency Room at St. Teressa's Hospital because of severe chest pains with possible MI. He was admitted to the hospital where he underwent an angiocardiography of the right heart with catheterization for contrast material. Findings were negative. Services were coded as 88.52 and the diagnosis was coded as 410.91.

3. This 24-year-old female on Medicaid was evaluated in the Emergency Room at University Hospital for severe generalized abdominal pains which had worsened over the past three days. The patient was examined and a KUB was performed with delayed films. Findings were negative and the patient was provided with a prescription for the pain. These services were coded as 74241 and 99201. The diagnosis was coded as 789.07.

4. This 3-year-old boy was seen in the hospital for removal of a stick from his left eye. This service was coded as 98.21-E1. The diagnosis was coded as 930.1.

5. This 71-year-old male was admitted to this ambulatory service center for a sling operation for stress incontinence which was coded as 57288.

6. Two resections involving the duodenum and jejunum were performed on this 42-year-old male with the use of a scope in the hospital. End-to-end anastomosis was performed at this time. This was coded as 45.62 and 45.91.

7. Saphenous vein bypass was performed in the hospital for this 58-year-old female for two coronary arteries. This was coded as 36.11.

8. This 43-year-old female was infused in the hospital with 1000 mg of fluorouracil. This was coded as 39.96 and J9190.

9. This 72-year-old diabetic male had fallen on some boards and a nail lacerated his left arm. A tetanus shot was provided in the hospital in addition to extensive debridement and 12 sutures. The wound became infected. This was coded as 250.00 and 958.3 with both linked to services coded as 99.56 and 86.59.

10. This 16-year-old 32-week pregnant girl was seen in the Emergency Room for complaints of severe abdominal pain and bleeding. She was diagnosed with detachment of the placenta. This was coded as 762.1.

11. This 13-year-old girl broke her tibia when playing basketball. She was seen in the emergency room where manipulation was performed and a short cast applied. This was coded as 27752 and 29405 with a diagnosis of 823.80.

12. In an ASC, this 46-year-old female had axillary lymph glands superficially removed which was coded as 38500 and 38740.

13. This 59-year-old male was seen in the outpatient clinic where an emergency endotracheal intubation was performed followed by a total laryngectomy. This was coded as 31360 and 31500-51.

14. In the hospital, an incision was made into the chest of this 73-year-old female and a thoracentesis completed. This was coded as 34.91 and 34.01.

15. In the hospital, a metal rod was removed from this 19-year-old male's brain after an injury at a construction site. This was coded as 01.24.

16. In an ASC, this 14-year-old female received a resection of two medial rectus muscles of the right eye. This was coded as 67311.

17. This 24-year-old male was involved in a motorcycle accident and received multiple lacerations on his face and both arms, three which required layered closure with the largest on the arm requiring debridement

Importance of Being a Coder 9

of the epidermis and dermis. The wound on his cheek was 2.3 cm; the three wounds on his arms were 1.7, 2.7 and 2.4 cm. This was coded as 12011 and 12002.

_____

_____

_____

18. A biopsy of the peritoneum was performed on this 61-year-old female in the hospital with a scope which was coded as 47561.

_____

_____

_____

19. This 48-year-old male received an ureterectomy while in the hospital with an end-to-end anastomosis. This was coded as 56.40 and 56.51.

_____

_____

_____

20. A biopsy of the ovary is obtained from this 23-year-old female in an ASC with the use of a scope. This was coded as 58900.

_____

_____

_____

COMPLIANCE

As a coder, it is important that you are up to date and familiar with regulations and regulatory agencies, so that you can keep your office compliant. Many organizations, such as the Joint Commission on Accreditation of Healthcare Organizations (JCAHO) and the Utilization Review Accreditation Commission (URAC), oversee compliance of procedures and regulations. The following legislation and guidelines are used in medical practices.

Health Insurance Portability and Accountability Act (HIPAA)

This new legislation, which has been debated for a long time, is designed to control fraud and abuse with special attention to privacy issues. New penalties and stricter enforcement of offenses have been limited concerning personal health information (PHI) of patients. When evaluating the medical care setting for HIPAA compliance, remember that a patient's privacy must be protected by ensuring that no unauthorized persons have access to patient information. This includes obvious settings as well as many innocuous settings, such as sign-in sheets, computer screens, office conversations, or medical records left outside of examining rooms. Patients passing by certain areas may hear or see the names of other patients, which would be a violation of HIPAA rules. Release of information also is an important HIPAA issue. Releases must be properly stated and signed by patients with specific instructions regarding the purpose of the release and to whom the information is to be released. In addition, HIPAA requires that patients be provided with an information sheet describing their right to privacy.

Fraud and Abuse

Regarding losses or failures, there are important distinctions that need to be made between fraud, abuse, omission and mistakes. Fraud implies purposeful intent to gain funds in an illegal manner. Fraud can include coding to a higher level of service than justifiable from the chart, billing services that were never provided, or listing diagnoses that are not valid to obtain payment.

Abuse is the failure to perform fair and reasonable billing practices or failing to provide other reasonable services. Examples of coding abuses include overbilling, improper billing, and billing for unnecessary services.

Acts of omission and mistakes are expected because nobody is perfect. However, when there are rules and regulations and a person is employed for the specific purpose of performing billing and coding duties, it is assumed that the person is knowledgeable in these areas. Significant patterns of mistakes or omissions can pose problems for a medical practice regardless of the good intentions of the physician or the office staff.

To detect fraud or abuse, profiles are created of practices nationwide. A profile consists of graphs of various codes and their usage. These graphs are used for national trending and indicate if a code is being used too often or too little in a particular practice when compared to other similar medical practices.

With the advent of several pieces of legislation, e.g., the False Claims Act (FCA), Stark Laws, and most recently the HIPAA, penalties and fines are levied for failure to properly bill, which includes coding for governmental and private claims. The OIG estimated that $11.9 billion was lost in improper payments in 2000. Penalties for inappropriate billing can include $10,000 per claim plus three times the claim amount as well as jail time and civil penalties, such as prohibiting practices from providing care for patients. In 2001, the federal government won or negotiated $1.7 billion from fraud cases. There were 445 criminal indictments and 3,756 participants were excluded from the right to provide health care to Medicare or Medicaid patients. Stark laws prohibit physicians or their relatives from referring patients to a facility, such as a laboratory, that has a financial relationship with the physician/relatives, such as a laboratory. The practice of professional courtesy can involve abuse since the practice reduces or eliminates associated health care costs, such as deductibles or copays. This is interpreted as a violation of anti-kickback laws, such as the FCA. Only charges to immediate family members of physicians can be waived legally. The consequences of failing to comply with rules and regulations demonstrate that coding is an important position within the medical field!

Prosecutions

The following examples are listed in the Office of Inspector General's reports which are available on their website (www.oig.gov).

- The Healthcare Corporation of America (HCA), the largest for-profit hospital chain in the United States, pled guilty to criminal conduct and paid over $840 million in fines and penalties for fraudulent billing, overbilling, billing for noncovered home health services, upcoding of pneumonia and other DRG and kickbacks.
- An Illinois business owner was sentenced to more than three years in jail and required to pay $6.7 million for submitting false claims for group psychotherapy sessions that either never occurred or were conducted by unlicensed individuals.
- Covenant Care agreed to pay $1.6 million for submitting false claims in which they had improperly allocated hours for services within distinct certified areas of its nursing facilities that were not conducted in those certified areas.
- LifeScan agreed to a $60 million settlement for marketing adulterated and misbranded medical devices.
- In Florida, the owner of a medical personnel staffing and home health management company was sentenced to 37 months in prison and ordered to pay $923,100 in restitution for receiving reimbursement for patient care costs that were actually spent on furniture, Christmas bonuses, and salaries.

- National Healthcare Corporation agreed to pay a $27 million settlement for inflating Medicare costs reports in which they overstated the number of hours spent by nursing staff in caring for patients, including billing for therapy services that were not provided.
- In Virginia, the president of a medical transport company was sentenced to 37 months in prison and required to pay $1.4 million in restitution for falsely charging for transportation of patients as wheelchair-bound who were not, in addition to billing for other services not rendered.
- In New York, a physician had to pay $529,000 in restitution to private insurance companies, forfeit $820,000 to the government and pay an additional $500,000 for a civil FCA case caused by billing for nerve block injections when he had actually performed a less expensive treatment.

ACTIVITY 2.2

Using the internet, access the OIG website (www.oig.gov) and provide summaries of five cases that were prosecuted by this office.

SUMMARY

Professional coders have become critical links in the proper functioning of the medical office, whether at a hospital, clinic or doctor's office. Coders help guarantee proper and full reimbursement for hospital services, which ensures that the hospital will be able to continue to provide proper patient services. Coders also facilitate compliance with government and private carrier contracts and regulations regarding issues other than coding including medical reports and charts, implementation of HIPAA, and proper use of Uniform Hospital Discharge Data Set (UHDDS), RBRVS (Resource-Based Relative Value System), and CCI (Correct Coding Initiative).

ACTIVITY 2.1 ANSWERS

In the following coding scenarios, costly facility errors have been made, either as inpatient or outpatient. There may be several errors in one scenario. Explain how these errors should be corrected. Some scenarios may have several missing or incorrect codes. Remember we are not coding for the physician, but for the facility.

1. A 32-year-old female was admitted to the hospital for an appendectomy. Services were coded as 44590 and the diagnosis was coded as 541.

ANSWER: This is an inpatient hospital admission, so a Volume 3 procedural code should have been listed instead of the CPT code 44590 for appendectomy.

2. This 45-year-old male came to the emergency room at St. Teressa's Hospital because of severe chest pains with a possible MI. He was admitted to the hospital where he underwent an angiocardiography of the right heart with catheterization for contrast material. Findings were negative. Services were coded as 88.52 and the diagnosis was coded as 410.9.

ANSWER: The contrast material is introduced via a catheter which must also be coded as directed in the coding book. Therefore, code 37.21 also should be listed as noted within the coding book. Also, the diagnostic code, 410.9 needs a fifth digit. The MI would not be coded as it was ruled out.

3. This 24-year-old female on Medicaid was evaluated in the emergency room at University Hospital for generalized abdominal pains that had worsened over the past three days. The patient was examined and a KUB was performed with delayed films. Findings were negative and the patient was provided with a prescription for the pain. These services were coded as 74241 and 99201. The diagnosis was coded as 789.07.

ANSWER: This is an emergency room visit, not an office or urgent care visit, so code 99201 is incorrect. With Medicaid, urgent care services require preauthorization or the hospital is not reimbursed; however, they would be reimbursed for an emergency room visit.

4. This 3-year-old boy was seen in the hospital for removal of a stick from his left eye. This service was coded as 98.21-E1. The diagnosis was coded as 930.1

ANSWER: The object was penetrating the eye and should have been coded as a penetrating object which would pay more. Code 98.21 is for removal of nonpenetrating objects. HCPCS modifiers, such as E1 for left eyelid, are not attached to ICD-9-CM procedural codes. In addition, the diagnostic code 930.1 is for a penetrating object in the eyelid, not the eye. Instead, code 871.6 is correct.

5. This 71-year-old male was admitted at this ASC for a sling operation for stress incontinence which was coded as 57288.

ANSWER: The patient is a male and code 57288 is for females only. Code 53440 should have been used instead.

6. Two resections involving the duodenum and jejunum were performed on this 42-year-old male with the use of a scope in the hospital. End-to-end anastomosis was performed at this time. This was coded as 45.62 and 45.91.

ANSWER: Only the resection is coded as the end-to-end anastomosis is included and there is an exclude note explaining this.

7. Saphenous vein bypass was performed in the hospital for this 58-year-old female for two coronary arteries. This was coded as 36.11.

ANSWER: The bypass was performed on two coronary arteries, so code 36.12 should have been used instead. In addition, there is a note stating that the cardiopulmonary bypass, otherwise known as heart-lung machine, should also have been coded, which is code 39.61.

8. This 43-year-old female was infused in the hospital with 1000 mg of fluorouracil. This was coded as 39.96 and J9190.

ANSWER: The HCPCS codes are not used in the hospital. Instead, Volume 3 of the ICD-9-CM directs coders to use 99.25 for infusion of the drug.

9. This 72-year-old diabetic male had fallen on some boards and a nail lacerated his left arm. A tetanus shot was provided in the hospital in addition to extensive debridement and 12 sutures. The wound became infected. This was coded as 250.00 and 958.3 with both linked to services coded as 99.56 and 86.59.

ANSWER: The administration of the tetanus shot is linked to the diabetes, which is incorrect. Instead, a diagnostic code should have been provided for the wound, which could be found under "wound, open, by site, complicated", which would be 884.1. The coding of the wound as complicated indicates that no code is needed for the infection, so code 9583 should not have been listed.

10. This 16-year-old 32-week pregnant girl was seen in the emergency room for complaints of severe abdominal pain and bleeding. She was diagnosed with detachment of the placenta. This was coded as 762.1.

ANSWER: Code 762.1 is for detachment of the placenta, but is for the fetus only. Because the services are being provided to the mother, the code instead should be 641.23.

11. This 13-year-old girl broke her tibia when playing basketball. She was seen in the emergency room where manipulation was performed and a short cast applied. This was coded as 27752 and 29405 with a diagnosis of 823.80.

ANSWER: Application of casts are included in the care of the fracture, so code 29405 should not have been used.

12. In an ASC, this 46-year-old female had axillary lymph glands superficially removed which was coded as 38500 and 38740.

ANSWER: This should have been coded as 38740 only as there is an exclude note stating that 38500 cannot be used with a select group of other codes of which this procedure is one.

13. This 59-year-old male was seen in the emergency room where an emergency endotracheal intubation was performed followed by a total laryngectomy. This was coded as 31360 and 31500-51.

ANSWER: The intubation, 31500, should not have a 51 modifier attached.

14. In the hospital, an incision was made into the chest of this 73-year-old female and a thoracentesis completed. This was coded as 34.91 and 34.01.

ANSWER: The 34.01 should not have been coded because it is included in the procedure for the thoracentesis.

15. In the hospital, a metal rod was removed from this 19-year-old male's brain after an injury at a construction site. This was coded as 01.24.

ANSWER: Incision into the brain to remove an object is coded as 01.39 as explained in the ICD-9-CM coding book.

16. In an ASC, this 14-year-old female received a resection of two medial rectus muscles of the right eye. This was coded as 67311.

ANSWER: There were two muscles resected, so code 67312 should have been used instead, because 67311 is for one muscle only.

17. This 24-year-old male was involved in a motorcycle accident and received multiple lacerations on his face and both arms, three of which required layered closure with the largest on the arm requiring debridement of the epidermis and dermis. The wound on his cheek was 2.3 cm; the three wounds on his arms were 1.7, 2.7 and 2.4 cm. This was coded as 12011 and 12002.

ANSWER: Three of the wounds required intermediate repair and should not have been coded from the simple repair codes. The simple repair on the arm should have been coded as 12001. The repair of the cheek was intermediate and should have been coded as 12051. The other two repairs of the arm as intermediate should be added together and coded as 12032.

18. A biopsy of the peritoneum was performed on this 61-year-old female with a scope which was coded as 47561.

ANSWER: This should have been coded from Volume 3 of ICD-9-CM as 54.24 because the surgery occurred in the hospital; instead it was coded from the CPTs.

19. This 48-year-old male received an ureterectomy while in the hospital with an end-to-end anastomosis. This was coded as 56.40 and 56.51.

ANSWER: The directions in the ICD-9-CM coding book state that anastomosis are to be coded in addition to the ureterectomy, but it is stated that the end-to-end is not coded.

20. A biopsy of the ovary is obtained from this 23-year-old female in an ASC with the use of a scope. This was coded as 58900.

ANSWER: There is a scope involved, so this should have been coded with 49321 instead.

Chapter 3 — PREPARING FOR THE EXAM

Objectives

(1) Complete and submit application packets correctly.

(2) Select the best coding books for the exam.

(3) Determine if you are ready for the next step in obtaining hospital coding certification.

(4) Emphasize the basics and minimize the effect of the changes.

(5) Be mentally prepared for the exam.

(6) Not be afraid to make educated guesses on the exam.

Key Terms

Assumptions—Something not done. Assumptions are deadly on the national exams. Do not assume anything! Check the details!

Applications—The necessary form from the national organization (either AHIMA or AAPC) which, when properly completed and submitted, insures your opportunity to take the exam.

Fear of failure—Not in our definitions.

Positive thinking—Success feeds off of positive thinking. It is critical to surround yourself with positive thinking, from yourself and others.

PREPARING FOR THE EXAM

Stay on top of the registration process! You will receive confirmation and testing information from the national organization, so if you have not heard from them one month after sending the application, call them! No communication can mean they did not receive your application and you will not be able to take the exam, and after months of preparation, that is a heartbreak! You must bring all admission and confirmation information with you when you take the test.

APPLICATION PACKETS

Make sure that you have thoroughly read the testing information provided in the application material. Read this information before testing day, and read it carefully when you begin the exam. Listen to the information provided by the proctor. Remember the teacher in third grade who gave you an F on a test because the directions said circle the right answer, but you underlined it!!! The same goes here; you must follow directions completely.

BOOKS

Choose your books wisely. Stay current with AHIMA's and AAPC's policies regarding books because they can change. You are allowed to use the current year CPT-4 and ICD-9-CM Volume 1, 2, and 3 for the exams. Make sure your ICD-9-CM coding book has a third volume, which is for coding hospital procedures. Remember that the books are allowed only on the second half of the exam (abstracting) with AHIMA, not the first part (multiple-choice) of AHIMA. AAPC allows the use of the books throughout the entire exam. Beware, however, for some versions or changes in the books may not be allowed, such as the *CPT-4 Expert*. No additions to the book, such as taping or word-for-word copying of information are allowed. AHIMA does allow the use of a medical dictionary on the second half of the exam.

There is a wide range of ICD-9-CM and CPT-4 books available, from basic to professional. Many features vary from book to book, such as indentations for indexes, extra information, drawings, and descriptions, so check the book carefully. All information is valuable when taking the exam, particularly anatomical figures. For example, some ICD-9-CM books have notations for fourth and fifth digits, which are very helpful. Some books may change the order of the volumes, such as placing Volume 1 first or second in the ICD-9-CM book. Books can be purchased as spiral bound, regular bound, or as a notebook. Much of this is preference, but you should be familiar and comfortable with your coding books. As you study, you can mark your books, although only coding-pertinent information is allowed, not excessive information, such as whole sections of complete medical terminology and definitions. Have a medical dictionary that contains a lot of information; don't limit yourself to a small one just because it is cute and handy.

READY?

The CPC-H and CCS exams are based on services provided by hospitals. The CPC-H exam is for outpatient hospital coding; the CCS exam is for inpatient and outpatient hospital coding. Evaluate your education and experience to determine if you are prepared to take the exam or if you need more time to study or work. Because hospital coding requires an advanced level of experience and learning, a wider range of topics is covered in-depth on these exams, including management and regulation of medical practices, and detailed disease processes and medical procedures.

V codes also are used on the exams. Modifiers are included on all parts of the hospital coding exams when CPT codes are used. HCPCS Level II is not used on the abstracting portion of the CCS exam, but is used on the CPC-H exam. E codes, morphology codes, and HCPCS Level III codes are not used on either exam except for E codes related to adverse effects of drugs. The exams cover all medical specialty areas, so if you have worked in a specialty office for many years, make sure you know all areas of the medical field, including pharmacology, laboratory, testing, disease processes, anatomy, physiology, hospital reimbursement, and regulations.

BASICS VERSUS CHANGES

One thing certain in coding is change. Although the secret to coding is in the details, understanding and interpreting those details is what matters, not the constant changes that occur every year. It is about the basics. *You must know the basics well to interpret and research the details!* It is impossible for any book to provide you with absolutes or every answer to the national test questions. A book can only provide tools that can help you achieve your goals.

There are plenty of good books about hospital billing and coding available. Sometimes this information is a chapter within a book or it can be an entire book. This particular book does not contain all the information necessary to answer every question on the national hospital coding exams and focuses on assessment and preparation. Both experience and knowledge are a must. Knowledge is found in many good books that focus on hospital billing and coding, management, governmental regulations, medical terminology, anatomy, and pharmacology. The American Hospital Association (AHA) publishes the well-respected *Coding Clinic,* which is invaluable on the job or when studying for the national coding exams. The American Medical Association publishes the *CPT Assistant,* which is also helpful.

If this book had all of the answers to the national exams, it would defeat the purpose of the exams, which is to test your ability to think and interpret medical hospital coding and reimbursement information, not memorize a multitude of numbers and facts that change constantly. The national exams should require thinking, or what is better known as "an educated guess."

ATTITUDE

Taking an exam is a mental activity which is true for the national coding exams, because they are based on thinking skills. The thinking skills are based on your understanding, not memorization. If you understand what you are studying when you study the exam material, then you are prepared. Study on a regular basis and do your best with the knowledge and understanding you have gained. The best attitude going into the exam is that you cannot know it all but will do your best.

TAKE THE LONG WAY, NOT SHORTCUTS! BEWARE OF SKIMMING!

There are no shortcuts when preparing for the national exams, whether you study by yourself or with an established program or group. The tests are designed for those who have worked for many years in the coding field, which means they probably have many more hours of learning than those whose only source of knowledge is what they have studied. Studying for the test must be taken seriously. Start studying at least six months before the exam for approximately ten hours per week. In the last few months before the exam, make the national exam your top priority. Try to study approximately two hours per night. Make sure to include actual practice scenarios and time tests as part of your studying. Study the coding books; they are an essential part of your study program!

DO NOT OVERLOOK THE SIMPLE!

Some information in this book is very simple, but people often miss the simple things when under stress. Do not fail the national exam because of a few simple mistakes! Familiarize yourself thoroughly with the basics so you cannot miss them.

OTHER AREAS OF STUDY

Your previous knowledge and experience may be limited to a few specialties. However, you must know the terminology, anatomy, pharmacology, laboratory, and medical tests from a variety of specialties. The fundamentals are a good starting point for these areas of study, gradually working your way up to more advanced topics. Basics should include combining forms (roots, suffixes, and prefixes), the most widely prescribed drugs, common tests and labs, and major anatomical sites. You should have completed a comprehensive medical terminology course, which includes anatomy and pharmacology, but you do not need to know everything. Remember, the medical field is extensive and complex, and you cannot know everything.

Abbreviations also are important and should be included when you study medical terminology. Although not every question will involve abbreviations, there will be some on the national exams. Medical terminology and abbreviations are used in the descriptions of coding scenarios and there may be medical terminology questions and abbreviations on the national exams. With abbreviations, make sure you use the proper meaning as indicated by the information contained in the medical reports and charts. Remember, there are no shortcuts to learning!

Although important for the physician-based coding exams, knowledge of anatomy/physiology, disease processes, pharmacology, laboratory, and tests are critical for the hospital-based coding exams.

PRACTICE!

Nothing can substitute for practice. For abstracting, practice both short descriptions and long reports. Practicing makes you feel more confident. You also must learn to pace yourself—take timed tests with short answers, but also with long abstracting scenarios. Practice taking long exams also, so you get used to both the fast pace and long hours required for the national exams.

DETAILS!

As mentioned before, coding is all about details! With hospital-based coding, remember that you are coding for the hospital services, either inpatient or outpatient. To code correctly, determine for which type of facility and department you are coding, then apply the proper codes, such as CPT versus ICD-9-CM procedures.

Inpatient and outpatient hospital services are not based on the same premises as used in physician billing, such as whether a patient is new or established as is the case with Evaluation and Management codes from the CPT coding book. For example, if a patient is seen in an emergency room or hospital and discharged but returns later, this does not constitute an established patient. The patient is once again coded as new.

BOOK NOTES AND GUIDELINES

Read through the ICD-9-CM and CPT-4 books and study the notes, guidelines, and information before you take the exam. The exams may ask specific questions about these sections, such as definitions. *Take your time to check things out!* When coding, it is critical that you take the extra time to read the notes and other notations. It also is advisable that you take extra time to scan before (above) and after (below) a selected code. Although the code may seem correct, if you take a few moments more to search for notes and to review other codes, you may find something more appropriate or an addition that may change the code. ICD-9-CM codes require that you check before and after a code to determine if there are extra digits or if there are notes that apply to the selected code. Sometimes these notes may be on another page, so take the time to check! This is particularly true with CPT-4 codes where the codes may be general and may cover a wide variety of procedures. One code may seem correct, but if you examine neighboring codes, you may find another code better describes the service.

This is true also of the alphabetic index. If several codes are listed in the alphabetic index, check all of them, even if one appears correct. Take the few extra moments to read notes and check other codes.

Make sure to take your time and check the documentation in a patient's file. Start with the document in the patient's file that contains the final diagnosis. Then examine the other parts of the report and other documentation.

If you find anything that can be coded, such as diagnoses, symptoms, procedures, and services, write those codes above the description. Looking up the codes at this stage and writing them down lets you know what you need to look for in the reports. You will not have to look them up again, if you should need to come back to the scenario later. This will save time and alleviate confusion because you will know what you are looking for and will not have to reread all of the reports

STUDY SCHEDULE

Do not procrastinate! Make a schedule of what you will study and when. Remember that you cannot study everything. To begin with, about six months before you take the exam, study one hour a day during the week and two or more hours on the weekends for several months. Then for the last couple of months, study at least two hours every night. By exam time, you should have studied more than 300 hours, particularly if you have minimal experience in the field.

STUDYING TECHNIQUES

STUDY GROUPS

Study groups are highly recommended as are traditional coding courses. Perhaps the most beneficial aspect of study groups is that you can practice your thinking skills. During study groups you can verbalize your thought processes, which the group can evaluate. If your thinking process is faulty, the group can help you correct your thinking. This forces you to discuss your thought process in a practiced setting with input from others. Be curious in study groups and ask questions! It is admirable to be inquisitive but realistic. Fine tuning your thinking processes with feedback from others will help you argue sensibly with yourself during the exam!

There are various coding courses available online and in seminar format. Check them out! They are not all the same; some are hands-on whereas others are lectures; some are 40 hours long and others may be more than 100. Ask questions, check with local employers and coders working in the field, and with national/state associations and chapters. There usually are fewer courses, textbooks, and seminars that focus on hospital coding than for physician coding.

There are many publications available, too many to read all of them. Fortunately, you don't have to—review with a study group one or two coding textbooks and supplementations. The *Coding Clinic* is one important publication you should review if possible as it provides the most up-to-date coding information. Coding courses and seminars also may be beneficial.

Do not forget to read the coding books themselves! Make notes and highlight important information that will be easily visible during the exam and can prompt your memory. Guidelines and notes (remember the details!) are critical on the exam. The coding books are perhaps the most important publications to study. The week before the exam, review the coding books once again.

Remember to rest! Studying too much will not make any more information go into your head. Following a study schedule, as noted above, is sufficient.

Sleep on it! Studies have found that a good night's sleep actually does help you retain more.

For the coding exams, you need to understand, not memorize, how coding is done. This may make you feel unsure of yourself, because coding tests, unlike traditional tests, focus on your ability to apply your knowledge, interpret information, and think.

Preparing for the coding exams is serious and a lot of work. It is recommended that you do not study too much the day before the exam. Sometimes your brain needs a rest. Instead, make sure that everything is organized and ready. Know where the testing site is. If you are staying in a hotel because your testing site is out of town, drive the day before and not the morning of the test. You need to be rested. The tests are long and tiring and, after the test, you may want to stay overnight. Be good to yourself for all of the hard work you have done.

MENTAL PREPARATION

POSITIVE THINKING

Surround yourself with people who will help you remain positive and study effectively.

SIMPLE BEGINNINGS

Start studying, even if it is not for as many hours as suggested or as you would like—just start even if only for half an hour. Procrastination is tempting and the work may seem insurmountable, but all you really have to do is start—anywhere! Things will come together over time. Sometimes your study plans will be disrupted. Minimize these disruptions as much as you can but expect a few.

Remember, you don't have to know everything to pass! In fact, you will *NOT* ever know everything! But this does not mean you cannot succeed. Do not become totally absorbed with minor details when studying or during the test. Keep it simple! Do not complicate things!

BE HARD ON YOURSELF!

The harder you push yourself before the exam, during your study time, the easier the exam will seem and the easier you will be able to handle the stress and still perform. Take long and tiring exams. Pick the hardest coding scenarios you can, such as cardiac catheterizations. Code long and complicated hospital stays. The more you practice, the better your chances of passing.

ANSWERS

Answer what you can. First answer the questions that you can do easily. With the difficult questions, answer what you can; if you miss a code or two, but get some codes right, you can still pass. Do not confuse and overwhelm yourself. Keep it simple and do what you can! And keep reminding yourself, to pass you are not expected to know everything.

Answering first what is easiest is important. Mentally, it is critical for you to answer several questions correctly at the beginning of the test. This will help you avoid mind block and give you confidence. Do not confuse or discourage yourself in the first few minutes of the exam. There will be difficult questions on the exam, you will not know everything, and it is okay to bypass the hard ones. (Just be sure to mark the questions you skip or do not complete in the column with a star or check so you remember you did not answer those questions, did not answer all the question, or need to recheck your answer.) Taking the test this way will help you avoid wasting precious time on a difficult question. As you take the exam, remember that it is better to answer ten questions rather than just one, so answer the easiest questions first and leave the hard questions until last. Mentally, the first few questions will seem the hardest. Once you get your brain working and have been steadily answering questions, it will seem easier. When you return to a question you originally skipped, it may make more sense especially after you have gained confidence and answered other questions. It is much easier to tackle hard questions, after you already have answered the easier questions.

Answer what you do know. If you are fairly sure of your answer, mark it. You can return to the question later and review it to ensure that your answer was correct, or make another selection if you decide otherwise. Do not waste your thought processes. Mark an answer if you are fairly confident it is correct. If the answer requires several answers, mark what you do know, and return to the question later to finish marking the other answers. If a question is very confusing, answer what you can. You will not be getting a 100% on the exam, so it is okay to miss a few. Keep in mind that in some test formats, providing too many answers is considered wrong.

ASSUMPTIONS

Be careful of skimming and assuming! Although, you will not have time to read and reread everything, particularly when abstracting, details are critical! If you skim or assume, you can miss the details. Take time to read the details!

FAILURE, NO FEAR

This exam is difficult for everyone. Many who take the exam think they have failed, but discover that they actually passed.

However, many people do fail the test. It is normal to fail; everyone fails at many things throughout their lives, yet become great successes. Failure does not define you, it is a learning process that everyone experiences and should help us become stronger. There are no losers; the only failure is to not try or to give up. With all the emphasis on coding certification, it often is forgotten that being able to do the coding job well is what matters.

Fear of failure stifles the mind and limits opportunities, which keeps us from being our best. It only takes one time to succeed to be a success. The good thing about trying, also, is that at least you know you tried. Do not regret having tried.

If you do not pass the exam, you are in great and intelligent company. You can take the exam again, and you can have a great career, with or without the certification, for, as any good employer will tell you, it is your attitude that counts the most!

On the day of the exam, take the exam, do your best and realize you cannot anticipate everything or know everything on the test. At exam time, if you have studied and worked hard, then you have done your best. Leave it at that. There is nothing more to be done at that time.

SUMMARY

Taking a test, especially a national exam, is an art. There are many steps you can take that will help you do your best on the exam. One of the first steps is determining if you are ready to take the test and if this is the right choice for you. Establishing a study plan and using study aides is important for your success. Study aides include study groups, textbooks, and building your self-confidence.

Chapter 4 TAKING THE EXAM

Objectives

(1) Establish time frames to pace yourself before and during the exam.

(2) Know what you need to bring and do before and after you get to the testing site.

(3) Ensure that the test is completed properly.

(4) Learn to check details!

(5) Know the rules of the testing process and how to follow instructions.

(6) Establish a thinking process that works to your benefit.

Key Terms

Educated guesses—What you must learn to do well since the exam is not predicated on what you have studied, but on what you can deduce from the wide variety of information that exists within the medical office field.

ID—Your ID is established at test time when you present several forms of identification, one of which must contain a picture. You cannot take the test without your ID.

Multiple-choice—Type of questions that comprise the entire AAPC exam and the first part of the AHIMA exam in which a question is asked and several possible answers are given.

Process of elimination—A useful process of eliminating multiple-choice answers by determining incorrect codes and then eliminating them.

Ticket—Tickets are issued by the national organizations no later than a week before the exam. Do not forget your ticket, which is your admission to the test. Without a ticket you cannot get in.

BE EARLY!

Arrive at least a half an hour early! Being early helps if miscommunication or problems occur; the extra time allows time for you to correct the problem. Hopefully, you have studied long and hard, and your efforts should not be for nothing. So come early to ensure that if there are complications, you have time to make other arrangements. It never hurts to be early, but being late can be devastating! The rules are strict!

BOOKS

Bring your books—ICD-9-CM and CPT-4—to the exams. The HCPCS book for AHIMA is not used. Make sure that your ICD-9-CM book includes Volume 3 for procedures. AHIMA does allow a medical dictionary, although it (and the other books) can only be used during the second half of the exam. There are limitations on what can be written or taped in your books. No extensive medical terminology, definitions, or pages of other books can be used. Additional information that could be used on the job is allowed in the coding book, but no stapled, glued, or taped information is allowed. No papers or other reference materials are allowed. You may bring a non-printing, battery-operated calculator. You can write on the exam, which is an excellent idea and suggestions for written notes are included in other chapters. For the computerized CCS exam, everything is answered online, although the cases are provided in paper form.

OTHER MATERIALS

You must bring two forms of **identification (ID)**. One must be a picture ID and one must have your signature. A social security card is not acceptable. You may wish to bring a watch to keep track of time in case no clock is available. You also should bring several #2 pencils. It is recommended that you also bring a pencil sharpener and good erasers.

EXTRAS

Bring something to eat and drink during breaks—you do not want your energy to run down. Snacks and beverages must be kept in your bag or backpack. A water bottle is usually allowed in the classroom.

PACE YOURSELF

Remember to pace yourself! Use a watch. There are 150 **multiple-choice** questions on the AAPC exam, which must be completed in five hours. There are 60 multiple-choice questions on the first part of the AHIMA exam, which must be completed in one hour. There are seven outpatient records and six inpatient records, which must be completed within three hours.

CODING

Every scenario must have at least one ICD-9-CM code and one service code (CPT/HCPCS or Volume 3 of ICD-9).

Read everything, especially the instructions in the coding books. Although you must be concerned with time, especially on the AAPC exam, if you fail to read the directions or questions carefully, you can miss points. *Success is in the details!*

Make sure that you read all of the questions and possible answers, particularly in the multiple-choice section. Do not read one choice and assume it is right without reading the rest of the choices.

If you finish early, go back and check your answers. Use all of your time if you can. Remember, your first answer has the greatest chance of being right. If you decide to change your first answer, make sure you have good justification. Sometimes the first choice represents your unconscious thought, which may recall something that your conscious mind does not.

Remember to check the information before codes to determine if there are any other criteria that apply to this code, such as "code also" or fifth digits.

If you finish the exam early, do not leave! It is always amazing and somewhat frightening to others taking an exam, when someone finishes up in a very short time and leaves; however, this is not a wise way to take tests. If you have time, always stay and recheck your answers carefully. People who finish early don't necessarily score better than those who stay. This is especially true with the national coding exams, because the questions are intricate and complicated, so it is easy to miss small details that affect the answers. Review your test if you have time; do not be in a hurry to leave!

ON LINE

Be careful! Select codes from the correct line in the coding books and put your answers in the right place on the answer sheet. This is important particularly with tables in the coding books and test forms that require bubbles to be filled in with pencil. It is easy to be off by just one line when marking your answer, which could result in many more answers being marked in the wrong answer line. Use a piece of paper or a ruler to mark where your answers go to ensure that you are marking the answer on the right line. Keep in mind that AHIMA now offers computerized tests.

AHIMA

CCS

The CCS exam is computerized. Although the multiple-choice questions are completely computerized, the cases are provided in paper format so you can examine the test pages, but you must submit your answers via the computer. Questions relative to ICD-9-CM Volumes 1, 2, and 3 and to CPT-4 are included on the exam. If you have worked in a specialty area for many years, be sure you study all other areas of the medical field. You are allowed to use a current edition of ICD-9-CM codes, CPT-4 codes, and a medical dictionary, but they cannot be used during the first portion (the multiple-choice section) of the exam.

There are tests in this book that are in the CCS exam format. Part I contains 60 multiple-choice questions with 60 minutes allowed for completion. Competencies for the test are:

I. HEALTH INFORMATION DOCUMENTATION

1. Interpret health record documentation using knowledge of anatomy, physiology, clinical disease processes, pharmacology, and medical terminology to identify codeable diagnoses and/or procedures
2. Determine when additional clinical information is needed to assign the diagnosis and/or procedure code(s)

3. Consult with physicians and other healthcare providers to obtain further clinical information to assist with code assignment
4. Consult reference materials to facilitate code assignment
5. Identify patient encounter type to assign codes (e.g., inpatient versus outpatient)
6. Identify the etiology and manifestation(s) of clinical conditions

II. DIAGNOSTIC CODING GUIDELINES

1. Select the diagnoses that require coding according to current coding and reporting requirements for inpatient services
2. Select the diagnoses that require coding according to current coding and reporting requirements for hospital-based outpatient services
3. Interpret conventions, formats, instructional notations, tables, and definitions of the classification system to select diagnoses, conditions, problems, or other reasons for the encounter that require coding
4. Sequence diagnoses and other reasons for encounter according to notations and conventions of the classification system and standard data set definitions [e.g., Uniform Hospital Discharge Data Set (UHDDS)]
5. Determine if signs, symptoms, or manifestations require separate code assignments
6. Determine if the diagnostic statement provided by the healthcare provider does not allow for more specific code assignment (e.g., fourth or fifth digit)
7. Recognize when the classification system does not provide a precise code for the condition documented (e.g., residual categories and/or non-classified syndromes)
8. Assign supplementary code(s) to indicate reasons for the healthcare encounter other than illness or injury
9. Assign supplementary code(s) to indicate factors other than illness or injury that influence the patient's health status
10. Assign supplementary code(s) to indicate the external cause of an injury, adverse effect, or poisoning

III. PROCEDURAL CODING GUIDELINES

1. Select the procedures that require coding according to current coding and reporting requirements for inpatient services
2. Select the procedures that require coding according to current coding and reporting requirements for hospital-based outpatient services
3. Interpret conventions, formats, instructional notations, and definitions of the classification system and/or nomenclature to select procedures/services that require coding
4. Sequence procedures according to notations and conventions of the classification system/nomenclature and standard data set definitions (e.g., UHDDS)
5. Determine if more than one code is necessary to fully describe the procedure/service performed
6. Determine if the procedural statement provided by the healthcare provider does not allow for a more specific code assignment
7. Recognize when the classification system/nomenclature does not provide a precise code for the procedure/service

IV. REGULATORY GUIDELINES AND REPORTING REQUIREMENTS FOR INPATIENT HOSPITALIZATIONS

1. Select the principal diagnosis, principal procedure, complications and comorbid conditions and other significant procedures that require coding according to UHDDS definitions and official coding guidelines
2. Evaluate the effect of code selection on Diagnosis Related Group (DRG) assignment
3. Verify DRG assignment based on Prospective Payment System (PPS) definitions

V. REGULATORY GUIDELINES AND REPORTING REQUIREMENTS FOR HOSPITAL-BASED OUTPATIENT SERVICES

1. Apply guidelines for bundling and unbundling of codes
2. Apply outpatient PPS reporting requirements:
   a. Modifiers
   b. CPT versus HCPCS II
   c. Medical necessity (i.e., linking diagnosis to procedure/service)
   d. Evaluation and Management code assignment
3. Select the reason for encounter, pertinent secondary conditions, primary procedure and other significant procedures that require coding
4. Verify APC assignment based on OPPS definitions

VI. DATA QUALITY

1. Assess the quality of coding from an array of data (e.g., reports)
2. Educate physicians and staff regarding reimbursement methodologies and documentation rules and regulations related to coding
3. Participate in the development of institutional coding policies to ensure compliance with official coding rules and guidelines
4. Analyze health record documentation for quality and completeness of coding (e.g., inclusion or exclusion of codes)
5. Review health record documentation to substantiate claims processing and appeals (e.g., codes, discharge disposition, patient type, charge codes)
6. Analyze edits from the Correct Coding Initiative (CCI) and Outpatient Code Editor (OCE)

VII. DATA MANAGEMENT

1. Manage accounts (e.g., unbilled, denied, suspended)
2. Recognize UB-92 data elements
3. Identify Charge Description Master (CDM) issues (e.g., revenue codes, units of service, CPT/HCPCS, text descriptions, modifiers)
4. Identify accounts subject to the 72-hour rule
5. Identify hospital-based outpatient accounts subject to the "to/from" dates of service edits
6. Identify cases needed for health record reviews (e.g., committee, clinical pertinence, research)
7. Analyze case mix index data

TOTAL

Reprint from AHIMA. 2005

The questions in the multiple choice section vary in content, but will include the wide variety of topics that relate to coding, including billing and regulations. Some questions will be short, but some will have billing information presented with several questions pertaining to the billing information provided.

Part II contains 13 medical abstracting cases—six inpatient and seven outpatient cases. With the outpatient cases, CPT-4 procedural codes should be used, including the use of modifiers. No HCPCS are to be listed. E codes are used only when external drug use influences patient care. For these cases, up to four diagnostic codes are allowed and up to four CPT-4 procedural codes.

The inpatient cases are coded from the ICD-9-CM procedural Volume 3. The cases consist of coding scenarios you might see in a hospital chart. These scenarios may contain a single report or various reports, such as operative, discharges, history and physicals, and lab/pathology. From these reports, determine the proper codes only after going through all of the reports. For inpatient hospital cases, you are allowed up to nine ICD-9-CM diagnostic codes and six procedural codes, although many scenarios will not require these many codes.

The principal diagnosis must be listed first for inpatient and outpatient visits. No E codes should be listed except those that are associated with the adverse effect of properly prescribed and administered drugs. Do not assign M codes. Use modifiers only with the outpatient ambulatory cases that involve CPT-4 coding and that are located in the first abstracting section on the test. Inpatient hospital cases require ICD-9-CM procedural codes and do not require modifiers. HCPCS Level II codes should not be listed and only codes are needed, not descriptions, for the cases. Code only as many codes as needed, which varies from case to case. Do not overcode for points will be deducted for overcoding. You cannot return to another portion of the exam once it has been completed.

Although books are not allowed on Part I, all needed information from the coding books is provided on the exam. An information box is available with the necessary information to answer a question, such as a list of modifiers and codes, which can be used to answer the question.

AAPC

The CPC-H exam is a paper exam. Bring a current edition of ICD-9-CM, CPT-4, and HCPCS. These can be used throughout the entire test. You can write information in your code books, except word-for-word descriptions from study materials, medical terminology and definitions, or any part of the AAPC study guide. Nothing can be loose, pasted, stapled, glued, or taped into your coding books.

You are given five hours to complete 150 multiple-choice questions for the CPC-H exam. Some of the questions will be direct, whereas others will involve coding scenarios. Coding scenarios will offer you four choices, which may contain a number of codes for each choice. There are tests in this book that reflect the AAPC exam format. With the CPC-H exam, time is a critical factor and it often is difficult to complete the exam within the allotted five hours.

The CPC-H exam is divided into three sections. Section one covers medical concepts, medical terminology, anatomy, coding concepts, payment methodologies, and compliance. Section two covers code assignment of codes from ICD-9-CM Volumes 1 and 2, CPT, and HCPCS. Section three covers coding applications, surgery, and modifiers.

There are two breaks allowed during the test, which are not counted as part of the five hours. If absolutely necessary, one person at a time is allowed to leave the exam for a short break, but generally, you will need each available second to complete the exam. During regular breaks, it is highly recommended that you remain on the premises and do not wander too far. Unfortunately, people have been in the middle of an exam, took a break, got lost and never made it back in time! You cannot afford to lose time or be denied entrance to the remainder of the exam.

DOCUMENTATION

Documentation is the basis of coding. "Not documented, not done" is the proverb of coders. Although this concept is frustrating to those who like to overanalyze, it is critical to remember for the exam, because in coding, if it is not documented, it is considered never to have happened and should not be coded.

Do not diagnose the patient! Only physicians can do this. You can code only from the documentation. If the documentation is inadequate, it can be addressed in the future by educating physicians and staff.

Code only from what is on the exam. There is nothing else! Do not include services that you know must have been provided if the physician does not state those services in the report. Lower level codes, such as 99201 as opposed to 99202, should be used if the physician does not provide enough information in the written report to warrant a higher-level code. This should make your job easier as a coder, and on the exam, because you do not need to diagnose the patient yourself, nor do you need to know more than the physician! Just code what is in the report.

Coding only what is documented simplifies the coding process. It means that coders do not need to know or anticipate everything, which is impossible to do anyway. It is critical that you remember this on the exam—code only what is documented, nothing else. *And do not assume anything!*

"Not documented, not done" also applies to illegible writing on reports. If the only documentation provided is illegible, then it cannot be coded. Keep in mind that, as a coder, you must be able to explain to another person why you selected the codes you did. This fact is particularly important for the exam but more so when coding on the job. Employers want to hire coders who think and can explain their method of thinking.

SPECIFIC

Be as specific as you can be! Sometimes you will have to select generalized codes, but always try to be as specific as you can. Do not use unspecified codes if possible. Usually these codes are marked with a 9 as the last digit.

SKIPPING

If you skip a question, only partially complete it, or are unsure of your answer, *mark it on the test in the outside column.* Use a star or other symbol so that you can easily find it when you recheck the exam. If you do not mark these questions, you miss finding one when you recheck. Or, worse, you could spend extra time looking for the questions you did not finish. Marking them makes everything easier and you are less likely to miss a question.

NOTES

You cannot write on the answer key, but you can write on the test booklet. Circle and underline all valuable information, so you do not have to constantly re-read everything. When you find a code, mark it next to the description in the test booklet. By doing so, you save time and lessen confusion, because you will not have to keep searching for the codes. Mark these codes next to the description, such as in the discharge diagnosis, as one of the first things that you do for each case. These codes should be done before reading the rest of the reports, as it gives you the information you need to know to select the proper code. For example, when selecting a code for diabetes, you must know whether it is Type I or Type II, whether it involves complications, etc. You may not know these selection criteria for the code until you see the code. Once you have seen the code, then you know what other factors to look for in the reports, which can save you time and minimize confusion.

MULTIPLE-CHOICE

When you are completing a multiple-choice section, through the **process of elimination**, cross off all answers you know are incorrect. The wrong answers often can tell you what the right answers are. Therefore, you do not have just one way to find the right answer, but several.

Sometimes you can determine the wrong codes in a multiple-choice question by examining just a few codes. Start with the first code listed in each multiple-choice answer. Check them and see if they are valid. If you find that they are not valid, such as the patient is insulin-dependent, but the code does not signify this, then cross off all the answers with codes that state insulin-dependent. You can then continue onto the next code listed, crossing off what you know is wrong until you have found the right answer. Usually you can be fairly certain about your answers using this process of elimination. For example, select the proper codes for the following scenario:

A 62-year-old man is seen in the hospital for acute systolic malignant hypertensive heart failure. In addition, he is diagnosed with chronic renal disease. Which are the correct codes?

   A. 404.03, 428.21

   B. 404.01

   C. 404.01, 428.21

   D. 428.21, 593.9

The heart failure is described as hypertensive, so it is coded as 404.01, hypertensive heart disease with the mention of heart failure. The fifth digit "1" is for heart failure. Because the heart failure is linked to hypertension, answer D is incorrect as this codes for heart failure but without it being linked to hypertension. There is chronic renal involvement, which is always coded as hypertensive renal disease, but this is not stated as failure, so answer A also is wrong, because it codes the renal disease as failure. The code also states that the type of heart failure also must be listed, so code 428.21 must also be used, which is for acute systolic heart failure. So, answer B is wrong, because there is no code for the heart failure. The correct answer, then through the process of elimination, is C—404.01 and 428.21.

Make **educated guesses**. The purpose of exams is to test your ability to abstract knowledge from your previous experience and apply it to new situations as given in the exam. The national exams are not tests of memorization but of problem-solving skills with the application and interpretation of coding knowledge to coding scenarios. Since you are relying upon educated guesses, it is best to stick with your first answer. Often when thinking hard, our brains vaguely remember something that we cannot consciously perceive at the time, particularly under stress; our first guess often reflects this unconscious knowledge, so your first guess is often the right guess! If you start constantly second-guessing yourself, you may be wrong more than you are right. It is good to review your test once you have finished, but wait before changing an answer—remember, there is strength of knowledge in your first answer. Have some solid information before you change your first educated guess. Often when people change their first answer, they guess wrong the second time. Be careful!

Get as close as you can to the diagnosis description or procedure! Sometimes, particularly when using the alphabetic index, you will not find the exact description of a diagnosis or procedure. Get as close as you can by finding related terminology, if possible. Try several different terms. Even when you think you have found something, take time to look a little further. This is particularly true when several codes are listed as possibilities under a term.

For example, transfer of a shoulder muscle lists several codes i.e., 23395-23397, 24301, 24320 in the alphabetic index. Briefly examine each one to determine which is correct. You may find more information and important differences. Although time is limited on exams, taking a few extra seconds to explore may help you make the right choice.

Even when you find a code that appears to be correct, check before and after the code. This is particularly important with fourth and fifth digits, since the additional digit code may be located directly underneath the main code or may be a few pages back. For example, diabetes mellitus lists the fifth digits at the beginning of the 250 section. Sometimes, a code listing in either Volume 1 or Volume 2 contains references to other codes that would be more appropriate. Essentially all answers are in front of you when coding from abstracts because they are in your coding books—they are in the details! Remember, these tests are not about memorization but about knowing how to research and think.

Be sure to read and follow the notes! If they cross-reference (e.g., see, see also) you to other information, check the information. If you are instructed to "see condition," the condition is the presenting problem, not the word "condition." A presenting problem/condition, for example, might be "failure" or "disease." Take the extra seconds or minutes to make sure your answer is correct by checking the cross-reference notations.

Details in the alphabetic index also are important! When the alphabetic index directs you to a specific section or code, follow the instructions precisely. This is particularly true with the ICD-9-CM coding book. For example, an acute slipped capital femoral epiphysis is coded as 820.01, and not 732.2, because the details in the index instruct you to go to "fracture, by site" for traumatic. Read the fine print!

SUMMARY

Proper attitude and preparation is important to a successful outcome on your exam! There are many steps you can take to ensure that you are not flustered before and during the exam. Basics, details, knowing how to use your time wisely, and being well prepared are critical.

Chapter 5   AFTER THE EXAM

Objectives

(1) Interpret the results of the national exams.

(2) Understand the certification process and requirements.

(3) Understand how to make an appeal.

(4) Identify the requirements for taking the exam again.

(5) Understand how to maintain certification and earn continuing education units (CEUs).

(6) Evaluate the future for employment and continuing education.

Key Terms

Appeals—An appeal is a process by which problems experienced during testing can be examined and possibly rectified by the national organization. It is important to follow through with appeals if you have a legitimate complaint.

CEUs—Continuing education units, which must be earned to maintain your certification. AAPC and AHIMA have both similarities and differences in how CEUs are earned and how certification is maintained.

Certification—The process of becoming a certified coder, which culminates in taking the AHIMA or AAPC national certification test. After a coder passes the national test, they are considered certified coders. Certification must be maintained by earning CEUs and completing self-assessment exams. Certification acknowledges a person's competency in coding and their achievement of all required qualifications.

Coding Edge—A publication of the American Association of Professional Coders that provides information on coding issues and resources available to coders and other interested professionals. The publication is sent to its members.

Lifelong education—A commitment to being your best by continually updating your skills and knowledge, which is critical in a constantly changing world.

RESULTS

You must pass all three parts of the CPC-H exam to become certified. Passing scores for the AHIMA exam are not an established set score; however, you must pass both parts of the exam to become certified. AHIMA's testing agency (Thomson Prometric) uses statistical scoring to ensure fairness for all candidates. AHIMA states that this process provides adjustment for the "fluctuations in difficulty across examination forms." After individual examinations are scored, each item is analyzed and reviewed by members of the Examination Construction Committee before final scoring. Therefore, a passing score cannot be established prior to testing.

CERTIFICATION

Every successful applicant who passes the exam receives a certificate identifying their credentials. AAPC examinees receive the distinction of Certified Professional Coder–Hospital (CPC-H) and AHIMA examinees receive the distinction of Certified Coding Specialist (CCS).

COMPLAINTS/APPEALS

Comments concerning the test are allowed on the exams during examination time and also on the evaluation form at the end of the testing period. There also is an appeal process. It is important that you take this seriously, as the national organizations do consider these complaints/**appeals**. If you can prove the validity of your complaints, these organizations may make accommodations. For AHIMA, the complaint/appeal must be submitted within 30 days after the completion of the exam and must be written. The address for mailing complaints/appeals is listed in the applicant handbook.

TESTING AGAIN

AHIMA allows you to retake the CCS exam 91 days after your last test, but you must pay again. With AAPC, you can retake the CPC-H examination one more time at no extra charge but no sooner than 30 days after taking the first exam and not more than one year later. You must resubmit your request to retake the exam.

RENEWING CERTIFICATION

AAPC and AHIMA will send you information regarding recertification and how to earn **Continuing Education Units (CEUs)**. You also can access this information at their websites (ahima.org and aapc.com). With AAPC, as part of the **certification** process, you must apply for membership. For AHIMA, membership is a separate process from certification maintenance.

AHIMA CEUs

Each AHIMA certification must be renewed every two years at a cost of $50 for nonmembers for one credential (CCS, CCS-P, Record Health Information Technician [RHIT], or Record Health Information Associate [RHIA]) or $10 for members, which is inclusive of all credentials earned. With AHIMA, you must renew your certification every two years by earning 20 CEUs or 30 CEUs if dual certified as CCS and CCS-P. As part of the renewal process, a self-assessment exam must be completed each year, which counts for 5 CEUs or a total of 10 CEUs for the two-year renewal period. The self-assessment exam costs members $25 each year and nonmembers $50 per year and can be accessed online at AHIMA's website (www.ahima.org). The self-assessment is composed of short-answer and multiple-choice questions with some medical scenarios. The exams are not graded, but, rather, the experience of taking the exam should provide an educational opportunity. Additional CEUs can be earned by attending pre-approved CEUs, including seminars, workshops, web-based training, audio conferences, subscriptions, local meetings, teaching, and more. Programs do not need to be pre-approved by AHIMA before attending them to earn CEUs, although full credit may not be given for unapproved programs. For most programs, two hours of seminar/conference/classroom time equals one CEU.

AAPC CEUs

AAPC certification must be renewed every year at a cost, in 2005, of $85. With AAPC, you must be earning CEUs throughout the year by attending coding seminars and preparing coding materials to meet renewal deadlines. For the CPC-H, you must demonstrate that you have earned 18 CEUs in a year. With AAPC, if you have dual certification in both CPC and CPC-H (physician and outpatient hospital), then you must submit documentation for 24 CEUs. To earn CEUs, you should attend AAPC-approved seminars, workshops, and presentations throughout the year. In addition, CEUs can be obtained from coding courses, attending local organizational meetings, publication or presentation of coding materials, development of coding scenarios, and other sources. Unapproved programs may be eligible as CEUs, but full credit may not be given per AAPC.

To become a fully certified coder, AAPC certified apprentices must complete their coding experience and submit proper documentation that demonstrates completion of one or two years of medical coding experience, as well as earning CEUs yearly.

ACTIVITIES THAT QUALIFY FOR CEU CREDIT

The national organization will assess the activities you present for CEU requirements and determine how many CEUs are awarded per activity. AHIMA and AAPC differ on what they will accept and how many CEUs are awarded per activity. The organization will send you information prior to recertification time that includes the CEU application and qualifying activities. Some activities will require prior approval by the national organization. These include participation in related educational programs, attendance at AHIMA meetings, national conventions, state, local, or regional meetings, visiting vendors at conferences, seminars, college or independent study programs, publication or presentation of coding seminars or materials, speaker or panel participant at a related event, approved quizzes, development of scenarios and written summaries of relevant publications.

NATIONAL RECOGNITION

National coder certification is recognized anywhere in the country. Some states and organizations require coders to be certified before then can obtain employment. Being a certified coder provides many new career opportunities and a professional status that earns respect from administrators and coworkers in government agencies, hospitals, regulatory agencies, and physician offices. This professionalism is enhanced by attending and networking at seminars, national and state organizational meetings, and in educational institutions. Networking is an important factor for furthering your career opportunities and for creating a source of information and feedback on coding issues. The certification is certainly hard-won through hard work and you should value it.

EMPLOYMENT OPPORTUNITIES

There are a wide variety of employment opportunities for certified coders in a rapidly growing and good-paying field. The medical field is rated as one of the fastest growing fields in the country. In a survey recently conducted by AAPC, certified coders were earning up to $18,000 more than the national average salary. The average salary of a certified coder was $39,046 annually which translates into $18.75 per hour. Noncertified coders also make more than the national average salary—up to $8000 more—but do not make as much as a certified coder.

Hospitals usually demand that their coders are certified. Coders usually are started as outpatient coders, and then transfer to inpatient coding later. Hospital coders tend to make significantly more than coders in physician offices or clinics, with inpatient coders making up to $20 per hour or more. There is a high demand for inpatient coders, especially in certain specialty areas such as cardiovascular. Because of the great demand, inpatient coders sometimes contract with a national company and travel around the country to perform coding at hospitals.

Salaries for certified coders rose by approximately $7000 in two years (1999 to 2001). These figures are dependent on where in the country the coder is employed, their educational level, and in what employment capacity. Coders in the central part of the country earned the least, and coders on the West Coast earned the most. Coding consultants earn the highest wages at an average of $52,075. A college degree can add $10,000 more to a coder's salary. More information regarding these statistics can be obtained from the Bureau of Labor Statistics at http://www.bls.gov. (American Academy of Professional Coders. *Coding Edge*, Volume V; Issue X:October 2003). Career opportunities depend on your ambition, which can be enhanced by additional education. Do not limit yourself unnecessarily. Earning national certification in coding is a great accomplishment and an indication of what else you can achieve.

In addition to coding, there are a multitude of employment possibilities available to coders. Employment opportunities can be found in a variety of settings, such as hospitals, clinics, physician offices, governmental agencies, health insurance companies, skilled nursing facilities, behavioral health facilities, consulting firms, rehabilitation hospitals, outpatient care facilities, home health care, vendors (such as drug companies), etc.

Hospitals are more demanding concerning coding experience and certification. Therefore, most coders beginning in the field will find employment with physician offices or clinics rather than starting with hospital coding. If you are just beginning, do not despair if you are not successful in getting a job involving only coding. Work your way up the career ladder by doing your best and learning as much as possible! Although hospital coders usually perform more coding-related duties, gaining experience in billing, medical records, and/or medical office management will make you well-rounded as a professional coder. Many jobs in the medical field are not limited strictly to coding but include other tasks, such as billing or medical records. You may have to take an entry-level job despite your certification, because there usually are more entry-level jobs available in an organization. Your additional knowledge and education should only help you get promoted quicker and easier, so the extra knowledge definitely is to your advantage.

Opportunities for coders include positions such as inpatient and outpatient hospital coding, billers, clinic coders, claim reviewers, supervisors, compliance specialists, analysts, auditors, consultants, educators, case managers, medical record technicians, and medical office specialists. There are a wide variety of employers for whom a coder can work including private and governmental agencies, hospitals, schools, insurance companies, physician offices, or clinics. Some positions may involve a great deal of patient contact, some may involve contact with physicians and insurance companies, and some may involve little contact with anyone. Other positions may combine both billing and coding into office duties and may even include patient checkout. Medical practices are very diverse in how they divide responsibilities among employees but learning as much as you can will always be to your benefit, so do not confine yourself strictly to coding. Except in hospitals, most positions include other duties, which may be because medical practices usually are smaller. Often in a medical practice, positions are either exclusive to one particular area or generalized. If a coding position is specialized, it may include only data entry of codes from a cover sheet, or a position could involve abstracting codes from actual files. Specialization may be confined to coding and billing for a single insurance company or a government agency such as Medicare. Keep the door of possibilities open in your search for employment, and remember that you have many skills that will help you in your job.

CONTINUING EDUCATION

There are many reasons to continue your education, a primary reason being the requirement to earn CEUs to maintain your coding certification. Because of new legislation and requirements, the coding field changes constantly making CEUs necessary to provide credibility and professionalism to the field. Continuing education offers protection to coders, since coders must be able to justify their coding choices and the office duties they perform.

Not only must you maintain and improve your education by using and reviewing current coding books and reference materials, you should also tap into AHIMA and AAPC resources and networking opportunities. Coders should also help develop coding compliance programs for medical practices.

PROMOTE YOUR EDUCATION CONTINUOUSLY!

In addition to CEUs, either before or after earning your national coding certification, you should consider becoming proficient in computers and medical software that includes basic concepts in scheduling, accounting, word processing and databases. Have a good working knowledge of general office procedures, including filing, use of calculators, faxing, and insurance terminology. Because rules and regulations change regularly and address particular specialties, you cannot know everything, but a good solid basic understanding of medical office practices, including billing and coding, will make you a valuable employee anywhere. Continuing education, therefore, is critical to the success of all coders.

Your continuing education also can include earning other certifications within the medical field. Additional education can include Certified in Healthcare Privacy (CHP), an AHIMA certification, Certified in Healthcare Security (sponsored by Healthcare Information and Management Systems Society [HIMSS], administered by AHIMA—CHS), Certified in Healthcare Privacy and Security (CHPS [co-sponsored with HIMSS]), RHIA, and RHIT. There are other organizations that offer certification in different areas. Always check the legitimacy of the organization and its national recognition before committing to a certification program. There are also higher educational institutions that offer degrees and courses in health information management, which may lead to certification, an associate degree or even a bachelor's degree. Your knowledge is yours to earn and own—value it!

Do not forget other opportunities provided to you by your experience and education. Beginning with your success as a coder, make education part of your career goals. Do not limit yourself! Earning your national coding certification, clearly demonstrates that you have the potential to be the best!

SUMMARY

After the exam is the time to evaluate your future—do you need to take the exam again; do you want to take the exam again; what will you do with your certification; what are your future employment and educational opportunities? There are many different opportunities for certified professional coders, as well as for medical office workers. Failure to pass the exam does not exclude you from opportunities in the medical office field. Also consider that there are other ways to approach your future. There are other quality opportunities available to people interested in bettering themselves and furthering their employment and educational opportunities.

Chapter 6   ICD-9 CODING

Objectives

(1) Understand the process of development and purpose of the ICD-9-CM book.

(2) Navigate the structure of the book to find correct codes.

(3) Interpret and utilize proper terminology within the ICD-9-CM book.

(4) Distinguish between tabular and alphabetic indices and use them properly.

(5) Understand how to properly use tables within the alphabetic index and appendices and how they link to the tabular index.

(6) Differentiate between three-, four-, and five-digit codes and how to assign each one.

(7) Distinguish the principal code.

(8) Select which codes should be listed.

(9) Become skilled at analyzing details and codes.

(10) Understand the flexibility involved when coding particularly specialty areas; know coding idiosyncrasies within specialties.

(11) Understand the importance of distinguishing symptoms from diagnoses so that the diagnoses—not the symptoms—are coded.

Key Terms

Alphabetic Index—Index contained within the ICD-9-CM coding book that lists diagnoses and conditions in alphabetic order to assist in the proper selection of codes.

Digits—The characters that compose the code. ICD-9-CM codes can be 3, 4, or 5 digits, counting from left to right. The digits, particularly the fourth and fifth digits, provide greater specificity among similar codes.

E codes—E codes are codes that list the *external causes* of a patient's conditions, diagnoses, and symptoms. They are not required for national exams.

ICD-9-CM—International Classification of Diseases, 9th Revision, Clinical Modification. ICD-9-CM is the codes and coding system used to code for diagnoses, symptoms, or other health issues to demonstrate medical necessity. They range from three digits to five digits and are numeric.

Hypertension Table—A table contained within the alphabetic index, which provides classification for hypertension and related conditions. This table is divided into malignant, benign, and unspecified categories.

Morphology codes—Morphology codes are used to classify neoplasms for statistical purposes. The statistics are maintained by Cancer Registries. These codes are not required for the national exams.

Neoplasm Table—A table contained within the alphabetic index that provides classification of neoplasms by anatomic site and eponyms. This table is divided into benign and malignant. Malignant is divided by primary, secondary, and *in situ*. There are additional categories for uncertain behavior and unspecified. *In situ* indicates that the cancer has not spread, but is confined or noninvasive. Metastasized means that the cancer has spread and indicates that there is both a primary and secondary site(s).

Surgical complications—Surgical complications include a wide variety of services that have occurred as a consequence of surgery, such as the need for replacement or repositioning of an implant or graft, and do not represent negligence necessarily.

Table of Drugs and Chemicals—This table contains four digit poisoning codes in addition to external causes for the use/abuse of drugs and chemicals, which result in medical problems. External causes are divided into accident, therapeutic use, suicide attempt, assault, and undetermined.

Tabular Index—The tabular index is a listing of the actual codes in numeric order with notations that assist with correct coding.

V codes—V codes are used for circumstances other than disease or injury, which are not typically listed in the regular tabular codes. These are generally classified into problems, services, and factual circumstances.

ICD-9-CM CODING BOOK

Medical codes are used not only for billing but also for statistical and research purposes.

ICD-9-CM codes are used for diagnoses, symptoms, or other health issues to demonstrate medical necessity. These codes range from three-digits to five-digits and are numeric. The proper diagnostic codes must be used and must be linked to the correct procedural codes, or medical necessity cannot be demonstrated, and the physician will not be reimbursed.

The ICD-9-CM coding book consists of Volume 1, which is the tabular list of codes (including V and **E codes**), Volume 2, which is the **alphabetic index**. The book also may contain Volume 3, which is for procedures used for hospital billing purposes. Volume 3 also has an alphabetic index. Additional sections of the book include Appendices, which contain **morphology codes**, a glossary of mental disorders, classification of drugs by American Hospital Formulary Service (AHFS) list, classification of industrial accidents according to agency, and a list of three-digit codes. There also is a **table of drugs and chemicals**. These sections comprise the entire ICD-9-CM book.

The ICD-9-CM coding book comes in a variety of shapes and styles. Some books use finger indents for each section, some use tabs, and some use different colors. Some have anatomical and procedural figures. Some books give explanations of procedures and terminology. Books vary in the conventions that are used and portray these in a variety book and page styles, such as coloring or symbols. Most books have symbols that indicate conventions, such as when codes are new or when fourth- and fifth-digits are required. Some books have Volume 2 located in the front of the book; others retain Volume 1 at the front.

TABULAR/ALPHABETIC

Volume 1 is the tabular list and contains the actual codes. Volume 2 contains the alphabetic index. Remember, never code directly from Volume 2 but also check Volume 1. Volume 3 is not used for physician coding and is not on the physician-based national coding exam.

Tabular Sections

The tabular section of the ICD-9-CM coding book is categorized by *Sections, Categories, Subcategories, and Subclassifications*. Sections are groups of common three-digit codes such as Chronic Obstructive Pulmonary Disease, 490–496. Categories are individual three-digit codes (chronic bronchitis, 491). Subcategories are four-digit codes (simple chronic bronchitis, 491.0), and subclassifications are five-digit codes (obstructive chronic bronchitis, with acute exacerbation, 491.21). Each lower level as indicated by an additional digit is contained with the higher level of code (code 491.0 is contained within 491).

DIGITS

It is critical that the proper number of **digits** be provided when coding with ICD-9-CM codes. Some codes are three digits only, some are four, and others are five. However, all the digits required by the code must be used for reimbursement purposes, otherwise there will be no payment for services rendered. Fourth and fifth digits are distinguished by many different criteria, such as anatomic site, state of consciousness, status of pregnancy, or degree of burn. A fourth or fifth digit of 8 usually indicates *other specified* conditions, whereas a digit of 9 usually indicates *unspecified*. An option for *multiple* often is provided within the fourth or fifth digits. Most ICD-9-CM coding books provide references within the tabular, and sometimes the alphabetic, index regarding the need for a fourth or fifth digit. This is a very helpful convention to have when taking the national exams.

INDENTATIONS

When moving between headings, be careful of indentations! They become very confusing in ICD-9-CM as some sections have many indented descriptions and codes. In addition, be sure you write down the correct code. The lines are small and the information voluminous, so it is easy to go to the wrong line when selecting a code. Use a piece of paper or ruler to ensure that you select the code you want and not the one above or below.

Details!

Do not miss fourth or fifth digits! Remember though that codes vary with some only three digits long, some four or five. This is an easy point lost or gained! Most coding books will have symbols indicating if a fourth or fifth digit is required. Code to the greatest specificity that you can. Unspecified codes usually marked with a 9 as the fourth or fifth digit, may be questioned or rejected by payors. Remember, code only what the doctor has documented.

A fifth digit of 8 indicates the *other* classification, which means the description of the service is not included in any other level of that code, but there is information provided regarding the service. A fifth digit of 9 indicates that no information was given and, therefore, the code is unspecified. Because it is best to be as specific in coding as you can be; do not use the fifth digit 9, unless you absolutely have no other information available. If there is not enough information, then you must select an unspecified code that meets the description of services provided. Sometimes *other* and *unspecified* will be contained with the same code.

ACTIVITY 6.1

Either answer the following questions or select the proper codes for the following descriptions.

1. In the alphabetical index, what other main terms can be checked for complications of the heart?

2. Under the category of *pregnancy* within the alphabetical index, is there a *due to* subcategory? If yes, under what category are they listed?

3. What is the code for burns over 31% of the body with 22% being third-degree?

4. How many digits are required to code bacterial pneumonia?

5. For epilepsy with petit mal status, what is the fifth digit?

6. Within the alphabetical index under *complications,* what is the code for infection of vascular catheter?

7. How many digits does the code for splinter in the eyelid have?

8. What other terms could *agenesis* be located under within the alphabetical index?

9. Are pallor and flushing coded together as one code? Why or why not?

10. Describe herniation of nucleus pulposus in other terms.

11. How many digits are required for code 551?

12. Utilizing only the alphabetical index, is V24.0 the correct code for admission for examination of a patient for postpartum status? Why or why not?

13. True or false: Within the alphabetical index, only codes for the pregnancy of the mother are listed.

14. Using the alphabetical index only, what is the fifth digit for the code for management of the mother's care due to fetal distress during delivery?

15. What additional information regarding the fifth digit is provided in the tabular list, which would assist in selecting the proper fifth digit for the code in the previous question?

16. What kind of congenital hiatal hernia is not included within the code 750.6?

17. What is the proper code for myocardial infarction of the basal-lateral wall?

18. What is the proper code for syphilis that is endemic and nonvenereal?

19. What is the proper code for hypertension exacerbated by anomaly of the renal artery?

20. What is the proper code for carcinoma of the connective tissue of the jaw?

TAKE YOUR TIME!

Be sure to read the details! Details refer to those items and words in the code book, for instance, *notes, includes, excludes, see also,* and *see condition*. These *notes* can be found either directly underneath the code, before it, or after it. They also can be found under the original heading so you must check backwards until you reach the main heading. Often, after these statements, you will be directed elsewhere to select the proper code. Some *notes,* such as *includes,* are not all inclusive, meaning that the *notes* listed may not include all conditions or situations when the code might be applied. *Excludes,* however, are decisive, such as when a code is excluded, it absolutely cannot be used under presenting circumstances. Sometimes codes will be described as *not allowable* when certain other codes are listed, which is known as *mutually exclusive codes*. It is important to read the *excludes,* because some conditions are not coded with similar conditions, but instead may have their own codes or be in another section. For example, pregnancy conditions have separate coding, which is found in its own section.

Being detailed does not mean that you must know every little detail. Consider yourself a researcher investigating the details contained within the code book information, such as notes and conditions. It is more important to understand the basics of reimbursement well than to attempt to know everything that is written about reimbursement, including coding. You cannot know it all! Learn the basics well, which means understanding the format of the code books, their terminology, their structure, their indexes, etc. Do not attempt to know everything that has ever been written. Codes, insurance contracts, and government regulations change constantly, and no one, including you, can know it all. As mentioned previously in this book, you do not need to know everything to pass the exam. You only need a passing score to pass!

In fact, there is the possibility that some of the questions or coding scenarios may actually be easy. Don't lose your cool if you actually find some questions easy!!

There also are statements and instructions that are critical to read! If you see statements such as *use additional code if desired* or *use additional code to further specify complication* before or after but within the coding category, you must do so.

You should rely upon the precise details of the alphabetic index. A good example of this is the requirement to use certain multiple codes together. When looking up a code, you may find it listed with another code that appears in slanted brackets. This means that you must list the two codes together in the order shown. Follow the ICD-9-CM's sequence for codes when listed. Another example is when you look up a term in the alphabetic index, which is referenced under another term. Sometimes when you reference a code in the code section of the book, it may appear inappropriate. It is at these times that you trust the alphabetic index.

In hospital-based coding, you must learn what is a symptom and what is a diagnosis, and then know what symptoms are a part of which diagnoses. The symptoms should not be coded in addition to the diagnosis that contains them. However, if symptoms are not a part of that diagnosis, then they would be coded in addition to the diagnosis code. This can become difficult, as sometimes a condition may or may not be coded in addition to another condition depending on their relationship and the structure of the coding book. Hospital coding also is difficult because terminology may further describe a diagnosis, in different terminology than the coding book. For example, status asthmaticus can be described as intractable or refractory.

Conventions

It is critical that you understand the conventions used in your book. To better understand your book, familiarize yourself with it before you take the exam. To achieve this, use your ICD-9-CM coding book when studying.

Most ICD-9-CM coding books now provide a notation next to a code informing coders if an additional digit, such as a fourth or fifth digit, is required. Some fourth and fifth digits, the complete code and its description may be provided immediately below the three- or four-digit code. However, the digits also may occur earlier in the listings, so be careful when checking for the extra digits.

Brackets have several different meanings depending on which volume of the ICD-9-CM book they appear in. Slanted brackets in the alphabetic index indicate that two codes listed together must be coded together in the exact order they appear. Regular brackets within the tabular volume can indicate which fifth digits are acceptable.

ACTIVITY 6.2

Provide all required codes for the following, or answer the question.

1. Why or why not would medical services not offered in the home, because the spouse is unable to provide the services, be coded as V63.1?

2. What is the proper code(s) for scoliosis due to rickets?

3. What is the proper code for calcinosis intervertebralis?

4. Bright's blindness should be coded as what?

5. Why would acute pharyngitis not be coded as 460?

6. Hardening of the arteries should be located within the alphabetical index under what terms?

7. For coding Paget's disease with osteosarcoma, where specifically would you have to look?

8. If a patient has pneumonia due to AIDS, would V08 be the correct code? Why or why not?

9. What else must be coded when using the code for disorders relating to extreme immaturity of a baby?

10. What definition is provided regarding the code for disorders relating to extreme immaturity of a baby?

11. What is the correct code(s) for parasitic conjunctivitis due to filariasis?

12. In addition to the code for complications, what additional code should be coded for cytomegolovirus (CMV) infection due to kidney transplant?

13. Patient has arthropathy due to erythema nodosum. What is the proper code(s)?

14. For the dependent personality disorders code, what is excluded?

15. What codes are mutually exclusive with 496?

16. What types of codes are excluded from the category for diffuse diseases of connective tissue?

17. Dislocation can include what other conditions?

18. If a patient has chronic renal failure and heart failure due to hypertension, what would be the code(s)?

19. If a patient has a dislocation of the proximal end of the ulna, this should be coded as what?

20. What is another description for Jadassohn-Lewandowski syndrome?

Alphabetic Index

The alphabetic index comprises Volume 2. Your book may have it located at the front, before Volume 1 (the tabular list of codes), instead of after Volume 1. The alphabetic index also has several tables located in it, such as hypertension and neoplasm, which are very useful in locating codes.

DO NOT CODE DIRECTLY FROM THE ALPHABETIC INDEX!

Although the ICD-9-CM alphabetic index is thorough and precise, you cannot code directly from it. You must use the tabular code listing to ensure that you have all the available information, such as excludes and digits, to make the proper code choice.

The ICD-9-CM alphabetic index is rich with precise and descriptive directions. You must follow these directions closely as the directions sometimes will specify order or other places to find correct codes.

Sometimes during your primary search in the alphabetic index, you may be unable to find the main diagnostic term you have selected. Select the next closest diagnostic term. If you still are not able to find the proper diagnosis, continue to select terms not listed in the diagnosis but associated with it, until you find a code that is a good description of the diagnosis. Sometimes, you may even need to use a term that is not correct but will put you in the area of coding where you may find the correct code.

Modifying Terms

Modifiers in ICD-9-CM are not the same as modifiers in the CPT. Modifiers in ICD-9-CM refer to terms that provide information to help code correctly. Modifiers can be nonessential in which case they usually are listed after the code in parenthesis and provide additional information about the code. These modifiers are not necessarily all inclusive; there may be additional words that describe the diagnosis which are not listed. It also means that if the physician's diagnosis does not include certain nonessential words, it cannot be used.

Although modifiers do not necessarily exclude certain codes from being selected, it should be noted that some modifiers, known as *connecting terms,* are critical for selecting the correct codes. Connecting terms include words such as *with, due to,* or *and.* These modifiers indicate that other codes may be needed to identify the condition with which the modifier is used.

Not otherwise specified (NOS) and *not elsewhere classified* (NEC) are used in the ICD-9-CM coding book. NOS is for classification of *unspecified* codes, where no information is specified regarding the code. NEC is for classifications where specific information is given, but the specific information is not listed in the index. They are usually indicated by a fourth or fifth digit of 9. It is preferable not to use these codes because they are so vague, and it is always best to be as specific as possible.

Follow directions. *See also* or *see* are known as *cross-references.* You should check all cross-reference listings.

Other phrases include *use additional code, if desired.* This phrase means you must provide the additional code, if enough coding information is available. *Code first underlying disease* means you should code the disease that caused the symptoms or diagnosis first. This is known as coding the *etiology of the condition.* In addition, you must code the manifestation that results from the etiology.

ACTIVITY 6.3

Answer the following questions or provide the correct code(s).

1. Which of the following codes are listed as NEC: V67.59, 528.5, 759.4, 062.8, and 312.0.

2. What is the code for vitamin deficiency NEC?

3. What are some modifying terms that describe pneumonopathy due to inhalation of other dust?

34   Chapter 6

4. What is the proper code(s) for benign ovarian neoplasm with hyperestrogenism?

5. What are two modifying terms for infarction of the lung?

6. Under what other term can Phagedena be found?

7. What conditions are included in code 722.11?

8. What is the code(s) for severe meningococcal sepsis with acute renal failure?

9. What is excluded with the code for adhesive middle ear disease?

10. What is the code(s) for drunkenness as a toxic effect of alcohol?

11. Fistula of the peritoneum also can be located within the alphabetical index under what term?

12. Infection resulting from a shunt can be located under what terms in the alphabetical index?

13. What are some modifying terms for polymyositis?

14. What is the code for acute rheumatism NEC?

15. What is the correct code(s) for paratyphoid fever with postdyseneteric arthropathy?

16. What is the code for lymphoma NOS of the lymph nodes of the axilla?

17. To what coding categories does the fifth digit of code 719 apply?

18. What is the proper code(s) for urinary tract infection due to *E. coli*?

19. What is the proper code(s) for psychogenic psoriasis?

20. Under what other term(s) can descriptions of *overactive* be found within the alphabetical index?

Nonspecific Terms

Some terms, such as condition, disease, or problem, may seem insignificant, broad, or general but they are listed in the alphabetic index and are used to refer to more specific terminology. If the notation, *See Disease* (or condition, etc), is used, then you must search using terminology that is more specific, that does not mean *condition* or *disease* necessarily. It means look for the definitive term, like arrest for cardiac arrest. When you are directed to go to another term, look for any specific information listed in the code. However, if no terms within the diagnostic statement are listed, simply pick the main general code. Other nonspecific terms that are used include due to, with, management, affected by, etc. Although a term may be classified under itself, some, such as *injury, cervix, with,* may be classified under broader terms first, which can be confusing.

Never underestimate or assume general terms. You may think a term is too general and therefore would not be located in the alphabetical index or would not be provided a code, but that may not be true. To locate terms, begin with the most specific major term, then proceed to more general and/or minor terms. Never assume, however, that general or broad terms would not be included in the alphabetical index. Always check it out!

ACTIVITY 6.4

Answer the following questions.

1. Pain in the ear is known as what?

2. What is the correct code(s) for a patient who has chronic and acute bronchitis with chronic obstructive pulmonary disease (COPD)?

3. What is the code(s) for incompetent veins with varicosity?

4. When selecting the proper code for cardiac failure by looking under *cardiac,* is it correct then to check under the term *condition* as directed in the alphabetical index? Why or why not?

5. Which of the following subcategories are listed within cesarean delivery in the alphabetical index: malpresentation, dystocia, bicornis, pre-eclampsia, metrorrhagia, fetal distress, incarceration of uterus, pelvic arrest, and premature labor.

6. Is airplane glue sniffing considered an *effect*? Why or why not?

7. How is *mass* specified within the alphabetical index?

8. Under what terms in the alphabetical index can you find the effects on the mother of a Down syndrome fetus?

9. What is the code for a sensory problem with the chest?

10. Codes for labor can also be found under what other term?

11. For fitting of a cardiac device (pacemaker), within the alphabetical index which terms would be considered general?

12. Is vomiting considered a *habit*? Why or why not?

13. Under what terms in the alphabetical index can you find a tear in the cervix due to a legal abortion, beginning with the word *tear*?

14. Under what terms would you find the code for crushing chest injury within the alphabetical index?

15. What would be the proper code(s) for the previous description?

16. What is Sheehan's disease?

17. Under what category(ies) is *laceration* listed within the category for *injury*?

18. Regarding the absence of the ulna with incomplete humerus, does this include complete absence of distal elements? Why or why not?

19. Under history in the alphabetical index, code V17.8 is listed for *musculoskeletal disease*. Is this the correct code for a patient with a history of musculoskeletal disease? Why or why not?

20. What is the code(s) for multiple fractures with hemorrhage?

LINKING

Diagnostic codes must support the services (ICD-9-CM Volume 3, CPT, and HCPCS codes) provided, which is achieved through proper code linkages Every coding scenario must have an ICD-9-CM code, which must be linked to the service provided.

PRINCIPAL AND PRIMARY DIAGNOSIS

A *principal diagnosis* is the main reason the patient receives inpatient medical care. For outpatients, the main reason a patient seeks medical treatment is known as a *primary diagnosis*. For inpatient services, the principal diagnosis is the main reason for admission after testing and evaluation. For outpatient services, the primary diagnosis is the main reason services are provided that day.

WHAT TO CODE

Some codes contain descriptions that include several diagnoses or conditions. List these codes only if all of the information is provided in the written report. Otherwise, use a code for each diagnoses and/or condition to ensure that a full description is provided.

Code to the highest degree of diagnostic certainty, especially with fourth and fifth digits. Specificity also refers to the diagnostic statement. The diagnostic code, which is the main reason the patient is receiving services, is the primary code and should be listed first. This does not mean, however, that this code is necessarily the patient's most serious condition. In addition, if a diagnosis is not stated, then you must code the symptoms. But if you have a diagnosis, then you do not code the symptoms. Also, particularly with inpatient stays, further workup may provide a more definitive diagnosis than a previous one. The most specific diagnosis should be coded instead.

According to the Uniform Hospital Discharge Data Set (UHDDS), the main reason a patient is admitted and seen is listed as the principal diagnosis for inpatients. For inpatient services, patients may be diagnosed with additional conditions, known as *incidental* or *secondary*, while in the hospital. But incidental or secondary conditions are not the main reason for admittance, so they are not coded as the principal diagnosis even though they may be a more serious threat. For outpatient services, the main reason for providing services is coded as the primary diagnosis. Do not code conditions that no longer exist, even if they are referred to in the documentation.

If there are no diagnoses or symptoms, **V codes** are used to indicate that there were no abnormal findings. For example, *worried well* is a V code that is used for visits that do not indicate a diagnosis or symptoms.

SECONDARY DIAGNOSES

Your first priority when coding diagnostic codes is to determine the main reason the patient was seen by the physician. Additional codes also may be listed for coexisting or ongoing conditions, such as chronic diseases, that are either part of the current care or may influence the outcome of the care. However, do not code symptoms if you have already coded the diagnosis to which the symptoms apply. Chronic conditions, including diabetes, COPD, and hypertension, usually should be coded, however, as they may influence a patient's care.

History codes should be coded only if they are relevant to the patient's care. For example, if a patient has a superficial laceration repaired in the office, there is little concern about anything else. However, if the patient is having a pacemaker implanted and had an angioplasty six months before, additional codes could be included for the postsurgical status of the angioplasty.

ACTIVITY 6.5

Answer the following questions. When providing codes, be sure to list them in proper order.

1. What is the code for the principal diagnosis for a patient who is admitted to the hospital for severe

abdominal pain? During hospitalization, the patient experiences a myocardial infarction (MI).

2. What is the code for the principal diagnosis for a patient who is admitted to the hospital for severe abdominal pain? Tests reveal that the patient has appendicitis. During hospitalization, the patient experiences an MI.

3. What is the code for the principal diagnosis for a patient who is seen at the outpatient clinic for severe abdominal pain? Tests reveal that the patient has appendicitis. During hospitalization, the patient experiences an MI.

4. What is the principal diagnosis for a patient who is seen in an emergency room for acute pulmonary edema? The patient is prepared to be transported to the hospital for possible pulmonary embolism with infarction?

5. What is the principal diagnosis for a patient who is seen in an emergency room after being injured in a car accident? Patient sustained a large laceration on his left arm and left leg, which required sutures. During this visit, patient complained of epigastric pain and tachycardia, which he states has been ongoing prior to the accident. Tests revealed the patient had a perforated ulcer for which he was treated.

6. What is the principal diagnosis for a patient seen in the hospital who sustained a minor head wound when hit by a car while on his bicycle? At this time, the patient complains of nausea and abdominal pain, which tests reveal is due to a perforated ulcer. Treatment is provided in the hospital at this time for the perforated ulcer.

7. What is the principal diagnosis for a patient who is admitted to the hospital for paresthesia in the fingers of his hand? After evaluation and tests, she is diagnosed with angina.

8. What is the correct code(s) for a patient who is seen today in the emergency room for chest pain and diaphoresis? The patient is diagnosed with obstructive cardiomyopathy. Patient's medical history indicates that she had a hysterectomy 3 years ago and 6 months ago received a left heart catheterization.

9. What is the correct code(s) for a patient who had a failed abortion and is now experiencing abdominal tenderness, fever, tachycardia, and vomiting? After testing and evaluation, the patient is diagnosed with pelvic inflammatory disease (PID) and sepsis.

10. In the alphabetical index, under what terms do you find the code from the previous question?

11. What is the correct code(s) for a patient who has a 30-year history of alcoholism and is seen as an outpatient for hepatomegaly and edema? Patient subsequently is diagnosed with cirrhosis?

12. What is the correct code(s) for a patient who is evaluated in the hospital after being transferred from urgent care with severe heartburn, difficulty eating, and possible hiatal hernia.

13. What is the correct code(s) for a patient seen in the emergency room for facial pain, multiple sclerosis, and nystagmus?

14. What is the correct code(s) for a patient, who, as an outpatient, had a hysterectomy a year ago for carcinoma of the uterus in addition to chemotherapy a year ago and is diagnosed now with metastatic carcinoma of the kidney?

15. What is the principal diagnosis for a patient admitted to the hospital for dizziness and palpitations? After evaluation and testing, the patient is diagnosed with renal artery stenosis. On the third day of hospitalization, the patient begins to demonstrate edema and is found to be in respiratory distress.

16. What is the principal diagnosis for a patient who was seen in an Emergency Room for dizziness and palpitations with possible renal artery stenosis but is found to be in respiratory distress. Care is provided for stabilization, and the patient then is transferred to the hospital.

17. What is the correct code(s) for a patient who is seen as an outpatient for paresthesia in the right arm in addition to anhidrosis and hematuria? The patient is diagnosed with peripheral neuropathy.

18. What is the correct code(s) for a patient who is seen in an Emergency Room for severe migraine headaches and vomiting?

19. What is the primary code for a patient who is seen as an outpatient with dysuria, fatigue, and back pain? Lab tests reveal cystitis due to Chlamydia. The patient is diagnosed at this time with carcinoma of the lungs as shown on x-rays.

20. What is the principal code for a patient who is seen in the hospital for dysuria, fatigue, and back pain? Lab tests reveal cystitits due to Chlamydia. The patient is diagnosed at this time with AIDS, for which the patient is provided treatment and counseling.

ANATOMIC SITES

Beware of body parts! Upper extremities are the arms; lower extremities are the legs. Some anatomic sites will be distinguished within the fourth and some within the fifth digit. Some codes are more inclusive and separately list more anatomic sites, like knees and wrists; others do not list them. Be particularly careful when coding for knees, elbows, wrists, and ankles, as they are sites where two other anatomic sites join. In addition, some categories of codes list further delineations of anatomic sites, such as shaft. *BEWARE!* Different parts of an anatomic site may be listed separately, like upper and lower ends or shaft. Often the codes are explanatory and the use of fourth or fifth digits is demonstrated easily. However, if not, the anatomic site can be located in the alphabetic index either under the condition or the site itself. For example, with fractures or knee problems, coding directions may be provided regarding the condition at that anatomic site, and if the problem is included as part of another anatomic site or by itself.

Sometimes a digit is provided for multiple sites, sometimes not. If a site does not include a digit for multiple sites and multiple sites need to be coded, then you must list the code several times with differing digits for each different anatomic site.

ACTIVITY 6.6

Answer the following questions:

1. If a patient has chronic osteomyelitis of the forearm and upper arm, what would the fifth digit be?

2. How is the anatomic site for carcinoma of the skin *in situ* applied within the codes?

3. What is the correct code(s) for carcinoma of the trachea with metastasis to the duodenum and peritoneum?

4. For dislocation of the wrist, is there one code available? Why or why not?

5. Which digit is used to denote the anatomic site for a foreign body within the eye that happened 3 months ago?

6. For cellulitis, which code and digit indicates the anatomic site?

7. What is the correct code(s) for fracture of the tibia and fibula?

8. Is aseptic necrosis of the bone categorized by anatomic sites for coding? Why or why not?

9. What is the correct code(s) for bilateral partial paralysis of the vocal cords?

10. Are V codes ever differentiated by anatomic sites? Why or why not?

11. What is the correct code(s) for ulcers of the heel and midfoot?

12. What anatomic sites does 441 describe?

13. What is the correct code(s) for a cerebral embolism?

14. Is the code for rotator cuff syndrome of the shoulder further defined by anatomic site(s)? Why or why not?

15. For fracture of the tibia, what digit indicates the anatomic site?

16. Do three digits ever specify anatomic sites? Why or why not?

17. What is the correct code(s) for lymphoma of the spleen and intra-abdominal lymph nodes?

18. Is the code for injury to the peripheral nerve of the upper limb further specified by anatomic site? Why or why not?

19. What is the correct code(s) for sphenoidal sinusitis?

20. What is the correct code(s) for laceration of the esophagus?

POSSIBLE, PROBABLE, RULE OUT

For outpatient coding, which is similar to physician coding, *possible, probable, rule out,* etc., do not exist and should not be coded as a definitive diagnosis. These are known as *qualified diagnosis. Versus* and *either/or* are coded in the same way. If a physician wants such statements coded as a definite diagnosis, explain to the physician that uncertain terms cannot be used; a definitive diagnosis must be stated as a certainty without terms such as possible or probable.

For inpatient coding, *rule out, possible, probable, suspected,* or other similar terms may be coded as a confirmed diagnosis because of the extensive nature of the testing, workup, and evaluation that is performed in the hospital. However, *ruled out* (that is with a "d") is not coded as a confirmed diagnosis because the past tense of *rule* indicates that it was determined that the diagnosis does not exist. In the case of ruled out, the symptoms should be listed instead. This same scenario applies to *either/or* or *vice-versa* coding, when two or more diagnoses are provided and both should be coded. The diagnosis that applies to the symptom that prompted the patient visit should be coded first.

However, in hospital coding, if there are suspected conditions, such as *versus* or *either/or,* but a symptom also is listed, the symptom is coded first, then the other comparative diagnoses coded, if possible.

If conditions are described as *threatened* or *impending*, then these should not be coded, unless the physician's report indicate that it is a definite diagnosis. If it is not reported as definite, then the condition prior to the threatened or impending diagnosis should be coded.

ABNORMAL LABS

Abnormal lab findings are not coded, unless the physician provides a conclusive description of their significance in the written report. Organisms must be coded if relevant to the diagnosis, as directed by the ICD-9-CM book. Some conditions will be coded with the organism as one code, but others (both the condition and the organism) will have to be listed separately.

Regular and V codes can be used to code abnormal findings if no other information or more definitive diagnosis is provided. These codes are located in sections such as 790 and V71.

ILL-DEFINED FINDINGS

Sometimes the patient's condition is not diagnosed for a variety of reasons, including the patient does not follow up with the physician, the patient leaves against medical advice (AMA) before a complete evaluation can be performed, patient expires, etc. There are specific codes used for findings that are not well-defined. Findings indicated by nondescript terminology include *abnormal, unspecified, positive, elevated*, or *generalized*. These nondescript terms should be coded only when there are no other definitive diagnoses. Check for more information before using these codes. Although you may find generalized codes throughout the coding book, there is a section of codes—780 to 799—specifically for these general codes.

FOLLOW-UP

Remember, a patient may be receiving physician services not for acute conditions but as follow up to prior services. Follow-up usually signifies that a patient has a medical history of a particular condition and should be coded using a V code. If the doctor indicates that a visit is for a follow-up exam, there cannot be a current diagnosis. Follow-up medical care implies that the patient was treated and the condition no longer exists.

History, however, does not always equate with follow-up. A patient may have a history of a condition such as cancer and still have cancer. This is the case with recurrences of conditions such as cancer or heart problems. Sometimes a physician may use terms improperly, such as *history* (for instance, history of diabetes); however, the patient's diabetes would still be ongoing. It is your duty as a coder to educate physicians and staff about proper wording and documentation to ensure that codes are properly selected. In a confusing scenario, such as with histories, you should check thoroughly through the reports or ask the physician. Keep in mind, you cannot code something extra that provides additional reimbursement if the physician did not document it in the original report. Clarifying confusing or omitted terminology that is absolutely critical to select the correct code is not the same as upcoding without proper documentation.

ACTIVITY 6.7

Answer the following questions.

1. What is the proper code(s) for a patient who is admitted to the hospital for tests and evaluation for complaints of lymphadenopathy, fatigue, and weight loss with possible enteritis?

2. What is the correct code(s) for a patient seen in the Emergency Room who is experiencing paresthesia of the left leg? The patient has a history of poliomyelitis.

3. For an injury to the blood vessels of the head, to what is the unspecified code referring and which digit refers to the unspecificity?

4. What is the correct code(s) for a patient who is admitted to the hospital for bronchitis complicated by emphysema? In addition, the patient has a history of COPD.

5. Under what terms is ischemia of the heart that occurred 7 weeks ago coded as?

6. What is the correct code for a patient who was injured while riding his bicycle? He sustained a fracture of his tibia. Patient is asymptomatic for HIV infection.

7. What is the correct code(s) for a patient who is admitted to the hospital with complaints of chest pain, tinnitus, and hearing loss? Patient was given diagnoses of possible otorrhea and Wegener's granulomatosis.

8. What is the correct code(s) for a patient who is seen in the Emergency Room for cramping and diarrhea, with suspected Crohn's disease?

9. What is the correct code(s) for a patient who is seen as an outpatient where he is diagnosed as having either cirrhosis or cholelithiasis?

10. What is the correct code(s) for a patient who is seen in the Emergency Room for fever, headache, and rigidity of the neck? Lab reveals presence of pneumococcus.

11. Describe the unspecific terms for the category that includes infectious diarrhea.

12. What is the correct code(s) for a patient who is admitted to the hospital for chest pain and coughing, which is diagnosed as pulmonary embolus versus heart failure?

13. What is the correct code(s) for arteriosclerotic vascular disease NOS?

14. What is the correct code(s) for a patient who is admitted to the hospital for Kaposi's sarcoma of the palate due to AIDS?

15. What is the correct code(s) for a patient who is seen in the hospital for suspected MI? EKG reveals abnormal readings.

16. What is the correct code(s) for a patient who is seen as an outpatient for suspicious nodular lumps of the arm, possibly carcinoma? Tests were inconclusive.

17. What is the correct code(s) for a patient who is seen in the Emergency Room for severe lower abdominal pain and dysuria? The patient has a history of a brain tumor and urinary calculi.

18. Name some unspecified malignant neoplasms of the histiocytic tissue.

19. What is the correct code(s) for a patient admitted to the hospital for fracture of his tibia when he fell from a roof? Patient has been diagnosed with AIDS but is asymptomatic at this time.

20. What is the correct code(s) for a patient who is seen in the hospital for shortness of breath due to carcinoma, but patient leaves AMA shortly after admission?

THERAPY CODES

If a patient is receiving therapy for an existing condition, usually the diagnosis is coded first, except with radiotherapy, chemotherapy or rehabilitation, which are coded first from the V codes and then the diagnosis is coded second.

ACUTE AND CHRONIC

Sometimes acute and chronic condition require the use of two codes, one code for each, and at other times one code includes both the acute and chronic conditions. *Read the details!* Remember, acute is a condition that is current. Chronic is an ongoing condition that has existed for an extended time period. This is an important distinction. As with all coding, you must code only what you know from the medical reports, what is occurring at that time, and what is relevant to patient care. Chronic conditions not relevant to patient care should not be coded. Keep in mind that some chronic conditions affect other conditions. For example, diabetes can affect the vascular system, eyes, heart, etc.

Acute and chronic are displayed in several ways in the ICD-9-CM coding book. Sometimes they are distinguished as acute and chronic in the alphabetic index, but not always. They may be distinguished as acute and chronic in the tabular lists of codes. Acute and chronic sometimes are listed in the modifiers immediately following the term for the condition in the alphabetic index; sometimes they will be listed later, underneath the code under the term *acute* or *chronic*. If the codes distinguish between acute and chronic, then the correct code needs to reflect the one specified in the report. If the condition is not specified as acute or chronic but this distinction is necessary to code properly, then select the code for *unspecified*. Read the details! It is not necessary to memorize anything, just read the details!

If both codes are needed, the acute code must be listed first and the chronic code listed second. Both codes must be listed if both acute and chronic are contained within the description and are not both contained within one code.

You must code as the physician has reported—acute or chronic. Do not diagnose the patient yourself! If you do not know whether it is acute or chronic from the reports, then code as unspecified. As mentioned above, if the codes are separate for acute and chronic, the acute code must be listed first. If acute or chronic are not distinguished, select the general code listed in which case, there will be no distinction for acute or chronic.

ACTIVITY 6.8

Answer the following questions.

1. What is the correct code(s) for a patient who is seen in the emergency room for acute pneumonia due to severe acute respiratory syndrome (SARS)-associated coronavirus?

2. What would be the code if the diagnosis in the previous question was chronic?

3. Is it true that the only categories for thyroiditis are chronic and acute? Why or why not?

4. What is the difference between acute and chronic diabetes?

5. What is the correct code(s) for a patient admitted to the hospital for chronic ischemic heart disease?

6. Is code 416.8 correct for chronic pulmonary heart disease? Why or why not?

7. What is the correct code(s) for a patient who is seen in outpatient for chemotherapy with diagnosis of pineoblastoma?

8. Is code 590.80 the correct code for acute pyonephrosis? Why or why not?

9. What is the correct code(s) for acute necrotizing ulcerative stomatitis?

10. Is code 743.00 the correct code for chronic anophthalmos? Why or why not?

11. What is the correct code(s) for acute and chronic mastoiditis?

12. What is the correct code(s) for a patient who is seen in outpatient for radiation treatment for malignant melanoma?

13. What is the correct code(s) for chronic appendicitis?

14. What is the correct code(s) for acute and chronic gastroenteritis?

15. What is the correct code(s) for a patient who received chemotherapy for Schwannoma and returns 3 months later, for follow-up?

16. What is the correct code(s) for chronic venous insufficiency?

17. What is the correct code(s) for recurrent cyclitis?

18. What is the correct code(s) for a patient admitted to the hospital with chronic meningitis due to Lyme's disease?

19. What is the correct code(s) for a patient admitted to the Emergency Room for chronic nephritis and systemic lupus erythematosus?

20. What is the correct code(s) for patient seen as an outpatient for chronic bronchitis with acute exacerbation and obstruction?

LATE EFFECTS

A *late effect* is the residual condition resulting from a previous condition. Late effect codes are useful, particularly for reimbursement, for these codes provide more information. Regarding late effects, a distinction must be made between what is being billed for currently and what is a consequence of a past diagnosis or condition that has been billed previously. The current diagnosis or condition is known as a *residual effect*. Remember, you cannot charge for a service that was billed for already, so you code for the residual effect. Residual or late effects are helpful in proving medical necessity for current charges. Terms that signify a late effect include *sequelae, residual,* or descriptions that signify elapsed time between the original diagnosis and the current one such as *malunion* or *scar tissue.*

The current condition should be listed first and the late effect code should be listed second. Select the late effect based on the previous condition, and not the current condition. For example, scar tissue and late effect of burn would be the correct codes for a coding scenario in which a patient had fractured his/her arm several months ago and then returned for correction of the malunion. You would look under fracture for the late effect. This also is located under *late (effect) of fracture*.

An ongoing condition such as a chronic condition cannot be coded as a late effect since it still exists. An example of such a condition is osteoarthritis.

Late effects can be found easily in the alphabetic index under *late,* although this section is not long or very comprehensive. Sometimes you may have to find the closest code making sure it says *late effect of* so that you are not billing for it at the present time. This type of code is a supplement to your main code. Do not select the codes in the parenthesis that are listed in the late effect section of the alphabetical index. The codes in the parenthesis do, however, provide you more information to ensure that you are selecting from the proper codes. *Remember*—late effects codes are not enclosed within the parenthesis. The codes within the parenthesis are the condition or diagnosis codes that should have been used when the condition first appeared. These codes would not be used now because late effects are being coded.

Cerebrovascular accident (CVA) late effects are coded differently from other late codes. They are coded from the 438 category and contain both the late effect and the cause of the late effect within the one code. This is in contrast to other conditions of the brain, such as trauma, that would have a 900 level late effect code along with the current symptom(s), such as 907.00 for intracranial injury with CVA late effect codes. CVA late effect codes include one code that is a description of the late effect, such as flaccid hemiplegia, which does not need to be coded additionally. One code covers both the residual effect (hemiplegia) and cause (CVA).

ACTIVITY 6.9

Answer the following questions.

1. What code would have been the original code used for a third-degree burn of the left arm that occurred 6 months ago?

2. What is the correct code(s) for a patient who is exhibiting facial droop due to a CVA 5 months ago?

3. What is the correct code(s) for a patient who is experiencing seizures due to viral encephalitis 6 months ago?

4. What is the correct code(s) for a patient with sequelae from an injury to the brachial plexus nerve?

5. What is the correct code(s) for repair of scar tissue on the face due to third-degree burns?

6. What is the correct code(s) for malunion of fracture of the ulna?

7. What original codes would have been used to code for a current condition that is now described as a late effect of a sprain of the supraspinatus tendon?

8. What is the correct code(s) for a patient who is diagnosed with dysphasia due to a CVA three months ago?

9. What is the correct code(s) for the late effect of surgical amputation?

10. What other term is used in the coding book to describe late effects of encephalitis?

11. What is the correct code(s) for a patient who is experiencing paresthesia in the toes after suffering a fracture of the fibular and tibia four months ago?

12. What is the correct code(s) for a patient who presents today with an infection after a cesarean delivery 2 months ago?

13. What original codes would have been used to code for a current condition that is now described as a late effect of colitis due to *Entamoeba histolytica*?

14. What is the correct code(s) for a patient who experienced a concussion with loss of consciousness for 45 minutes a month ago and is now experiencing severe headaches?

15. What is the correct code(s) for a patient who is experiencing phantom limb pain?

16. What is the correct code(s) for a patient who is experiencing effects from a skull and face fracture two months ago?

17. What original codes would have been used to code for a current condition that is now described as a late effect of a gunshot wound to the left hand?

18. What original codes would have been used to code for a current condition that is now described as a late effect from a laceration of the lung due to being crushed at a concert?

19. What is the correct code(s) for a patient who has experienced hydrocephalus due to an intracranial abscess?

20. What is the current code(s) for a patient who has experienced sequelae due to an injury to the axillary nerve of the shoulder?

MULTIPLE INJURIES

Code the most severe injury first. Beware of codes that include several sites or contiguous sites (adjacent). These may be specified as one code for *multiple sites*. Other codes may list each site separately. If sites are listed separately, then you must code each one separately. Watch those fourth and fifth digits!

The conjunction, *with*, indicates involvement of all listed sites. *And* indicates that either or both sites can be involved; in other words, *or* can be included with *and* (i.e., and/or). Whereas only one site may need to be listed to use a code defined with the word *and*, all sites must be listed in the diagnoses or conditions to use a code that has the word *with*.

COMBINATION VERSUS MULTIPLE CODING

Don't get combination codes confused with multiple coding. They are definitely different! A combination code includes descriptions for several diagnoses or conditions in one code. It is important to read the notes (such as exclusions and inclusions) when selecting combination codes, since reference may be provided. Other terms to watch for that indicate combination coding are *complicated by, due to,* and *with*.

Remember, if, in its description, a code includes all of the information for more than one diagnosis and/or condition, then you do not need to code anything more.

With multiple coding, more than one code must be coded so all of a patient's conditions are included. Order is important with these codes. Sometimes these codes are listed in the alphabetical index in their specified order. The ICD-9-CM alphabetic index and tabular list of codes are precise, and the rules need to be followed closely. Some codes require the use of other codes to further distinguish the patient's condition. For example a patient with AIDS who is hospitalized with pneumonia would have both AIDS and the pneumonia coded. Keep in mind that with -inclusive codes such as AIDS, only ongoing conditions influencing the current care are coded in addition to the AIDS code (042).

Other multiple codes can be determined by notes such as *code first, and, code also,* or *use additional code* in the ICD-9-CM coding book. Read your notes! Although some terminology might be the same for both combination coding and multiple coding (such as due to), the difference lies in how a code is listed in the code book. If there is a relationship between two conditions, but there is no one code that includes both, then multiple coding would be used to ensure that both conditions were included. Etiology, or the underlying causes, should be listed first and the manifestation, or result, should be listed second. Within the alphabetic index, multiple codes also may be indicated by brackets, which should be coded in the order they are listed.

At this point it is a good idea to examine the term *with* more thoroughly as its use can be particularly confusing, and it can be used improperly. In the code book, *with* means that there is a relationship between two codes, as discussed in the previous paragraphs. If one combination code can be used, then it must be used to include both conditions described by *with*. For instance, find the word *disease,* then the word *heart*. Immediately below heart is *with*. Now find *hypertensive*. Notice that the adjective hypertensive is listed, and not *hypertension*. This is because *with* implies a relationship between the hypertension and the heart disease. As you look further, you will find the term *arteriosclerotic*. Once again, the adjective is listed as arteriosclerotic. This, too, implies a relationship between the arteriosclerosis and the heart disease. However, if the diagnoses do not list the hypertension or the arteriosclerosis as having a relationship with the heart disease, then these categories for codes cannot be used, and the arteriosclerosis or hypertension would be listed separately in addition to the code for heart disease. Be careful and do not code conditions in a causal relationship unless the physician clearly indicates that they are causal. If not clearly stated by the physician, the conditions are not considered causally related. Equally confusing is the use of the word *and,* which can sometimes imply a causal relationship of *with,* which is indicated by multiple coding statements.

ACTIVITY 6.10

1. What is the correct code(s) for gastroenteritis due to Salmonella?

2. Does a patient have to have both heat stroke and sunstroke to qualify for code 992.0?

3. Does a patient have to have hyperosmolality and hypernatremia to qualify for code 276.0?

4. What is the correct code(s) for a patient diagnosed with subacute sclerosis and pernicious anemia?

5. What is the correct code(s) for a six-month pregnant patient diagnosed with urinary tract infection (UTI)?

6. What is the correct code(s) for a patient diagnosed with cardiopathy due to beriberi?

7. Must all digestive organs and spleen be diagnosed with neoplasm in order to use code 197.8? Why or why not?

8. What is the correct code(s) for a patient diagnosed with for eosinophilic meningitis?

9. What is the correct code(s) for a patient diagnosed with retinal microangiopathy?

10. What is the correct code(s) for a diabetic patient with retinal microangiopathy?

11. What is the correct code(s) for a patient diagnosed with pericarditis who has TB and diaphoresis?

12. What is the correct code(s) for a patient diagnosed with hypertensive malignant heart disease with congestive heart failure?

13. What is the correct code(s) for a patient diagnosed with malignant hypertensive heart disease with diastolic heart failure?

14. Do both lymphosarcoma and reticulosarcoma have to be coded together in order to use code 200?

15. What is the correct code(s) for a patient diagnosed with septicemia due to Pseudomonas?

16. What is the correct code(s) for a patient diagnosed with gynecomastia and hemianopia with a pituitary tumor?

17. What is the correct code(s) for a patient diagnosed with galactosemia and cataracts?

18. What is the correct code(s) for a patient diagnosed with cirrhosis due to xanthomatosis?

19. What is the correct code(s) for a patient diagnosed with hematoma due to open wound of the liver?

20. What is the correct code(s) for a patient diagnosed with vitamin B12 deficiency who is experiencing neuropathy and glossitis?

Hypertension Table

Hypertension is displayed in a table format in the ICD-9-CM alphabetic index and listed under Hypertension. Hypertension usually is diagnosed for systolic pressures over 140 and diastolic pressures over 90; however, the physician must determine the diagnosis.

Do not code hypertension as benign or malignant, unless the physician documents it. Benign hypertension is characterized by slow development but can become as severe as malignant hypertension, just not as rapidly. While benign indicates a good prognosis, malignant does not. If the physician does not document hypertension as benign or malignant, code as unspecified.

Hypertension is classified as essential or secondary. If it has no known cause, it is termed essential or idiopathic. If the hypertension is associated with a disease process, such as renal disease, then it is termed secondary.

Hypertension can be the cause of cardiovascular problems, which is coded as hypertensive heart disease. It is important to note, however, that hypertensive heart disease is not coded the same as heart disease with hypertension; it is coded as *hypertensive heart disease*.

With chronic renal failure or disease, a relationship is usually assumed with hypertension, so each one is coded as hypertensive renal disease. Renal failure or disease can be acute and not related to hypertension and would then be coded separately. Be careful, therefore, when coding acute renal disease or failure to make sure that any hypertension mentioned is related to the renal conditions before coding as hypertensive renal disease or failure. If the acute renal disease or failure is not related to the hypertension, then both the renal disease or failure and the hypertension must be coded separately. Keep in mind that secondary hypertension requires two codes, the hypertension and the etiology.

Transient hypertension and elevated blood pressure are not the same as hypertension because the physician has classified them as a temporary conditions. Transient is found in the **hypertension table** but is not listed as benign or malignant. Both transient and elevated blood pressures are coded with 796.2, which are not standard hypertension codes.

Review the information regarding hypertension within the circulatory section of this book.

ACTIVITY 6.11

Answer the following questions.

1. What is the proper code(s) for a patient diagnosed with transient hypertension?

2. What three categories are there for hypertension?

3. What is the proper code(s) for renal failure and malignant hypertension?

4. Which of the following are not modifying terms for hypertension: vascular, idiopathic, secondary, etiological, diffuse, arterial, adaptive, subacute, primary, and paroxysmal?

5. What is the correct code(s) for transient hypertension in a pregnant woman?

6. What is the correct code(s) for secondary benign hypertension due to an aneurysm of the renal artery?

7. What does secondary hypertension instruct you to code?

8. What is the correct code(s) for hypertensive heart disease?

9. What is the correct code(s) for malignant hypertension due to polycythemia?

10. What is the proper code(s) for elevated blood pressure?

11. What is the correct code(s) for pre-eclampsia with essential hypertension that is malignant?

12. What is the correct code(s) for a patient diagnosed with diabetes with peripheral angiopathy and hypertension?

13. Is 401.9 the correct code for hypertension following surgery?

14. What is wrong with the following scenario: patient with hemiparesis due to a CVA three months ago with hypertension, which is coded as 436, 402.90, and 342.9.

15. What does idiopathic mean?

16. Who are you coding for if the codes for chronic hypertension are coded as 760.0?

17. What is the correct code(s) for patient with edema, inflammation, and chronic hypertension due to deep vein thrombosis?

18. Which classification of hypertension did you select for the previous question (benign, malignant, or unspecified)?

19. What is the code for malignant hypertension for a patient who has heart disease?

20. What are the diastolic and systolic pressures that usually are considered high enough to qualify as hypertension?

Neoplasm Table

Neoplasms are abnormal growths that must be identified correctly to be coded. Before beginning with the **Neoplasm Table**, remember that if the terminology of a diagnosis is unknown, it can be checked in the alphabetic index before proceeding using the Neoplasm Table. There also are neoplasms that are not found in the Neoplasm Table and must be located in the alphabetic index. This is important to keep in mind, because neoplasms have many names and can occur at the same site but in different structures, such as connective tissue, organs, or bone. Neoplasms are known as sarcoma, adenoma, carcinoma, etc., and may be solely or all-inclusively benign or malignant, in situ, etc. Schwannoma is a neoplasm of the connective tissue, for example, and can be benign or malignant as indicated in the alphabetic index. If you use the alphabetic index to further define the neoplasm, it will direct you to see the Neoplasm Table, but will not direct you to an anatomic site. It is understood that you will go to the anatomic site whenever you select a code from this table.

Like other coding categories in the ICD-9-CM book, beware of indentations from a wide variety of modifying terms such as *connective tissue, skin NEC,* or *brain NEC*. It is every easy to select the wrong code as you move across the table.

To use the table, first determine if the neoplasm is malignant, benign, or unspecified. Benign indicates that the neoplasm is not cancerous. In situ is an additional category for malignancies and indicates that the malignancy has been confined within a certain area without invasion to other areas of the body. An additional category is *uncertain behavior,* which is used when the behavior of the neoplasm cannot be determined and may exhibit both malignant and benign characteristics. Unspecified indicates that the behavior of the neoplasm is not specified in the documentation.

If the neoplasm is malignant, the next step is to determine whether it is primary, secondary, or in situ. Primary indicates that the malignancy began in the specified area. Secondary refers to areas where the malignancy has spread. The word *metastasis* indicates a secondary malignancy. Use the word *spread* in place of metastasis if you are having trouble understanding which malignancy is primary and which is secondary.

The primary site usually is listed first; however, if the secondary site is the primary reason for the care, then it should be listed first. This is true particularly if the primary site has been successfully treated and no longer is being treated. A history V code should be used to indicate a neoplasm that previously has been treated and, through excision or other treatment, no longer exists. If the malignancy is secondary according to the medical report, but no primary is given, then use a general code to indicate that there is an unknown primary site (199.1). However, if the primary neoplasm no longer exists because it has been eradicated, then use a history code from the V codes to list the past history of the primary site. If two or more sites are secondary, you must code all of them.

You can select more than one neoplasm code for a scenario such as if you have a primary site and a malignancy at a secondary site.

Rehabilitation services, chemotherapy, or radiotherapy should be coded first. Side effects such as nausea and vomiting can be coded in addition to the neoplasm code.

Morphology codes are used to classify neoplasms for statistical purposes within tumor registries, but these codes are not used for reimbursement purposes and are not included on the national exams.

ACTIVITY 6.12

Answer the following questions.

1. If a malignant carcinoma has not invaded other adjacent structures, how would this be coded?

2. Which of the following terms do not refer to neoplasms: gliomas, nodular, leukemias, hirsutism, sarcomas, astigmastism, seminoma, paronychium, and adenomas?

3. What is the correct code(s) for serous cystadenocarcinoma of the ovaries?

4. How did you find the code for the previous question?

5. Provide descriptive terms for cystadenocarcinoma.

6. How would you code for cystadenocarcinoma that is benign?

7. What is the correct code(s) for a patient who is receiving chemotherapy for leiomyoma of the uterus?

8. What did you look under in the Neoplasm Table to find the previous code?

9. Is oat cell carcinoma benign, malignant, or *in situ*?

10. What is the correct code(s) for oat cell carcinoma of the left lung?

11. What is the correct code(s) for Hodgkin's lymphoma of spleen and axillary lymph nodes?

12. Sarcomas can be found in which of the following anatomic sites: cardiac, muscle, glands, small intestine, bone marrow, uterus, bone, and fat?

13. What is the correct code(s) for Schwannoma of the thigh?

14. What is the correct code(s) for a patient who is diagnosed with bone metastasis? A year ago she had a mastectomy for carcinoma of the left breast and completed chemotherapy 5 months ago.

15. What is the correct code(s) for patient receiving chemotherapy for rhabdomyosarcoma of the arm?

16. What anatomic site does rhabdomyosarcoma refer to?

17. What is the correct code(s) for rhabdomyoma of the arm? Compare this to rhabdomyosarcoma with reference to coding.

18. What anatomic sites are included in leiomyosarcoma?

19. What is the correct code(s) for a patient who had a mastectomy a year ago and then was treated with chemotherapy and returns today for a follow-up exam?

20. What is the correct code(s) for ovarian adenocarcinoma that invaded the gallbladder and peritoneum?

SPECIFIC SYSTEMS

INTEGUMENTARY

The integumentary system includes the skin and its accessories, including hair, nails, and glands. Integumentary and related diagnoses can be coded from a wide variety of codes, including burns, infections, wounds, and neoplasms. The most severe injury is coded first. Minor integumentary injuries, such as abrasions, are not coded if there is a more severe injury at the same site. More extensive injuries, such as injuries to nerves and vessels, would be coded additionally.

A nevus is a benign skin neoplasm. List the diagnosis code only once. If, for instance, the nevus is on the back, breast, scalp, and temple, only code 216.5 is used.

Infections can be coded to 958.3. However, if cellulitis is diagnosed, then it should be coded as cellulitis (681–682), unless ICD-9-CM specifies otherwise. Cellulitis is inflammation of cellular or connective tissue. If cellulitis is secondary to an injury or ulcer, then code the injury or ulcer first. Cellulitis that involves gangrene is coded as gangrene (785.4).

Ulcers occur for many reasons, such as from diabetes or from wearing a cast. Ulcers caused by diabetes can be either related to neuropathy, infection, or vascular disease. Be sure to code the etiology, if known, in addition to the code for the ulcer. Gangrene can occur with integumentary conditions and should be coded additionally if not already contained within the code.

Wounds

Coding of wounds often is deferred to other injuries, such as burns and fractures. Read the details listed in the ICD-9-CM book concerning when not to use the wound codes. Open wounds are coded from 870 to 897.

Burns

Burns are coded by site, degree, and/or percentage of body burned depending on what information is documented in the patient's record. The 948 codes are used if the burn percentage is known. For example, if a patient is diagnosed with 30%, third-degree burns, then the percentage can be coded in addition to the anatomic site, because that amount of third-degree burns could effect the patient's recovery.

Burns include first-, second- and third-degree. Third-degree burns are the most serious type of burn. These burns involve all three layers of skin and can include muscles and tendons. Second-degree burns involve the first two layers of skin—the dermis and epidermis—and include blisters. A first-degree burn involves the first layer of skin, the epidermis, and does not include blistering. Only the highest degree burn for a specific area should be coded.

Use code 948 when the burn site is not given and the percentage of body burned is known. Percentage of burned area is determined usually by the Rule of Nines for adults and Lung-Browder for children. Rule of nines calculate sections of the adult body as: 9% for the front or back of thorax, abdomen, one leg, both arms, and only 4.5% for the head. With a child, the sections of the body are calculated as: 18% for entire head, chest/abdomen (front or back), 14% for each leg, and 9% for each arm. With code 948, the fifth digit indicates the percentage of surface burned that was third-degree. Third-degree burns greatly affect the outcome of the patient's care. When the amount of third-degree burns are not known or are less than ten percent, use 0 as the fifth digit. The fourth digit signifies the total percentage of surface burned.

The code for multiple burns sites should only be used if the specific sites are not known; otherwise, code for each grouping of sites as indicated in the code book, which is more precise.

Necrosis is considered a non-healing burn, which is coded as acute burns. Escharotomy is the removal of *eschar*, which is the tissue that develops in place of burned skin and debridement is the removal of foreign material from a wound. Like escharotomy, debridement is critical for burn care because it helps restore circulation, removes devitalized tissue, prevents infection, eliminates constriction caused by scar tissue, and prepares the site for grafting.

Scars are considered late effects of burns and should be coded using a 906 code in addition to the code for scars. A patient can be coded with both acute burns and late effects at the same time because burns heal differently. A first-degree burn may have healed, but a more serious third-degree still may be receiving treatment.

Infections from burns also should be coded using code 958.3.

ACTIVITY 6.13

1. What is the correct code(s) for a patient diagnosed with third-degree burns of the shoulder and chest areas?

2. What is the correct code(s) for a patient diagnosed with ulcer of ankle resulting from postphlebitis that is inflamed?

3. What is the correct code(s) for a patient diagnosed with infection from a hand wound sustained while bicycling?

4. What is the correct code(s) for a patient diagnosed with infection of the hand from a wound that was repaired with sutures and debridement and was sustained while bicycling a month ago?

5. What is the correct code(s) for a patient diagnosed with infection resulting from a hand wound with injury of the tendons?

6. What is the correct code(s) for a patient diagnosed with infection resulting from a hand wound with injury to the tendons that occurred a month ago?

7. What is the correct code(s) for a patient diagnosed with eight skin tags?

8. What is the correct code(s) for a patient diagnosed with endothelial sarcoma of the forearm?

9. What is the correct code(s) for a patient diagnosed with multiple lacerations of both knees and legs?

10. What is the correct code(s) for a patient diagnosed with lacerations of the left arm and fracture of the radius requiring sutures and application of cast?

11. What is the correct code(s) for a patient diagnosed with an infection from diffuse cellulitis of the left cheek?

12. What is the correct code(s) for a patient diagnosed with infection and cellulitis at the colostomy site?

13. What is the correct code(s) for a patient diagnosed with burns over 43% of his body with 17% described as third-degree?

14. What is the correct code(s) for a patient diagnosed with cellulitis of the big toe of the right foot?

15. What is the correct code(s) for a patient diagnosed with cellulitis of the big toe of the right foot, which is gangrenous?

16. What is the correct code(s) for a diabetic patient diagnosed with cellulitis of the big toe of the right foot, which is gangrenous?

17. What is the correct code(s) for a patient diagnosed with atherosclerosis who has cellulitis of the big toe of the right foot, which is gangrenous?

18. What is the correct code(s) for a patient diagnosed with scar tissue from burns to the left arm three months ago?

19. What is the correct code(s) for a patient who was bitten in the nose during a fight?

20. Under what terms in the alphabetical index is the code found for the injury in the previous question?

MUSCULOSKELETAL

As listed in the ICD-9-CM book, many musculoskeletal codes require the manifestation code also be identified. Be sure to read the details!

Arthropathies and arthritis are classified together (711–714). Underlying diseases and organisms—the etiologies—need to be coded as directed in the ICD-9-CM book. Osteoarthritis usually involves the bones and cartilage of the hips, spine, knees and joints in the foot. *Ankylosing spondylitis* is rheumatic inflammation of the vertebrae and joints.

Fractures

The terminology used for determining diagnoses for fractures differs in definition from that used to determine procedures for fractures. Specifically, the terms *open* and *closed* as used to determine diagnoses (ICD-9-CM) are unrelated to how these terms are used to determine procedures (CPT). In diagnostic terminology, *open* means the skin is broken as indicated by terms such as:

compound

infected

puncture

foreign body

missile

Closed means the skin is not broken and bone is not protruding as indicated by terms such as:

impacted

spiral

linear

comminuted

fissured

greenstick

simple

spiral

An easy way to remember what terms apply to closed or open is to imagine the open terms as indicating a break in the skin. For example a foreign object indicates something that enters the body site only by creating an opening. If you remember the terms for open, then you should not have to memorize the closed terms because everything other than the open terms would indicate closed. However, make sure you are familiar with the closed terms.

Dislocations also are coded as open or closed. Although dislocations are coded as dislocations, a fracture-dislocation is coded only as a fracture.

When there is no description indicating closed or open, a fracture should be coded as closed.

Separate fracture sites should be coded separately. This includes separate sites within the same bone as distinguished by a different fourth digit.

Sometimes there are terms that need to be distinguished. With fractures, when a disease process, which is known as *pathological,* causes the fracture, it should be noted within the code. Spontaneous fractures are considered pathological as the word spontaneous denotes that the fracture occurred without an external force and not as a result of trauma. In addition, the cause for the fracture should be noted, such as a disease. Underlying disease processes that cause fractures include osteoporosis, osteomyelitis, cancer, atrophy, or congenital. Compression fractures can be either pathological or not, so do not assume these fractures are pathological.

Malunion of a fracture occurs when the fracture site does not heal correctly in alignment. Nonunion occurs when the fracture site does not heal completely.

ACTIVITY 6.14

1. What is the correct code(s) for a patient diagnosed with malunion of a pathological fracture of the left arm?

2. According to the alphabetical index, what is the difference between arthritis and arthropathies for coding purposes?

3. What is the correct code(s) for a patient diagnosed with fracture of the skull with subarachnoid hemorrhage?

4. What is the correct code(s) for a patient diagnosed with spontaneous fracture of the radius?

5. Under what description in the alphabetical index is the code for spontaneous located?

6. What is the correct code(s) for a patient diagnosed with viral enteritis and arthritis?

7. What is the correct code(s) for a patient diagnosed with dislocation of the acetabulum?

8. What is the correct code(s) for a patient diagnosed with greenstick fracture of the lower epiphysis of the ulna?

9. What other anatomic sites are included in the code for the previous question?

10. What is the correct code(s) for a patient diagnosed with infected fracture/dislocation of olecranon?

11. What is the correct code(s) for a patient diagnosed with transient arthritis of hands and forearms?

12. What is the correct code(s) for a patient diagnosed with arthropathy with Behcet syndrome of the pelvis and thigh?

13. Is code 828.0 the correct code for impacted fractures of ribs, sternum, and the left arm?

14. What is the correct code(s) for a patient diagnosed with linear dislocation of the wrist?

15. What is the correct code(s) for a patient diagnosed with arthritis in the bones in both hands and forearms?

16. What is the correct code(s) for a patient diagnosed with pathological fracture of the fibula?

17. What is the correct code(s) for a patient diagnosed with serum sickness and hypersensitive arthropathy?

18. What is the correct code(s) for a patient diagnosed with splenadenomegaly and leucopenia including rheumatoid arthritis?

19. What is the eponym for the diagnosis in the previous question?

20. What is the correct code(s) for a patient diagnosed with collapsed fracture?

CIRCULATORY

There are many codes for circulatory conditions, and there are many details that must be reported within the codes. Medical terminology is very important when coding circulatory conditions. Many circulatory codes are categorized by anatomic sites, such as vessels involved, so know the circulatory system well to code it correctly. The circulatory section also can include grafts, which can be *autologous* (from self) or *nonautologous* (from an outside source or man-made). Gortex is an example of a commonly used man-made material.

Circulatory conditions can be caused by other conditions, which are etiological, such as rheumatic heart disease. These etiological conditions sometimes are included within the circulatory code, but often are not. Remember, if the code includes everything, and there are no notes indicating additional codes are needed, then list only one code. If the code does not include everything necessary, then you must use both the circulatory condition code and its etiological code.

Remember, do not diagnose the patient—that is the role of the physician. If it is not in the report, you cannot code it, whether it is a code or a linkage of codes. Also, keep in mind—not documented, not done!

Symptoms

Beware of coding symptoms instead of a diagnosis. Remember! You do not need to code the symptoms if you have the diagnosis. For example, angina is heart discomfort and is not coded when the primary code is a MI, because it is an inherent part of the infarction. However, angina is coded with other diagnoses that are not necessarily associated with the angina, such as arteriosclerosis. Another example is pleural effusion and edema. Both of these conditions are common symptoms of congestive heart failure and, therefore, are not coded in addition to heart failure or heart disease codes. Both pleural effusion and edema can be associated with many other conditions, including renal disease, pneumonia, and carcinoma. However, if the pleural effusion or edema is reported as not involved with another present condition, but was treated, code the effusion or edema separately, as there are many other reasons these conditions could exist besides the one currently being treated.

Myocardial Infarction

An MI is better known as a heart attack. The MI would be coded as 414.8 as instructed in the alphabetic index of the ICD-9-CM coding book. The majority of MIs occur in the left ventricle. Pulmonary edema is included in heart failure diagnosis as stated in the coding description. Right heart failure is secondary to left heart failure. Since the right follows the left, only one code is necessary, either right (inclusive of the left) or the left. Ischemia and/or occlusions may occur without the diagnosis of infarction and are coded with a different code than the infarction. Laboratory data and EKGs usually are administered to test for an infarction.

The fifth digit of the MI codes are based on initial and subsequent episode of care. These terms differ from those terms used to code hospital visits. With MIs, *initial* refers to the occurrence of an MI, so there may be several services/visits coded as initial, up until the time of discharge, if they are related to the original episode of the MI. Hospital transfers would, therefore, be coded as initial. Subsequent refers to all visits/services provided after a patient is discharged and returns for treatment and follow-up, including other MIs that may occur within eight weeks following the initial MI. With the initial or subsequent codes, symptoms are not coded. However, if a symptom increases or alters patient care, such as an arrhythmia, then both the MI and arrhythmia can be coded.

Congestive Heart Failure

Congestive heart failure occurs when the heart is unable to pump enough blood through causing fluids to accumulate. The heart's ability to make adequate changes that maintain adequate pumping is known as *compensation,* with the opposite, *decompensation,* representing the heart's inability to do so. Congestive heart failure usually begins on the left side with dominance on either side.

Angina

Angina pectoris is severe constriction and/or pain about the heart. Variant angina is angina pectoris in which constriction and pain occur during rest. With stable angina, there is no change in episodes within a previous six-week period. With unstable angina, there is change in episodes, which usually results in hospitalization. For angina, vasodilators are used to widen the lumen of the vessels, beta blocker are used to reduce heart rate and blood pressure, and calcium blockers are used to prevent coronary spasms.

Hypertension

Hypertension often is associated with circulatory problems and should be coded as reported—either as part of a circulatory condition or as a separate condition. Hypertension usually is listed as a secondary code. For more information, see Tables, Hypertension, which is discussed later in this book.

Arteriosclerotic Heart Disease

Arteriosclerotic heart disease (ASHD) frequently is coded as part of circulatory conditions, sometimes by itself, and sometimes contained within a combination code. Arteriosclerosis is the end stage of atherosclerosis. An important distinction is the physician's diagnosis indicating that the heart disease is caused by the arteriosclerosis, which would use one code for arteriosclerotic heart disease as opposed to two codes for heart disease and arteriosclerosis. This also applies to hypertension, which is coded separately unless the physician indicates it is related, such as hypertensive heart disease. However, with the kidneys, it is assumed that hypertension is an inherent part of the chronic condition therefore chronic renal failure and hypertension is coded as hypertensive chronic renal failure.

Be sure to check before and after a code to ensure that you have selected the proper code. A good example of this is an old myocardial infarction, which has its own code, 412, and is not listed in previous codes.

Cerebrovascular Accidents

CVAs can manifest other conditions, which may be ongoing for extended periods of time. Code the CVA as acute with codes 430–434 only when it is a condition that is currently being diagnosed and treated, not when it is old and has produced manifestations, which may be referred to as *residuals*. If the manifestation is acute or chronic, then code it in addition to the acute cerebrovascular disease code, unless the manifestation contained within the cerebrovascular disease code, is required by note to be coded, or is a late effect. Residual indicates that the effect is a late effect of the associated CVA, therefore the CVA is not acute. Use 438 codes for these descriptions. If the manifestation occurred at an extended time after the cerebrovascular disease, use a late code for the cerebrovascular disease in addition to the manifestation code. If a history of CVA is described but there are no residual effects, code from the V12.5 codes.

Transient ischemic attacks (TIAs) are similar to CVAs but are temporary and sudden focal neurological deficits that resolve within 24 hours. They can be caused by loss of cerebral blood flow and may indicate the possibility of a future CVA. They are coded with code 435.

Other Sites

Within the circulatory codes, other areas of the vascular system can be affected in addition to the heart and the brain. Common diagnoses and conditions include atherosclerosis, peripheral vascular disease, embolisms, thrombosis, phlebitis, varicose veins, hemorrhoids, and varices.

Hemorrhoids are categorized as internal or external. External hemorrhoids protrude to the outside. Hemorrhoids also are classified by complications, such as bleeding, strangulation, prolapse, and ulcerations. Strangulated means the blood supply is compromised.

Organisms

Organism can produce circulatory conditions. Rheumatic heart disease (390–398) is caused by rheumatic fever, which sometimes occurs after a streptococcal infection. Though occurring more often in children, it can occur in adults. There are various forms of cardiac involvement, but damage to the mitral valve is quite common with this type of infection.

Dysrhythmias

Cardiac dysrhythmias come in many forms and can be symptoms of other diagnoses. Be careful in coding dysrhythmias because they may be included in another diagnostic codes, such as MIs. To be coded in addition to other cardiovascular codes, the dysrhythmia must result in additional care or alteration of current care for the patient. Fibrillations (atrial or ventricular) are incomplete and rapid contractions of the heart. Atrial flutter occurs when the contractions of the heart are very rapid. Tachycardia is when the heart beats rapidly. Atrioventricular (AV) heart blocks interfere with the normal conduction of electric impulses of the heart from the atrium to the ventricles, which can result from MIs, heart disease, or drugs. Possible symptoms include syncope, palpitations, dyspnea, and hypotension. AV blocks can be first degree (delayed impulses in the ventricles), second degree (nonconduction of impulses from atrium to the ventricles, known as Wenckebach or Mobitz type I or II), or third degree (no supraventricular impulses, known as complete heart block). In addition, there are right and left bundle branch blocks. These can be first-, second- or third-degree. First-degree blocks are associated with valvular disease or atrial septal defects. Second-degree are classified as Mobitz I or Mobitz II. A third-degree block is known as complete blockage of the heart. If no other information is provided, use code 427.9.

ACTIVITY 6.15

1. What is the correct code(s) for a patient diagnosed with CVA?

2. What is the correct code(s) for a patient diagnosed with hemiparesis due to a CVA 7 weeks ago?

3. Why or why not was the CVA coded in the previous question?

4. What is the correct code(s) for a patient being treated for apraxia resulting from a CVA a week ago?

5. Why or why not was the CVA coded in the previous question?

6. What is the correct code(s) for a patient diagnosed, upon admission to the hospital, with monoplegia of the left leg and then is treated for acute CVA?

7. Why or why not did you code the monoplegia in the previous question?

8. Why or why not did you use a late effect code for the vertigo in the previous question?

9. What is the correct code(s) for a patient diagnosed with facial droop due to cerebral thrombosis two weeks ago?

10. What is the correct code(s) for a patient diagnosed with transient neurological signs who is admitted to the hospital for care of transient ischemic attack?

11. What is the correct code(s) for a patient diagnosed with varicose veins of the rectum with bleeding?

12. What term is the description in the previous questions commonly referred to as?

13. What is the correct code(s) for a patient diagnosed with phlebitis due to an elective abortion three weeks ago?

14. In the previous question, did you or did you not select a fifth digit? If yes, how did you select it?

15. What is the correct code(s) for a patient diagnosed with malignant heart disease due to hypertension?

16. What is the correct code(s) for a patient diagnosed with benign heart disease and hypertension?

17. What is the correct code(s) for a patient diagnosed with a long history of malignant hypertension who is diagnosed with renal and heart disease with edema?

18. What is the correct code(s) for a patient diagnosed with transient hypertension?

19. What is the correct code(s) for a patient diagnosed with Wenckebach AV block?

20. What is the correct code(s) for a patient diagnosed with progressive muscular dystrophy that has resulted in heart disease?

ANEMIAS

A patient can be suffering from anemia for many reasons, such as diseases, trauma, or deficiencies. Each type has its own codes and associated conditions, which must be coded as directed by the ICD-9-CM coding book.

Deficiencies (280) occur when the blood does not absorb iron adequately. When the anemia is due to another condition, such as from an ulcer, code the condition as well.

Hemolytic anemia results from excessive destruction of red blood cells combined with the inability of the bone marrow to produce enough red blood cells. Hemolytic anemia is classified by hereditary (282) or acquired (283). Aplastic anemia (284) occurs from the lack of production of red blood cells by the bone marrow. Aplastic anemia may occur from chemotherapy or radiotherapy.

Coagulation problems (286), such as hemophilia, also include coagulation problems from other conditions such as side effects from Coumadin and other similar drugs.

SEPSIS

Sepsis is an infection, whether from toxins or organisms, which is not coded as itself, but must be coded according to its location within the body. The specific nature of the sepsis must be coded, that is whether it is located within the blood, the urinary system, or skin. Sepsis can occur from natural circumstances or from medical services, such as implants or devices. Septicemia is the most common form of sepsis and is an infection of the blood, which can result in death if not controlled. Bacteremia and sapremia are forms of septicemia in which bacteria or toxins are the source of infection. Urosepsis refers to sepsis within the urinary tract. Puerperal sepsis refers to sepsis following childbirth. Any sepsis before, during, or after pregnancy must have a code specifically from the pregnancy codes in Chapter 11.

If the specific diagnosis of the manifestation of sepsis is known, such as cellulitis or lymphangitis, this should be coded. Septic shock also should be coded if associated with the septicemia. When the primary infectious agent is not known, it is referred to as cryptogenic. If the cause is known, such as an organism, it should be coded also, although usually the sepsis is included in the code used for the organism. Symptoms of sepsis that would not be coded separately because they commonly are associated with it include fever, and respiratory/cardiovascular problems, such as tachycardia.

SIRS is an abbreviation for systemic inflammatory response syndrome. SIRS refers to sepsis in which there is an overwhelming infection of the bloodstream by toxin-producing bacteria. It can be located in the gallbladder, liver, kidneys, bowel, skin (known as cellulitis), and lungs (known as bacterial pneumonia). SIRS can result in other severe illnesses, including meningitis and septic shock.

ORGANISMS

There are many types of organisms that cause diseases and conditions in humans. Organisms can be coded in two ways: either as a separate code but in addition to a code for the specific condition or as one code that includes both the organism and the condition. The latter, organism and condition included in one code, is usually coded from Chapter 1 codes. Be sure to read the coding notes for directions on specific coding scenarios. Organisms can include bacteria, fungi, viruses, or parasites. *E. coli* and rotavirus infections are responsible for a majority of gastrointestinal infections such as diarrhea.

Viral hepatitis is the effect of viruses on the liver. The most common types of hepatitis are infectious and caused by the A strain, serum hepatitis, which is caused by the B strain, and hepatitis C, which results from blood transfusions. Hepatitis A is mild in comparison to hepatitis B, which patients may have for a long time and usually is transmitted parenterally through blood products, sexual contact, or needles. Additional hepatitis cases can occur from hepatitis D and E.

There are several ways to code HIV/AIDS. If the patient has symptoms associated with the HIV/AIDS, then code 042 should be coded first and then code the symptoms. Usually, the HIV/AIDS code 042, should be listed first; however, if the patient has a condition that is unrelated to the HIV/AIDS, then the condition should be listed first with the appropriate HIV/AIDS code listed second. An exception to this coding sequencing is if the patient is pregnant, which should then list 647.8x as the principal diagnosis. If the patient does not exhibit symptoms but has been positively diagnosed with HIV/AIDS, then code V08 should be used. An additional code, 795.71, is for patients whose tests have not definitely demonstrated HIV. Once a patient has been diagnosed as positively exhibiting symptoms of HIV, then the patient must be coded always with 042 and never with V08 or 795.71. Code V01.7 is for a newborn whose mother has HIV and tests on the newborn show HIV is present in the newborn.

The most common conditions caused by the HIV infection are Kaposi's sarcoma and pneumonia caused by *pneumocystis carinii*. Each separate anatomic site that is affected by the Kaposi's sarcoma should be coded separately.

ACTIVITY 6.16

1. What are the two ways hemolytic anemia can occur?

2. What is the correct code(s) for a patient who is diagnosed with puerperal septicemia one week after delivery?

3. What is the correct code(s) for a patient diagnosed with systemic sepsis due to anthrax?

4. What is the correct code(s) for newborn diagnosed with sepsis due to herpes?

5. What is the correct code(s) for a patient diagnosed with acquired pancytopenia?

6. What are the descriptions of the codes under which acquired pancytopenia is coded?

7. What is the correct code(s) for a patient who is diagnosed with anemia due to acute blood loss from injuries sustained during a car accident?

8. How is the anemia in the previous question differentiated within the codes? Explain how this affected the choice of codes.

9. Within the codes, is buccal sepsis specifically described as sepsis? Explain.

10. What is the correct code(s) for a patient diagnosed with sepsis of the urinary tract with septicemia due to Salmonella?

11. Was there another choice of codes for the previous question? If so, why did you pick the code you did?

12. What is the correct code(s) for a patient diagnosed with cryptogenic septicemia?

13. What is the correct code(s) for a patient diagnosed with bacteremia due to Strep B?

14. What is the correct code(s) for a patient diagnosed with bacteremia with sepsis due to Clostridium?

15. What is the correct code(s) for a patient diagnosed with bacteremia due to RSV?

16. What is the correct code(s) for a patient diagnosed with pneumococcal septicemia with SIRS and hepatic failure?

17. What is the correct code(s) for a patient diagnosed with end-stage renal disease (ESRD) with anemia?

18. What is the correct code(s) for a patient diagnosed with chronic hypertensive uremia?

19. What is the correct code(s) for newborn with anemia due to Rh incompatibility?

20. What is the correct code(s) for a patient diagnosed with hyperchromic anemia due to acute blood loss?

RESPIRATORY

COPD (496) is generally an obstruction of the airways, which, when exacerbated by other conditions, is coded by the nature of the exacerbation. If known, the condition causing the exacerbation can be listed first and COPD listed as secondary. However, there are codes that cannot be coded in addition to COPD. These codes (for example, codes 491–493) are listed in the coding guidelines in the CPT book under excludes for COPD. These codes, by their nature, are inclusive of COPD. So, beware when coding COPD. Read the details in the coding book!

Respiratory insufficiency is not coded separately from COPD, as it is normally assumed to be a condition of COPD. Other conditions such as asthmatic bronchitis also are inclusive of COPD and would not be coded separately.

Asthma is a condition of hypersensitivity in which breathing is difficult because of spasms of the bronchial tubes and/or swelling of their membranes, which can be characterized by bronchospasms. Intrinsic asthma is caused by the body's own system and extrinsic asthma is caused by an external factor. Status asthmaticus (also referred to as intractable, severe, or refractory), which is included in the fifth digit categorization of asthma, is defined as an acute asthmatic attack in which normal treatment does not alleviate the condition. This type of asthma includes symptoms such as hypercapnia (excess carbon dioxide in the blood).

Pneumonia can be caused by bacteria, virus, mycoplasma, or by aspiration of a substance into the lungs. Organisms can include pneumoccocal, streptococcal, staphylococcal, or Haemophilus influenzae.

Pneumothorax results from air leaking into the pleural sac around the lungs. A pneumothorax is categorized as spontaneous, iatrogenic, or traumatic. Spontaneous pneumothorax (512) occurs without known trauma and air leaks into the pleural sac. Iatrogenic pneumothoraces (512) occur from previous medical treatment. Traumatic pneumothorax (860) is caused by a traumatic injury and is coded from the 860 codes not from the 512 codes. They are categorized further by the existence of an open wound.

Pulmonary edema can occur from cardiac or noncardiac problems. Whereas pulmonary edema caused by cardiac problems is coded with 428 codes, noncardiac pulmonary edema can be coded from the 500 codes, such as 508 or 518. Be careful because some codes may include edema as a symptom and should not be coded.

Tuberculosis usually is considered pulmonary; however, it can infect other body sites, such as the kidneys and lymph nodes.

Whereas upper respiratory infections (URIs) are coded specifically with 465, there is no specific code for lower respiratory infections (LRIs). If the exact diagnosis is not known for an LRI, use code 519.8, (other diseases of respiratory system not elsewhere classified).

Tonsillitis and adenoiditis can be coded together when they occur at the same time, or coded separately if only one condition exists. They also can be specified as chronic or acute. The same concepts of coding apply to other respiratory sites, such as sinusitis, pharyngitis, or rhinitis.

Remember that if a patient is diagnosed with both an acute and chronic condition at the same time, one code that includes both the acute and chronic condition is used and no other code is necessary. However, if there is not one code that includes both the acute and the chronic condition, you must code separately for each condition.

ACTIVITY 6.17

1. Asthma due to internal bodily conditions is known as what?

2. If normal treatment does not alleviate asthma, this can be referred to as what?

3. What is the correct code(s) for a patient diagnosed with COPD and who is status asthmaticus?

4. What is the correct code(s) for a patient diagnosed with COPD and pneumonia due to strep pneumoniae?

5. What are the three types of pneumothoraces?

6. What is the correct code(s) for a patient diagnosed with chronic sinusitis?

7. What is the correct code(s) for a patient diagnosed with acute and chronic sinusitis?

8. What is the correct code(s) for a patient diagnosed with upper respiratory infection?

9. What is the correct code(s) for a patient diagnosed with lower respiratory infection?

10. What is the correct code(s) for a patient diagnosed with acute pulmonary edema?

11. What is the correct code(s) for a patient diagnosed with edema due to congestive heart failure as associated with hypertension?

12. What is the correct code(s) for a patient diagnosed with TB associated with a pneumothorax?

13. What is the correct code(s) for a patient diagnosed with Rostan's asthma?

14. What exactly is the condition in the previous question described as?

15. What is the correct code(s) for a patient diagnosed with open sucking chest wound due to car accident?

16. What is the correct code(s) for a patient diagnosed with chronic tonsillitis and adenoiditis?

17. What is the correct code(s) for a patient diagnosed with edema of the glottis?

18. What is the correct code(s) for a patient diagnosed with allergic bronchitis?

19. What is the correct code(s) for a patient diagnosed with atelectasis?

20. What is the correct code(s) for a patient diagnosed with pleural effusion due to staph infection?

GASTROINTESTINAL

The gastrointestinal (GI) tract begins with the mouth and extends to the anus. Its functions include digestion, absorption, and elimination of nutritional substances. For coding GI diagnoses and procedures, it is critical you know your medical terminology well.

Diabetes

Diabetes mellitus and insipidus are not the same. Diabetes mellitus is caused by the improper secretion and utilization of insulin as produced by the pancreas. Symptoms include hyperglycemia and glycosuria caused by the improper utilization of sugars in the body from insulin dysfunction. Urine is notably sweet from the sugar content. Symptoms also include polyuria and polydipsia. Most references to diabetes refer to mellitus. The urine in diabetes insipidus, in contrast, is watery and tasteless. It is caused by the body's failure to adequately utilize antidiuretic hormone (ADH), which results in the kidney's inability to properly reabsorb water into the bloodstream, which is then excreted. Symptoms include polydipsia and polyuria.

When coding, the fifth digits with diabetes can be confusing, in part because of the descriptions. A fifth digit of "0" indicates that the diabetes is Type II diabetes and non-insulin dependent, Although a Type II diabetic patient may be prescribed insulin to correct a momentary problem, they are not on a regular maintenance and, therefore, are not insulin-dependent. Therefore, there may be reasons other than confirmed diabetes for a patient to take insulin. Do not assume, therefore that because a patient is taking insulin that patient is insulin-dependent. Characteristically, patients with Type II do not exhibit ketoacidosis from ketones in their urine, in contrast to Type I diabetes. Type II diabetes usually is treated by diet and by drugs, such as Diabeta and Micronase.

Type II is the most common type of diabetes and usually is adult-onset, but not always. A fifth digit of "1" indicates that the patient has Type I diabetes meaning there is a failure to produce insulin. These patients do require insulin. This type of diabetes occurs most often in young people, but can occur at any age, so the notation *juvenile type* does not exclude older people from being classified as Type I. Ketoacidosis, in which the body produces too many ketones, is a familiar indication of this type of diabetes.

Fifth digits, "2" and "3," indicate that the Type I or Type II is uncontrolled, which must be documented by the physician. Do not diagnose the patient as having Type I, Type II, or uncontrolled diabetes; only the physician can do this.

There are many system complications that are associated with diabetes including neurological, ophthalomological, and renal. Complications can include renal disease, neuropathy, amputation, and blindness. The codes for diabetes now include some manifestations, so beware! Despite the manifestation within the diabetes code, additional manifestation codes may be required as indicated in the coding book. Be sure that complications are complications from the diabetes, since some conditions may occur that are not diabetes-related and, therefore, should not be coded as such. Read the documentation to ensure that the physician has confirmed that the diabetes caused a complication.

In addition, there are several conditions that are related to diabetes, but are not diagnosed as diabetes. These include hypoglycemia and gestational diabetes, which can occur when a woman is pregnant, and impaired glucose tolerance, which is a mild impairment of the body's ability to metabolize glucose so the patient has a high blood glucose level after eating. Temporary treatment may be necessary for these conditions, but usually the body returns to normal functioning.

Gestational diabetes must include a code from Chapter 11, the 640 codes, for pregnancy. Impaired glucose tolerance is coded as 790.2. Hypoglycemia is coded with 251 codes. In addition, a baby may present with symptoms of diabetes as manifested from the mother, which is coded with 775.0. If, however, the baby exhibits hyperglycemia, this would be coded as 775.1 and may require temporary treatment with insulin.

Renal

Chronic renal failure (CRF) is the progressive development of kidney failure over many years. The later stages of CRF are known as end-stage renal disease (ESRD); therefore, they are both coded as renal failure, code 585. Causes of renal failure include nephritis, calculi, tumors, hypertension, drugs, glomerulonephritis, or disease. While some of the causes of renal failure can be treated, some can not and will continue to worsen.

It is important to note the connection between renal disease and hypertension, for hypertension produces chronic renal disease or failure. However, this relationship is not necessarily true for acute renal disease or failure, because a patient can have an acute event, such as trauma, which aggravates the renal disease or failure but may not be related to the hypertension. In the later stages of failure, hemodialysis or peritoneal dialysis are used for treatment, accessed through either a fistula or catheter. Catheters are for short-term use and fistulas are for long-term use.

Urinary Tract Infection

Do not code urinary tract infections (UTI, 599.0), unless the physician diagnosed the UTI. Even when laboratory tests reveal an organism and there are typical symptoms, without a diagnosis of UTI by the physician, a UTI code is not acceptable. As a coder, you are not to determine diagnoses.

Hernias

Hernias (550) are protrusions of any part of tissues and/or organs through an abnormal break in the muscular wall. A hernia of the digestive system is known as a *hiatal hernia,* which, specifically, is the protrusion of the stomach into the diaphragm. An irreducible hernia cannot be returned to its original placement. A strangulated hernia is when the blood supply is compromised because of constriction. Incarcerated hernias occur when the hernia completely obstructs the bowels. Both incarcerated and strangulated hernias are considered to be obstructive. Incomplete hernias have not gone through the break completely as opposed to complete hernias, which have broken completely through.

A twisting of the bowels is known as a *volvulus* and can result in infarction. It can be caused by obstructions, such as a tumor or adhesions.

Varices/Diverticulum

Varices are veins that are twisted and can occur in many locations, including the gastrointestinal tract. Varices may burst and result in bleeding and possible death. Often with GI varices, alcoholism is involved.

Diverticulum are sacs or pouches in the walls of an organ. Inflammation caused by the diverticulum is known as diverticulitis, which can be acute or chronic. This is seen frequently in the intestinal tract, particularly the colon, and can result in gangrene, abscesses, inflammation of the peritoneum, and even perforation. Diverticulosis is the presence of diverticula, but without symptoms.

ACTIVITY 6.18

1. If the urine of a patient is tasteless and watery, what type of diabetes do they have?

2. What is the correct code(s) for a patient diagnosed with volvulus of the duodenum?

3. What is the correct code(s) for a patient diagnosed with hiatal volvulus gangrenous obstructive hernia?

4. What description did you look under in the alphabetical index for the previous question?

5. What is the difference between CRF and ESRD?

6. What is the correct code(s) for a patient diagnosed with bronzed diabetes?

7. A hernia that completely obstructs the bowels is known as what?

8. What is the correct code(s) for a patient diagnosed with retinal edema as a complication of insulin-dependent diabetes?

9. What is the correct code(s) for a patient diagnosed with sliding incarcerated hernia?

10. What is the correct code(s) for a patient diagnosed with acute renal failure?

11. What is the correct code(s) for a patient diagnosed with chronic uremia with neuropathy?

12. What is the correct code(s) for a patient diagnosed with stones in the urinary bladder?

13. What is the correct code(s) for a patient diagnosed with UTI due to *E. coli*?

14. What is the correct code(s) for a patient diagnosed with hypoglycemia due to diabetes?

15. What is the correct code(s) for a patient diagnosed with hormone-induced hypopituitarism?

16. What is the correct code(s) for a patient diagnosed with diverticulosis with hemorrhage and diverticulitis of the ileum?

17. What is the correct code(s) for a patient diagnosed with diverticulosis of the ileum with hemorrhage?

18. What is the correct code(s) for a patient diagnosed with staghorn calculus?

19. Where is a staghorn calculus located?

20. What is the correct code(s) for a patient diagnosed with cholecystocolic fistula?

NEUROLOGICAL SYSTEMS

Hemiplegia/Hemiparesis

Hemiplegia and hemiparesis (342) can be spastic, flaccid, other specified, or unspecified as indicated by the fourth digit. Flaccid refers to the loss of muscle reflex and tone. Spastic refers to sudden involuntary reactions of the reflexes. The fifth digit denotes which side of the body is affected, categorized as dominant or nondominant, the dominant side being the side of the body that the patient favors. For example if a right-handed person's right side is affected by the hemiplegia or hemiparesis this is considered their dominant side. Remember that hemiplegia/hemparesis associated with CVAs are coded with their own distinct code.

Epilepsy

Epilepsy and seizures are not synonymous diagnoses. Seizures may occur without the diagnosis of epilepsy.

Epilepsy (345) is categorized by several different means. It can be distinguished by either convulsive or nonconvulsive seizures, by petit mal or grand mal, with mention of impairment of consciousness or without, etc. The fifth digit denotes the presence or absence of intractability. Intractability indicates that the epilepsy is not manageable.

The two main categories for epilepsy are idiopathic and psychomotor. Idiopathic epilepsy may occur during birth or from trauma, whereas brain lesions that develop cause psychomotor epilepsy. Idiopathic epilepsy can be either grand mal or petit mal. Grand mal seizures can last for several minutes, whereas petit mal seizures last only for ten to twenty seconds.

DRUG ABUSE

Drug abuse, dependence, and withdrawal are not defined the same. Drug abuse can be specified as not dependent when a patient has abused a certain drug at one time but has not abused the drug on a regular basis in the past. Code 305 would be used for this scenario. Code 304 represents drug dependence, which signifies ongoing abuse of drug(s) in the past, although this may be intermittent, in remission, or ongoing at the time current services are provided. Drug withdrawal is coded with codes 292.0 and indicates cessation of drugs and does not distinguish between abuse and dependence.

OTITIS

Otitis externa (380) is an infection of the external auditory canal. Otitis media (381) is an infection of the middle ear and may be further classified as suppurative (producing pus) or secretory (producing a secretion). Serous otitis media refers to the excretion of a serous fluid.

ACTIVITY 6.19

1. Within the codes, which of the following terms apply to hemiplegia: dominant, primary, suppurative, flaccid, and hysterical?

2. If a patient presents with a long history of alcohol abuse, can they be diagnosed with acute episode of alcoholism? Explain.

3. What is the correct code(s) for a patient diagnosed with acute iridocyclitis that is recurrent?

4. What is the correct code(s) for a patient diagnosed with continuous abuse of Dexedrine?

5. What is the correct code(s) for a patient diagnosed with otitis externa due to impetigo?

6. What is the correct code(s) for a patient diagnosed with epileptic seizure?

7. What is the correct code(s) for a patient diagnosed with sarcoidosis with meningitis?

8. What is the correct code(s) for a patient diagnosed with drug-induced myelopathy?

9. What is the correct code(s) for a patient diagnosed with myelopathy with cervical spondylosis?

10. What is the correct code(s) for a patient diagnosed with hemiplegia due to a CVA three weeks ago?

11. What is the correct code(s) for a patient diagnosed with vision loss due to senile punctate cataracts?

12. What is the correct code(s) for a patient diagnosed with visual impairment of the left eye at 20/250?

13. What is the correct code(s) for a patient diagnosed with intractable clonic epilepsy?

14. What is the correct code(s) for a patient who is a complete quadriplegic at C5–C7?

15. What is the correct code(s) for a patient diagnosed with exophthalmos due to hyperthyroidism?

16. What is the correct code(s) for a patient diagnosed with delirium tremens due to alcohol withdrawal?

17. What is the correct code(s) for a patient diagnosed with serous otitis media?

18. What is the correct code(s) for a left-handed patient diagnosed with spastic hemiparesis affecting the right side?

19. What is the correct code(s) for a patient diagnosed with migraines with characteristic auras?

20. What is the correct code(s) for a patient diagnosed with collapse of external ear canal due to inflammation?

PREGNANCY

Pregnancy codes can be tricky. There are several places in the alphabetic index where you can find the correct code. These places include the term itself (e.g., hemorrhoids), pregnancy, labor, or delivery. Be very careful when locating codes, for there are many indentations and a wide variety of main terms under which the condition may be listed. Main terms can include *management affected by, due to, damage from,* etc.

If the condition is listed under pregnancy/labor/delivery, then you must code from these codes first. If the condition is not listed in these sections and is not related to pregnancy, code only the condition and include V22.2 for incidental pregnancy. Be sure to read the notes—if the code indicates, you must code the condition also.

With pregnancy codes, the alphabetic index will often list similar codes for both mother and baby, so remember for whom you are coding.

Complications

Coding for pregnancy is difficult because there are many things that can happen in pregnancy, and the codes are very specific. Remember to differentiate! Anything that is important to the outcome of the care should be coded, such as multiparity, high-risk pregnancy, and early labor.

With pregnancies, be careful of time periods, because distinct codes can be necessary for specific time periods:

Preterm is for deliveries prior to 37 completed weeks of gestation.

Term is for deliveries between 38 and 40 completed weeks of gestation.

Post-term is for deliveries between 40 and 42 completed weeks.

Prolonged is for deliveries after 42 completed weeks.

Early onset of delivery is delivery before 37 completed weeks gestation.

Puerperium is associated with the term, *postpartum*. Puerperium refers to the time period required for the uterus to return to normal size and begins immediately after delivery and extends for six weeks.

Gravida refers to the number of pregnancies. Para refers to the number of viable pregnancies (i.e., approximately 22 weeks' gestation). Gestation refers to the weeks of pregnancy. For example, this may be stated as G2, P1, Ab 1, which means two pregnancies, one birth, and one abortion.

Pre-eclampsia, a condition developed by the mother during pregnancy, is toxemia characterized by hypertension, headaches, albuminuria, and edema of the lower extremities. Albuminuria and/or edema must be present to use this code.

Be careful to code for the proper patient! If you are coding for the mother, use the codes applicable to the mother, and not the baby. There are double codes for pregnancy complications, one for the mother and one for the baby. Code for the patient, whether it is the mother or the child, but not for both. Pregnancy codes for the mother only can be found within the alphabetic index under Pregnancy.

Pregnancy codes include a wide variety of codes and associated terminology, which is important to recognize, particularly within the alphabetic index. The terminology, such as *management affected by* and *complicated by* will help you locate the proper code.

Pregnancy codes also need to be broken down into antepartum, delivery, or postpartum visits. This will determine, not only the code, but also the fifth digit. Postpartum or antepartum conditions may exist with a delivery code. Postpartum begins immediately after delivery and continues for six weeks. Fifth digits 1 and 2 indicate that the delivery occurred during the current visit. Digits 3 and 4 indicate that delivery did not occur during the current visit. The fifth digits then separate out with 1 and 3 for antepartum or unspecified conditions; 2 and 4 are for postpartum conditions associated with the pregnancy.

Normal delivery code, 650, is used only for normal delivery including episiotomy but not much else. Even a normal delivery with a previous caesarean section is not coded as a normal delivery. No other pregnancy/delivery condition codes can be used with 650 as stated in the coding book.

If delivery occurred during the visit, then you must use delivery codes plus birth codes (V27) as stated in the codes. V27.0 is for a single liveborn. When the type of birth is unspecified, a fifth digit of 9 should be used.

Weeks matter with pregnancy codes as listed within the codes, such as when coding early or late onset of labor. For example, normal delivery at 36 weeks with single liveborn would include code 644.21 with V27.1 as the outcome of delivery code.

Abortion Codes

Abortions can be induced legally or naturally (known as spontaneous). Natural abortions can include miscarriages. Ectopic and molar pregnancies can result in abortions and have their own codes. The fifth digit for abortion codes is classified as "1" for incomplete, "2" for complete, and "0" for unspecified abortion. Complications associated with abortions also must be coded and are referenced by the fourth digit. A fourth digit of 7 indicates there are other specified complications, 8 indicates the complication was unspecified, and 9 indicates that there was no mention of a complication. It is important to understand the differences between these codes. *Other* means information was provided regarding complications, but it was specifically listed in the codes. *Unspecified* means that there was a complication but it was never specifically mentioned.

Complications of abortions can be coded additionally from codes 640–648 and 651–657 if not already stated in the abortion code, but never use complication codes, 660–669 for abortion complications because these codes are for live births. Pregnancy codes are very specific! Remember to differentiate!

Abortions can be missed, threatened, or recurrent. A missed abortion (632) occurs when a fetus dies before 22 weeks' gestation and remains in the uterus for more than four weeks. After six weeks, it is considered a dead fetus syndrome (641.3x) or dead fetus (656.4x), which can be coded only if the physician states that it is, in fact, dead fetus syndrome or dead fetus. Threatened abortion is indicated by intrauterine bleeding before 22 weeks' gestation. Recurrent (habitual) abortion (646.3 and 629.9) is three or more consecutive pregnancies that have aborted spontaneously at or near the same gestational age.

If a spontaneous naturally occurring abortion results in a live birth, it is not coded as an abortion but as a live birth. Code 644.21 (early onset of delivery) is used if the abortion results in a live birth. If a legally-induced abortion results in a live birth, it is coded as a failed abortion (638). The 639 codes specify complications from abortions that need to be separately indicated, such as an abortion that later resulted in complications.

ACTIVITY 6.20

1. What does G3, P2, Ab1 mean?

2. What is the correct code(s) for a patient who has retained a dead fetus at 23 weeks gestation?

3. What is the correct code(s) for a pregnant patient with malpositioning of the fetus, which was a successful version with delivery of a single liveborn?

4. Prolonged delivery means delivery after how many weeks of gestation?

5. What is the definition of a habitual aborter?

6. What is the correct code(s) for a newborn who is affected by the mother who is diagnosed with pre-eclampsia and edema?

7. What is the correct code(s) for a patient who delivered a single liveborn three days ago and returns today with deep vein thrombosis?

8. What is the correct code(s) for a 20-week pregnant patient who is diagnosed with mild hypertension with albuminuria?

9. What is the correct code(s) for a 24-week pregnant patient who is diagnosed with diabetes mellitus?

10. What is the correct code(s) for a 22-week pregnant patient who is diagnosed with carcinoma of the fibular bone?

11. What is the correct code(s) for a 28-week pregnant patient who is seen today for abnormal fetal heart rate?

12. What is the correct code(s) for a 21-week pregnant patient who is diagnosed with rubella, which is suspected to have affected the fetus?

13. What is the correct code(s) for a patient who delivered a single liveborn with the use of forceps?

14. What is the correct code(s) for a patient with corneal ectopic pregnancy?

15. What is the correct code(s) for a 13-week pregnant patient who miscarried with shock?

16. What is the correct code(s) for a 36-week pregnant patient who delivered a single liveborn?

17. What is the correct code(s) for a patient who delivered a single stillborn after rupture of amniotic sac 30 hours before?

18. What is the correct code(s) for a 29-week pregnant patient who is diagnosed with rubella?

19. What is the correct code(s) for a patient who delivered a single liveborn at 32 weeks with a history of premature births and who sustained obstetrical laceration of the cervix?

20. What is the correct code(s) for a patient who was previously treated for tubal pregnancy and who returns today for oliguria?

SURGICAL COMPLICATIONS

Complications, as discussed in this section, are conditions that result as a consequence of medical or surgical care. While many of these complications are listed in the 990 codes, some are listed elsewhere. Additional codes may be required to denote organisms or processes that contribute to the complication. Complications can include not only surgical procedures, but also services performed on implants and other devices. Be careful, when you are coding a complication as medically or surgically caused, to ensure that the complication is a result of the care, for it may not be related. It simply may be an expected postoperative condition, which is not coded as a complication.

Medical care complication codes can be found in the alphabetic index, usually under *Complications* but also under other conditional words, such as *inflammation* or *failure*. If a complication is not a normal foreseeable aftereffect of medical or surgical care, then it must be documented in the file before it can be coded as a complication. Complications can include surgical or mechanical aftereffects or malfunctions. Infections are coded as 996.7.

Burns that result from procedures are coded with regular burn codes and not surgical complication codes.

Warning: Some postoperative conditions may be expected and foreseeable, such as fever or pain, and should not be coded as complications of surgery or surgical care. Also, physicians may describe a condition as being postoperative, but it may not be caused by any previous surgical or medical care. It may have occurred after care had been provided and, therefore, is not considered a surgical complication.

If a surgical complication is coded and a history exists of a condition that influences the patient's care, then the history should be coded.

ACTIVITY 6.21

1. What is the correct code(s) for a patient who has rejected a kidney transplant?

2. What is the correct code(s) for a patient who is diagnosed with failure of an artificial skin graft?

3. What is the correct code(s) for a patient who is diagnosed with stertorous respirations following endotracheal surgery?

4. What is the correct code(s) for a patient who is diagnosed with shock occurring during delivery?

5. What is the correct code(s) for a patient who is diagnosed with an ulcer of the ankle due to a cast for a fractured tibia and fibula?

6. What is the correct code(s) for a patient who is diagnosed with phlebitis following chemotherapeutic infusion?

7. What is the correct code(s) for a patient who is diagnosed with phlebitis due to inserted arteriovenous fistula?

8. What is the correct code(s) for a patient who is diagnosed with second-degree burns of the chest from phototherapy?

9. What is the correct code(s) for a patient who is diagnosed with traumatic mydriasis after surgery for exotropia?

10. What is the correct code(s) for a patient who is diagnosed with continuing cardiac insufficiency after implantation of a pacemaker a year ago?

11. What is the correct code(s) for a patient who is diagnosed with phantom limb syndrome?

12. What is the correct code(s) for a patient who is diagnosed with malfunction of enterostomy?

13. What is the correct code(s) for a patient who is diagnosed with thrombus of graft?

14. What is the correct code(s) for a patient who is diagnosed with apraxia following hospitalization for a CVA three weeks ago?

15. What is the correct code(s) for a patient who is diagnosed with bradycardia following angioplasty two days ago?

16. What is the correct code(s) for a patient who is diagnosed with a nonhealing surgical wound?

17. What is the correct code(s) for a patient who is diagnosed with accidental laceration of the left lung by a catheter following surgery?

18. What is the correct code(s) for a patient who is diagnosed with elephantiasis following mastectomy?

19. By what other medical terms is the condition in the previous question known as?

20. What is the correct code(s) for a patient who is diagnosed with meningitis due to vaccination?

DRUGS AND CHEMICALS

Drugs are displayed in a table format at the end of the ICD-9-CM code alphabetic index. The table is titled "Table of Drugs and Chemicals." Effects of drugs are listed as either poisonings or adverse effects. The poisoning code is a five-digit code, whereas the adverse effects are coded using E codes. Poisoning codes are caused by the improper use of a prescribed drug such as when taken not according to a physician's orders. This can include medication taken by the wrong person, taking the wrong dose of a prescribed medication, or when someone else, including a nurse or physician, mistakenly gives medicine to a patient. Poisonings occur because of errors by hospital staff who may administer the wrong drug. Poisonings also can include alcohol or over-the-counter medications when taken without notifying the physician.

The E codes listed in the Table of Drugs and Chemicals are listed by sources of external cause. These causes consist of Accident, Therapeutic Use, Suicide Attempt, Assault, and Undetermined. If a drug is taken properly and the patient experiences symptoms or diagnoses, then the symptom or diagnosis should be listed first, and then the proper E code for the drug, which is found in the therapeutic use column in the table.

Remember, E codes selected from the therapeutic use category cannot be used in conjunction with poisoning codes and instead should be listed with the effect, such as a symptom or diagnosis. If, however, a drug is taken under circumstances that constitute a poisoning as previously mentioned, then select the proper poisoning code first. The symptom or diagnosis then would be listed, followed by the E code that specifies the conditions under which the drug was ingested. It is assumed that the poisoning was accidental unless otherwise stated.

If several drugs were taken, they should all be coded. If a prescription drug was taken in addition to alcohol and/or an over-the-counter drug that the patient takes without notifying their physician, then the prescription, alcohol, and/or over-the-counter drug all are coded as poisonings, as the physician did not prescribe the combination of drugs.

If there are no symptoms or diagnoses stated, then use code 995.2 (unspecified adverse effect of drug, medicinal and biological substances). This may occur if the physician reports that the patient has effects or sequelae but does not provide any additional information.

If you are coding for drug complications and you cannot find the drug listed in the Table of Drugs and Chemicals, the drugs are listed by type, such as coagulants or diuretics, in Appendix C, in addition to the proper poisoning code. Or you can use any drug reference book, such as the *Physicians' Desk Reference (PDR),* which will list alternative names for each drug.

There are many other substances listed in the Table of Drugs and Chemicals, such as insect bites, chemical exposures, household products, vaccines, and minerals.

ACTIVITY 6.22

1. What is the correct code(s) for a patient who is experiencing severe stomach pains after ingestion of one full bottle of Clonidine in a suicide attempt?

2. If a person takes another person's medication, is this considered a poisoning? Explain.

3. What is the correct code(s) for a 3-year-old child who ingested petroleum jelly?

4. What is the correct code(s) for a patient who was stung by a sea anemone?

5. What is the correct code(s) for a patient who is seen today for gum swelling due to use of Coumadin?

6. What is the correct code(s) for a patient who was admitted after taking Medrol during a drinking binge?

7. What is the correct code(s) for a patient who developed fever due to the use of Lidex?

8. What is the correct code(s) for a patient who developed numerous side effects from ingestion of Sinequan?

9. What is the correct code(s) for a patient who experienced erythema while taking Dilantin in addition to alcohol?

10. What is the correct code(s) for a patient seen today for anaphylactic shock due to accidental ingestion of Solfoton?

11. What is the correct code(s) for a patient who had an allergic reaction to the Rocky Mountain spotted fever vaccine?

12. What is the correct code(s) for a patient who was seen for side effects from ingestion of thorazine?

13. What is the correct code(s) for a patient who was exposed to cupric sulfate fumes while at work and fainted?

14. What is the correct code(s) for a patient who is admitted for extreme fatigue who has been taking digitalis glycoside?

15. What is the correct code(s) for a patient who has polyneuropathy due to use of Elavil?

16. What is the correct code(s) for a patient who ingested over-the-counter antihistamines in addition to Coumadin?

17. What is the correct code(s) for a patient who took a friend's Ativan and developed dyskinesia?

18. What is the correct code(s) for a patient who developed side effects after proper ingestion of Tylenol with codeine?

19. What is the correct code(s) for a 2-year-old child who was forced to ingest cigarette lighter fluid and experienced burns of the throat?

20. What is the correct code(s) for a patient who experienced seizures after ingestion of ethyl alcohol?

NONSPECIFIC CONDITIONS

Sometimes well-defined diagnoses are not always possible to code from the documentation, so nonspecific codes need to be used. Chapter 16 (780–799) of the ICD-9-CM coding book lists conditions that are not well-defined diagnoses including signs, symptoms, and ill-defined conditions, such as abnormal, elevated, decreases/increases, and abnormal findings. Like a fourth or fifth digit of 9, which is nonspecific, the 780–799 codes should not be used if a more definitive diagnosis is available.

V CODES

V codes are used for circumstances other than disease or injury that typically are not listed in the regular tabular codes. Beware, for there are many conditions and situations listed in the V codes. Some are unusual and not easily recognizable as circumstances that need to be coded such as medical care for a homeless person, which can be coded using V codes. These are generally classified into problems, services, and factual circumstances. They include personal and family histories, visits that result in no diagnosis, postsurgical care, newborn, routine exams, screening tests, follow-up exams, exposure to, donors, contact with, contraception, carrier of, and attention provided for certain conditions. V codes often are secondary codes, but can be primary codes such as therapy or rehabilitation services.

Services may be provided when the patient does not have a condition or problem, such as tests that result in no determination of a diagnosis. *But every service provided (CPT or ICD-9-CM procedure) must have an ICD-9-CM diagnostic code!*

Factual V codes denote circumstances unrelated to actual medical conditions, such as carriers of communicable diseases, immunizations, and living conditions. If a patient is a carrier or is exhibiting sequelae of a disease, this is referred to as *status*. *Screenings* indicate that a patient is receiving services to test for a condition. Some of these screenings are performed routinely for certain populations for various reasons, such as mammograms for older women. When a patient presents with symptoms, this is not a screening but rather a diagnostic exam.

History codes indicate that a condition no longer exists, and can include personal, family, health, or surgical histories. History within a medical report may be indicated by the term *follow-up*. Histories denote that the condition no longer exists, although the patient occasionally may receive treatment as a preventive measure, such as for tuberculosis and other communicable diseases or chemotherapies.

Individuals may be carriers of a disease, but do not exhibit signs and symptoms of the disease. If so, these individuals can be coded from the V codes (V02). Individuals may also not be exhibiting signs or symptoms of a disease but have been exposed to a highly contagious disease. These can be coded using V01 codes. This is an important code particularly for those populations that are highly susceptible, such as the elderly, sick, and infants.

There are two groups of similar V codes. One group of codes is for follow-up care after surgery and the other group is for postsurgical status. Follow-up care is part of the ongoing care after a recent surgery and usually would be considered as part of the global care. Postsurgical status begins after the global period has ended, with the surgery and its aftercare having already been completed, but the patient returns some time later for other treatment. The postsurgical status code should be included when the previous completed surgical care may affect the outcome of the current care. For example, if a patient is receiving an angioplasty, but had cardiac bypass surgery a year ago, postsurgical status for the bypass should be coded.

Only use V codes when necessary in physician billing. *Do not* overcode—only conditions affecting the current treatment of the patient, such as history or health status, should be coded.

ACTIVITY 6.23

1. What is the correct code(s) for a patient who received tetanus antitoxin?

2. What is the correct code(s) for a patient who was given a blood test for employment?

3. What is the correct code(s) for a patient who had to be fitted with another cast three months after removal of a previous cast?

4. What is the correct code(s) for a patient who is seen for severe headaches due to a brain shunt inserted one month before?

5. What is the correct code(s) for a patient who received vaccinations for tetanus and diphtheria?

6. What is the correct code(s) for a pregnant patient whose fetus is diagnosed with spina bifida?

7. What is the correct code(s) for the birth of a premature newborn weighing 1525 grams whose mother delivered with forceps?

8. What is the correct code(s) for a patient who is seen for fatigue with a family history of leukemia?

9. What is the correct code(s) for a patient who sustained a concussion with loss of consciousness from an attack that was religiously motivated?

10. What is the correct code(s) for a patient who has developmental dyslexia?

11. What is the correct code(s) for a patient who worried well?

12. What is the correct code(s) for a patient who requires intermittent renal dialysis?

13. What is the correct code(s) for a homeless patient who was admitted for heat stroke?

14. What is the correct code(s) for a patient who has episodic abuse of cocaine?

15. What is the correct code(s) for a patient who was seen for a PAP smear?

16. What is the correct code(s) for a patient who is diagnosed with osteosarcoma of the fibula who had a mastectomy a year ago?

17. What is the correct code(s) for a patient who has claustrophobia?

18. What is the correct code(s) for a patient who receives implantation of an artificial eye?

19. What is the correct code(s) for a single liveborn baby born in the hospital before the mother was admitted?

20. What is the correct code(s) for a patient who is admitted today for sterilization who gave birth to triplets a year ago?

E CODES

E codes are codes listing the *external causes* of patient's conditions, diagnoses, and symptoms. They contain a wide variety of subjects, including accidents, place of occurrence, environmental conditions, and drug effects. E codes have their own alphabetic index. They are not used in the national exams.

MORPHOLOGY CODES

Morphology codes are used to classify neoplasms for statistical purposes. These statistics are maintained by Cancer Registries. The morphology codes consist of the capital letter M followed by a four-digit number, which represents the histological state, such as epithelial or adenomas. This number is followed then by a slash and a fifth digit, which indicates the behavior of the neoplasm. This fifth digit can be 0 for benign, 1 for uncertain behavior, 2 for *in situ,* 3 for primary site of malignancy, 6 for secondary site of malignancy, and 9 for uncertain status of malignancy. The M codes are located in the Appendix or within the regular ICD-9-CM alphabetic index under the cited neoplasm. M codes are not used on the national exams and are not used for billing purposes.

SUMMARY

All reimbursement forms must have at least one ICD-9-CM code listed to ensure proper payment. Proper use of ICD-9-CM codes will ensure that the codes are properly linked to the appropriate service provided, i.e., CPTs and HCPCS. The ICD-9-CM book has a definite structure, which allows the coder to navigate through the book and locate correct codes. The code books have a thorough and complete alphabetic index, which is helpful in locating diagnoses, but the actual selection of the code must always be selected from the tabular list of codes to ensure that proper formats of codes are maintained, for example, fifth digits are included where required. There are tables for hypertension and neoplasms located within the alphabetic index, which are useful in locating correct codes for these conditions.

In addition to the actual alphabetical and tabular listing of codes, there also are Appendices, which are helpful in locating codes and researching changes. A Table of Drugs and Chemicals exists for coding issues involving the proper or improper use of drugs and chemicals. There also is a tabular and alphabetical index for E codes, but these are not required on the national exams.

The ICD-9-CM coding book is extensive and requires knowledge and experience to use it properly and fully.

ANSWERS

ACTIVITY 6.1

1. In the alphabetical index, what other category can be checked for complications of the heart?

ANSWER: Heart disease. If the alphabetical index is checked for complications of the heart, there is a reference also to check disease of the heart.

2. Under the category of *pregnancy* within the alphabetical index, are there any subheadings described as *due to*? If yes, under what category are they listed?

ANSWER: Yes, there is a *due to* subcategory listed under the category *hemorrhage*.

3. What is the code for burns over 31% of the body with 22% being third-degree.

ANSWER: 948.32; The fifth digit is for the percentage of burn that is third-degree, whereas the fourth digit is for the total percentage of body burned by all types of burns.

4. How many digits are required to code bacterial pneumonia?

ANSWER: The type of bacteria is not specified, so only four digits are required. The correct code is 482.9, which is not the same as 482 codes that specify the type of bacteria that caused the pneumonia, 482.81–482.89.

5. For epilepsy with petit mal status, what is the fifth digit?

ANSWER: There is no fifth digit for this category. Although fifth digits for epilepsy are located prior to this code, they do not apply to this code as indicated in the coding notes.

6. Within the alphabetical index under "complications," what is the code for infection of vascular catheter?

ANSWER: 996.62; be careful of the indentations under "complications," then under "infection," then "catheter."

7. How many digits does the code for splinter in the eyelid have?

ANSWER: Four. There may be no mark indicating that there is a fourth digit for the splinter in the external eye, which is indicated as code 930; however, upon further reading of the codes, the fourth digit classifications do have an unspecified site listed. So, for example, code 930.9 would be correct.

8. What other terms could *agenesis* be located under within the alphabetical index?

ANSWER: Absence, by site, and congenital as indicated within the alphabetical index.

9. Are pallor and flushing coded together as one code? Why or why not?

ANSWER: No, although they are listed within the same category code as 782.6, the fifth digit distinguishes them, so two codes must be listed.

10. Describe herniation of nucleus pulposus in other terms.

ANSWER: Displacement of intervertebral disc. Within the code 722.2—displacement of intervertebral disc—herniation of nucleus pulposus is listed. Within the alphabetical index, herniation of nucleus pulposus will direct you to displacement.

11. How many digits are required for code 551?

ANSWER: There can be four or five digits required, as some of the 551 codes have four digits and some have five.

12. Utilizing only the alphabetical index, is V24.0 the correct code for admission for examination of a patient for postpartum status? Why or why not?

ANSWER: No, V24.0 is for observation only, not a checkup. The correct code would be V24.2.

13. True or false: Within the alphabetical index, the only codes associated with the status of pregnancy are for the mother.

ANSWER: False, within the pregnancy codes in the alphabetical index, there are notations redirecting the coder to codes for the fetus if the fetus is the patient and not the mother.

14. Using the alphabetical index only, what is the fifth digit for the code for management of the mother's care due to fetal distress during delivery?

ANSWER: 1. The fifth digits are indicated within the alphabetical index.

15. What additional information regarding the fifth digit is provided within the tabular list, which would assist in selecting the proper fifth digit for the code in the previous question?

ANSWER: The **tabular index** indicates that only a 0, 1, or 3 could be selected as a fifth digit and since delivery is mentioned, only 1 can be selected as the proper fifth digit.

16. What kind of congenital hiatal hernia is not included within the code 750.6?

ANSWER: Congenital diaphragmatic hernia. This is indicated as an exclude within the tabular index.

17. What is the proper code for myocardial infarction of the basal-lateral wall?

ANSWER: 410.50; The episode of care is not specified, so the fifth digit must be a 0.

18. What is the proper code for syphilis that is endemic and nonvenereal?

ANSWER: 104.0; This diagnosis is not coded from the syphilis codes of 090 through 099 as indicated by the exclude notes within the tabular index.

19. What is the proper code for hypertension exacerbated by anomaly of the renal artery?

ANSWER: 405.91; This can be found within the hypertension table under renal (artery). It does not specify whether it is benign or malignant, so unspecified must be selected.

20. What is the proper code for carcinoma of the connective tissue of the jaw?

ANSWER: 143.9; Within the neoplasm table, this code must be found under connective tissue, not jaw. Be careful of indentations, particularly when there are page changes.

ACTIVITY 6.2

1. Why or why not would medical services not offered in the home, because the spouse is unable to provide the services, be coded as V63.1?

ANSWER: Code V63.1 excludes lack of care within the home because no household member is able to provide patient care. Instead, this is coded as V60.4.

2. What is the proper code(s) for scoliosis due to rickets?

ANSWER: 737.43 and 268.1. The scoliosis is a late effect of rickets and so 268.1 for late effect must be coded in addition to the scoliosis. In the tabular index, there is a note specifying that the cause also should be coded.

3. What is the proper code for calcinosis intervertebralis?

ANSWER: 275.49 and 722.90. Both of these must be coded as indicated within the alphabetical index. Brackets are used in the index to indicate the requirement for coding both. This listing in the alphabetical index also indicates the order for the codes.

4. Bright's blindness should be coded as what?

ANSWER: Uremia; The alphabetical index indicates that it should be coded as uremia.

5. Why would acute pharyngitis not be coded as 460?

ANSWER: Code 460 excludes pharyngitis, which should be coded as 462 as the tabular index indicates.

6. Hardening of the arteries should be located within the alphabetical index under what terms?

ANSWER: Arteriosclerosis as indicated within the alphabetical index.

7. For coding Paget's disease with osteosarcoma, where specifically would you have to look?

ANSWER: Neoplasm, bone, and malignant as specified within the alphabetical index.

8. If a patient has pneumonia due to AIDS, would V08 be the correct code? Why or why not?

ANSWER: No, As directed within the tabular index, V08 cannot be used because V08 is for patients with AIDS who exhibit no symptoms or who have no conditions present. The patient has pneumonia, so this code cannot be used. Once the patient exhibits symptoms associated with AIDS, then code 042 is always used.

9. What else must be coded when using the code for disorders relating to extreme immaturity of a baby?

ANSWER: Weeks of gestation must be coded from codes 765.20–765.29 as indicated in the tabular index.

10. What definition is provided regarding the code for disorders relating to extreme immaturity of a baby?

ANSWER: Extreme immaturity usually implies a birth weight of less than 1000 grams.

11. What is the correct code(s) for parasitic conjunctivitis due to filariasis?

ANSWER: 125.9 and 372.15. The tabular index, under conjunctivitis, specifies that the underlying condition also must be coded first.

12. In addition to the code for complications, what additional code should be coded for CMV infection due to kidney transplant?

ANSWER: 078.5 for the CMV infection as directed in the tabular index under code 996.8.

13. Patient has arthropathy due to erythema nodosum. What is the proper code(s)?

ANSWER: 695.2 and 713.4. The tabular and alphabetical index indicates that the erythema nodosum should be coded first in addition to the arthropathy.

14. For the dependent personality disorders code, what is excluded?

ANSWER: Neurasthenia and passive-aggressive personality.

15. What codes are mutually exclusive with 496?

ANSWER: Code 491 through 493 cannot be coded with 496 as indicated in the tabular index.

16. What types of codes are excluded from the category for diffuse diseases of connective tissue?

ANSWER: Those affecting mainly the cardiovascular systems as indicated within the tabular index.

17. Dislocation can include what other conditions?

ANSWER: Displacement and subluxation as indicated within the *includes* for codes 830 through 839.

18. If a patient has chronic renal failure and heart failure due to hypertension, what would be the code(s)?

ANSWER: 404.93

19. If patient has a dislocation of the proximal end of the ulna, this should be coded as what?

ANSWER: Dislocation of the elbow as indicated within the alphabetical index.

20. What is another description for Jadassohn-Lewandowski syndrome?

ANSWER: Pachyonychia congenital, code 757.5, as indicated within the alphabetical index.

ACTIVITY 6.3

1. Which of the following codes are listed as NEC: V67.59, 528.5, 759.4, 062.8, and 312.0.

ANSWER: V67.59, 528.5, and 062.8 are listed as NEC in the alphabetical index.

2. What is the code for vitamin deficiency NEC?

ANSWER: 269.2 is listed in the alphabetical index as NEC although it is described as *unspecified* in the tabular index.

3. What are some modifying terms that describe pneumonopathy due to inhalation of other dust?

ANSWER: Byssinosis, cannabinosis, and Flax-dressers' disease, as listed under code 504.

4. What is the proper code(s) for benign ovarian neoplasm with hyperestrogenism?

ANSWER: 220 and 256.0; Code 220 states that an additional code should be used for any functional activity within codes 256.0 through 256.1.

5. What are two modifying terms for infarction of the lung?

ANSWER: Embolic and thrombotic, which are listed in parenthesis in the alphabetical index.

6. Under what other term can Phagedena be found?

ANSWER: Gangrene; the alphabetical index states *see also gangrene*.

7. What conditions are included in code 722.11?

ANSWER: Any condition classifiable to 722.2 of thoracic intervertebral disc.

8. What is the code(s) for severe meningococcal sepsis with acute renal failure?

ANSWER: 036.2, 584.9, and 995.92; Code 995.92 for severe sepsis in the kidneys states that an additional code must be used to specify organ dysfunction. Also, the underlying systemic infection should be coded first.

9. What is excluded with the code for adhesive middle ear disease?

ANSWER: Glue ear as indicated by the exclude for this code, 385.1.

10. What is the code(s) for a patient who is drunk due to alcohol with severe abdominal pains?

ANSWER: 980.0, 305.0, and 789.00; The tabular index indicates that an additional code must be used for drunkenness.

11. Fistula of the peritoneum can also be located within the alphabetical index under what term?

ANSWER: Peritonitis; in the alphabetical index, *see also peritonitis* is listed under *fistula, peritoneum*.

12. Infection resulting from a shunt, can be located under what terms in the alphabetical index?

ANSWER: Under *see complications, infection and inflammation, due to (presence of) any device, implant or graft* classified to 996.0-996.5 NEC.

13. What are some modifying terms for polymyositis?

ANSWER: Acute, chronic, or hemorrhage as listed, within parentheses, in the alphabetical index.

14. What is the code for acute rheumatism NEC?

ANSWER: 729.0; This is listed in the alphabetical index under rheumatism NEC and in the tabular index as unspecified.

15. What is the correct code(s) for paratyphoid fever with postdyseneteric arthropathy?

ANSWER: 002.9 and 711.30; The tabular index for code 711.30 states that the underlying disease must be listed first. The part of the body is unspecified, therefore, the fifth digit of the 711.30 code must be a 0 for site unspecified.

16. What is the code for lymphoma NOS of the lymph nodes of the axilla?

ANSWER: 202.84; The tabular index lists 202.8 as the proper code for lymphoma NOS. The fifth digit is determined by the anatomic site, which is the lymph nodes of the axilla.

17. To what coding categories does the fifth digit of code 719 apply?

ANSWER: 719.0–719.6; 719.8–719.9; These codes are listed in the tabular index just prior to the listing of the fifth digits and after the heading for code 719.

18. What is the proper code(s) for urinary tract infection due to *E. coli*?

ANSWER: 599.0 and 041.4; In the coding book under 599.0, there is a note stating *use* additional code to identify organism, such as *Escherichia coli*, therefore, two codes must be listed.

19. What is the proper code(s) for psychogenic psoriasis?

ANSWER: 316 and 696.1; The alphabetical index lists two codes that are required to be used together as indicated by the brackets, as well as a note within the tabular list.

20. Under what other term(s) can a description of *overactive* be found within the alphabetical index?

ANSWER: Hyperfunction as stated within the alphabetical index under *overactive*.

ACTIVITY 6.4

1. Pain in the ear is known as what?

ANSWER: Otalgia as listed within the alphabetical index.

2. What is the correct code(s) for a patient who has chronic and acute bronchitis with COPD?

ANSWER: It is coded as 491.21 and 491.22. Both codes must be listed as there is acute and chronic bronchitis, not acute exacerbation of the chronic bronchitis. Also, the COPD is included in the code for the chronic bronchitis, so it is not coded.

3. What is the code(s) for incompetent veins with varicosity?

ANSWER: 454.9; Incompetent varicose veins can be found in the alphabetical index under *incompetence, vein,* and *varicose*.

4. When selecting the proper code for cardiac failure by looking under *cardiac,* is it correct then to check under the term *condition* as directed in the alphabetical index? Why or why not?

ANSWER: No, *see also condition* as listed in the alphabetical index does not mean *condition,* it means the actual condition. In this case, the condition is the failure.

5. Which of the following subcategories are listed within cesarean delivery in the alphabetical index: malpresentation, dystocia, bicornis, pre-eclampsia, metrorrhagia, fetal distress, incarceration of uterus, pelvic arrest, and premature labor.

ANSWER: Bicornis, malpresentation, fetal distress, preeclampsia, and incarceration of uterus. Be very careful in following the indentations within the alphabetical index. Use the alphabetic letter order to recognize when alphabetical changes occur and when categories and subcategories change.

6. Is airplane glue sniffing considered an *effect*? Why or why not?

ANSWER: Yes, airplane glue sniffing can be located under the term *effect* in the alphabetical index.

7. How is *mass* specified within the alphabetical index?

ANSWER: By anatomic site as listed within the alphabetical index.

8. Under what terms in the alphabetical index can you find the effects on the mother of a Down Syndrome fetus?

ANSWER: Down Syndrome is considered a fetal genetic abnormality and the concern is how this affects the mother, not the baby. This can be located under the terms of *pregnancy,* then *management affected by.*

9. What is the code for a sensory problem with the chest?

ANSWER: V48.5; If you look under *problem,* in the alphabetical index, you will not find chest. Another term for chest can be trunk, which does list *sensory.* No other terms are listed elsewhere, so this is the only way to find this code.

10. Codes for labor can also be found under what other term?

ANSWER: Delivery as listed within the alphabetical index.

11. For fitting of a cardiac device (pacemaker), within the alphabetical index which terms would be considered general?

ANSWER: Fitting and device. Within the alphabetical index, you can locate the code for this description by locating *fitting,* then *device,* and then *cardiac.*

12. Is vomiting considered a *habit*? Why or why not?

ANSWER: Yes, vomiting is listed under *habit* within the alphabetical index.

13. Under what terms in the alphabetical index can you find a tear in the cervix due to a legal abortion, beginning with the word *tear*?

ANSWER: Abortion, legal, with damage to pelvic organs. Under *tear,* in the alphabetical index, *cervix* can be located and includes a *with* statement that specifies *abortion.* However, this directs us to *Abortion, by type with damage to pelvic organs.* If you look under *by type,* you will not find this, for *type* refers to terms such as *legal.* When you select *legal,* however, be sure to look under *damage to pelvic organs* because the exact instructions from the heading *tear* indicate this.

14. Under what terms would you find the code for crushing chest injury within the alphabetical index?

ANSWER: Injury, internal, intrathoracic organs NEC. If you check under *crush* in the alphabetical index, you then can look under *trunk,* then *chest.* This entry instructs you to *see Injury, internal, intrathoracic organs NEC.*

15. What would be the proper code(s) for the previous description?

ANSWER: 862.8; You must follow the precise instructions of the alphabetical index, which instructed us to look under *injury, internal, intrathoracic organs NEC.*

16. What is Sheehan's disease?

ANSWER: postpartum pituitary necrosis.

17. Under what category(ies) is *laceration* listed within the category for *injury*?

ANSWER: Intracranial and liver as listed within the alphabetical index under injury.

18. Regarding the absence of the ulna with incomplete humerus, does this include complete absence of distal elements? Why or why not?

ANSWER: No, the humerus code is distinct from the complete absence of distal elements, although it is hard to distinguish by the indentations.

19. Under history in the alphabetical index, code V17.8 is listed for *musculoskeletal disease.* Is this the correct code for a patient with a history of musculoskeletal disease? Why or why not?

ANSWER: No. this code is for family history of musculoskeletal disease.

20. What is the code(s) for multiple fractures with hemorrhage?

ANSWER: 804.3; Be careful when selecting this code as there are several indentations that go into other columns in addition to the generalized terms *with, multiple,* and *fractures* being used.

ACTIVITY 6.5

1. What is the code for the principal diagnosis for a patient who is admitted to the hospital for severe abdominal pain? During hospitalization, the patient experiences an MI and is treated for both.

ANSWER: The abdominal pain, 789.00, is coded as the principal diagnosis as there is no more definitive diagnosis for the pain, which is why the patient was admitted. Although the MI, 410.91, is more serious, it was not the reason the patient was admitted, so it is not the principal diagnosis.

2. What is the code for the principal diagnosis for a patient who is admitted to the hospital for severe abdominal pain? Tests reveal that the patient has appendicitis. During hospitalization, the patient experiences an MI.

ANSWER: The appendicitis, 541, is coded as the principal diagnosis. Although the patient was admitted for abdominal pain, there was a more definitive diagnosis, which is the appendicitis. The MI, 410.91, is more serious, but it was not the reason the patient was admitted, so it is not the principal diagnosis.

3. What is the code for the primary diagnosis for a patient who is seen at the outpatient clinic for severe abdominal pain? The patient experiences an MI and is treated for this.

ANSWER: The abdominal pain is not listed as the primary diagnosis because the major services were provided for the MI, 410.91.

4. What is the primary diagnosis for a patient who is seen in an emergency room for acute pulmonary edema? The patient is prepared to be transported to the hospital for possible pulmonary embolism with infarction?

ANSWER: 518.4; Code only the edema as the embolism and infarction are described as possible, which is not coded with outpatient coding.

5. What is the principal diagnosis for a patient who is seen in an emergency room after being injured in a car accident? Patient sustained a large laceration on his left arm and left leg, which required sutures. During this visit, patient complained of epigastric pain and tachycardia, which he states has been ongoing prior to the accident. Tests revealed the patient had a perforated ulcer for which he was treated.

ANSWER: 533.5; The perforated ulcer is more severe and required the greatest services, so it is coded as the principal diagnosis. Even though the wounds were the main reason for admission, they did not require the greatest service, so they are not listed as the principal diagnosis.

6. What is the principal diagnosis for a patient seen in the hospital who sustained a minor head wound when hit by a car while on his bicycle? At this time, the patient complains of nausea and abdominal pain, which tests reveal is due to a perforated ulcer. Treatment is provided in the hospital at this time for the perforated ulcer.

ANSWER: 873.8; The main reason the patient was admitted is the head wound, which would be coded as the principal diagnosis despite the fact that the perforated ulcer is more severe and required more services.

7. What is the principal diagnosis for a patient who is admitted to the hospital for paresthesia in the fingers of his hand? After evaluation and tests, she is diagnosed with angina.

ANSWER: 413.9; The paresthesia is a symptom of angina, which was diagnosed after evaluation and testing; therefore, the angina should be selected as the principal diagnosis.

8. What is the correct code(s) for a patient who is seen today in the hospital for chest pain and diaphoresis? The patient is diagnosed with obstructive cardiomyopathy. Patient's medical history indicates that she had a hysterectomy 3 years ago and 6 months ago received a left heart catheterization.

ANSWER: 425.4 and V15.1; The history for the catheterization should be coded as it is significant for the cardiomyopathy admission. Chest pain and diaphoresis are symptoms of the cardiomyopathy and should, therefore, not be coded.

9. What is the correct code(s) for a patient who had a failed abortion and is now experiencing abdominal tenderness, fever, tachycardia, and vomiting? After testing and evaluation, the patient is diagnosed with PID and sepsis.

ANSWER: 638.7; The PID is coded from the abortion codes as directed within the alphabetical index. The symptoms of abdominal tenderness, fever, and vomiting are not coded as these are symptomatic of the PID. The tachycardia, however, is not a symptom of the PID and should be coded.

10. In the alphabetical index, under what terms do you find the conditions in the previous question?

ANSWER: See *abortion, by type, with sepsis*. You must look under *disease,* then *inflammatory*. There is a listing for PID for a female, resulting from an abortion. The listing instructs you to *see abortion, by type, with sepsis*.

11. What is the correct code(s) for a patient who has a 30-year history of alcoholism and is seen as an outpatient for hepatomegaly and edema? Patient subsequently is diagnosed with cirrhosis?

ANSWER: 571.2 and V11.3; The history of alcoholism is pertinent to the treatment of this patient since the alcoholism is a contributor to the cirrhosis. The hepatomegaly and edema are not coded as they are symptomatic of cirrhosis.

12. What is the correct code(s) for a patient who is evaluated in the hospital after being transferred from urgent care with severe heartburn, difficulty eating, and possible hiatal hernia.

ANSWER: 553.3; Only the hiatal hernia is coded as the heartburn and difficulty eating are symptomatic of a hernia. For hospital coding, a statement of possible qualifies as a diagnosis that can be coded.

13. What is the correct code(s) for a patient seen as an outpatient for facial pain, multiple sclerosis, and nystagmus?

ANSWER: 340; Both facial pain and nystagmus are symptomatic of multiple sclerosis and are not coded.

14. What is the correct code(s) for a patient, who, as an outpatient had a hysterectomy a year ago for carcinoma of the uterus in addition to chemotherapy a year ago and is diagnosed now with metastatic carcinoma of the kidney?

ANSWER: 198.0 and V10.41; The carcinoma of the kidney is considered secondary as the cancer spread. There is no code for chemotherapy as this was given a year ago. The carcinoma of the uterus must be coded as history because it was excised.

15. What is the principal diagnosis for a patient admitted to the hospital for dizziness and palpitations? After evaluation and testing, the patient is diagnosed with renal artery stenosis. On the third day of hospitalization, the patient begins to demonstrate edema and is found to be in respiratory distress.

ANSWER: 440.1; The renal artery stenosis was determined to be the main reason for admission. Although the respiratory distress is significant, it is not related to the main reason for admission.

16. What is the primary diagnosis for a patient who was seen in an Emergency Room for dizziness and palpitations with possible renal artery stenosis, but is found to be in respiratory distress. Care is provided for stabilization, and the patient is transferred to the hospital.

ANSWER: 518.82; The main services provided are related to the respiratory distress, so this is the primary diagnosis. The possibility of renal artery stenosis is not coded as this is outpatient service, and possible diagnoses are not coded.

17. What is the correct code(s) for a patient who is seen as an outpatient for paresthesia in the right arm in addition to anhidrosis and hematuria? The patient is diagnosed with peripheral neuropathy.

ANSWER: 354.9 and 599.7; The peripheral neuropathy and hematuria would both be coded as they are not related to each other; however, the paresthesia and anhidrosis would not be coded as they are related to the peripheral neuropathy.

18. What is the correct code(s) for a patient who is seen in an Emergency Room for severe migraine headaches and vomiting?

ANSWER: 346.90; The vomiting is not coded as it is symptomatic of a migraine headache.

19. What is the primary code for a patient who is seen as an outpatient with dysuria, fatigue, and back pain? Lab tests reveal cystitis due to Chlamydia. The patient is diagnosed at this time with carcinoma of the lungs as shown on x-rays.

ANSWER: 162.9; The carcinoma is coded first as it prompted the most significant service at this time.

20. What is the principal code for a patient who is seen in the hospital for dysuria, fatigue, and back pain? Lab tests reveal cystititis due to Chlamydia. The patient is diagnosed at this time with AIDS, for which the patient is provided treatment and counseling.

ANSWER: 099.53; The AIDS is a more significant code, but it is not the reason for admission.

ACTIVITY 6.6

1. If a patient has chronic osteomyelitis of the forearm and upper arm, what would the fifth digit be?

ANSWER: 9; The fifth digit of 9 is used for multiple sites. Forearm and upper arm are specified as separate sites within the fifth digit classifications, so although they are both arm sites, they are considered multiple sites.

2. How is the anatomic site for carcinoma of the skin *in situ* applied within the codes?

ANSWER: The fourth digit indicates that anatomic site, not a fifth digit.

3. What is the correct code(s) for carcinoma of the trachea with metastasis to the duodenum and peritoneum?

ANSWER: 162.0, 197.4, and 197.6; Notice that the anatomic sites are specified within the fifth digit of the codes for the secondary neoplasms.

4. For dislocation of the wrist, is there one code available? Why or why not?

ANSWER: There are multiple codes available. First they are distinguished by open or closed, then they are distinguished by which part of the wrist is dislocated.

5. Which digit is used to denote the anatomic site for a foreign body within the eye that happened 3 months ago?

ANSWER: While the code 360 indicates the eye, the fifth digit indicates the anatomic part of the eye.

6. For cellulitis, which code and digit indicates the anatomic site?

ANSWER: The fourth digit as indicated within the 682 codes.

7. What is the correct code(s) for fracture of the tibia and fibula?

ANSWER: 823.82; Notice that the fifth digit includes both the fibula and tibia, so a second code is not necessary.

8. Is aseptic necrosis of the bone categorized by anatomic sites for coding? Why or why not?

ANSWER: Yes, but not typical anatomic sites, like legs and arms. It is classified by the parts of the bone, such as head and condyle.

9. What is the correct code(s) for bilateral partial paralysis of the vocal cords?

ANSWER: 478.33; Notice that the bilateral specifies additional information about the anatomic sites since both sides are included.

10. Are V codes ever differentiated by anatomic sites? Why or why not?

ANSWER: Yes, codes such as histories, surgeries, and exams can refer to anatomic sites, such as with acquired absence of organs (V45.7).

11. What is the correct code(s) for ulcers of the heel and midfoot?

ANSWER: 707.14; Note that multiple sites are included in this code.

12. How, for anatomic sites, is code 441 described?

ANSWER: By the specific location of an aneurysm.

13. What is the correct code(s) for a cerebral embolism?

ANSWER: 434.0; Notice that the anatomic site is specified by artery.

14. Is the code for rotator cuff syndrome of the shoulder further defined by anatomic site(s)? Why or why not?

ANSWER: No, the only reference to an anatomic site is in the fourth digit, not the fifth.

15. For fracture of the tibia, what digit indicates the anatomic site?

ANSWER: Both the fourth and fifth digits indicate an anatomic site, with the fifth digit indicating whether it is tibia and/or fibula and the fourth digit indicating which part of the tibia and/or fibula.

16. Do three digits ever specify anatomic sites? Why or why not?

ANSWER: Yes, anatomic sites can be specified by any digit, but often by the fourth or fifth. For example, 231 refers to carcinoma of the respiratory system.

17. What is the correct code(s) for lymphoma of the spleen and intra-abdominal lymph nodes?

ANSWER: 200.87 and 200.83; Notice that two codes must be used, although there is a fifth digit for multiple sites; however, this fifth digit is for multiple sites for lymph nodes only, which does not include the spleen.

18. Is the code for injury to the peripheral nerve of the upper limb further specified by anatomic site? Why or why not?

ANSWER: Yes, the fourth digit specifies what type of nerve it is, e.g., ulnar, median, etc.

19. What is the correct code(s) for sphenoidal sinusitis?

ANSWER: 461.3; Note that the fourth digit specifies anatomic structures of the sinuses.

20. What is the correct code(s) for laceration of the esophagus?

ANSWER: 862.32; Notice that the fourth and fifth digits distinguish the anatomic sites.

ACTIVITY 6.7

1. What is the proper code(s) for a patient who is admitted to the hospital for tests and evaluation for complaints of lymphadenopathy, fatigue, and weight loss with possible enteritis?

ANSWER: 558.9; Although the enteritis is described as possible, since the patient is in the hospital, the possible diagnosis is coded.

2. What is the correct code(s) for a patient seen in the Emergency Room who is experiencing paresthesia of the left leg? The patient has a history of poliomyelitis.

ANSWER: 782.0 and V12.02; This is not a late effect. The physician, however, has noted that the patient has a history of polio, but has not stated that the paralysis is due to the polio. Because it may be significant, the history is coded.

3. For an injury to the blood vessels of the head, to what is the unspecified code referring and which digit refers to the unspecificity?

ANSWER: Blood vessel of head and neck, as listed in code 900.9 of which the fourth digit refers to the unspecificity.

4. What is the correct code(s) for a patient who is admitted to the hospital for bronchitis complicated by emphysema? In addition, the patient has a history of COPD.

ANSWER: 491.20; Although COPD, particularly with bronchitis and emphysema, would not be considered a history, it is not coded because codes from categories 491 through 493 cannot be coded in addition to the COPD (496).

5. Under what terms is ischemia of the heart 7 weeks ago coded as?

ANSWER: Infarct, myocardium as listed in the alphabetic index under ischemia.

6. What is the correct code for a patient who was injured while riding his bicycle? He sustained a fracture of his tibia. Patient is asymptomatic for HIV infection.

ANSWER: 823.80 and V08; The diagnosis of AIDS is not the etiology for the fracture, so the AIDS is not listed first. The patient does not have any symptoms at this time for AIDS, so it is coded as V08.

7. What is the correct code(s) for a patient who is admitted to the hospital with complaints of chest pain, tinnitus, and hearing loss? Patient was given diagnoses of possible otorrhea and Wegener's granulomatosis.

ANSWER: 446.4; With inpatient coding, diagnostic statements described as possible are coded; the otorrhea, as well as the other symptoms, are all symptoms of Wegener's granulomatosis, so only this is coded.

8. What is the correct code(s) for a patient who is seen in the Emergency Room for cramping and diarrhea, with suspected Crohn's disease?

ANSWER: 729.82 and 787.91; Only the cramping and diarrhea are coded, as the Crohn's disease is described as suspected and, therefore, not coded for outpatient coding.

9. What is the correct code(s) for a patient who is seen as an outpatient where he is diagnosed as having either cirrhosis or cholelithiasis?

ANSWER: There are no known codes at this time because provisional statements, such as possible or either/or, are not coded as diagnoses; instead, the symptoms would be coded, but no symptoms are provided. The coder should discuss this with the physician.

10. What is the correct code(s) for a patient who is seen in the Emergency Room for fever, headache, and rigidity of the neck? Lab reveals presence of pneumococcus.

ANSWER: 780.6 and 784.0 and 781.6; Although the lab reveals presence of pneumocoocccus, which is an indicator of meningitis and which the symptoms confirm, it is not the responsibility of the coder to interpret the abnormal lab results, so only the symptoms are coded.

11. Describe the unspecific terms for the category that includes infectious diarrhea.

ANSWER: Ill-defined intestinal infection, codes 009.

12. What is the correct code(s) for a patient who is admitted to the hospital for chest pain and coughing, which is diagnosed as pulmonary embolus versus heart failure?

ANSWER: 428.9 and 415.1; With inpatient coding, provisional statements, such as versus or possible, are coded due to the definiteness of evaluation and testing that occur in the hospital. The chest pain and coughing, therefore, do not need to be coded, but both diagnoses of embolus and heart failure should be coded.

13. What is the correct code(s) for arteriosclerotic vascular disease NOS?

ANSWER: 440.9, which is described as generalized and unspecified atherosclerosis.

14. What is the correct code(s) for a patient who is admitted to the hospital for Kaposi's sarcoma of the palate due to AIDS?

ANSWER: 042 and 176.2; AIDS is related to the Kaposi's sarcoma, so the AIDS is listed first.

15. What is the correct code(s) for a patient who is seen in the hospital for suspected MI? EKG reveals abnormal readings.

ANSWER: 410.91; The MI is listed as suspected, so it is coded. The abnormal lab does not influence this decision to code the infarction, since coders cannot use lab results to make their own patient diagnosis.

16. What is the correct code(s) for a patient who seen as an outpatient for suspicious nodular lumps on the neck, possibly carcinoma? Tests were inconclusive.

ANSWER: 784.2; The only known complaint of the patient is the lumps, so this is coded as a mass as directed in the alphabetical index. The carcinoma is stated to be possible and since this is outpatient, it is not coded.

17. What is the correct code(s) for a patient who is seen in the Emergency Room for severe lower abdominal pain and dysuria? The patient has a history of a brain tumor and urinary calculi.

ANSWER: 788.1, 789.0, and V13.01; The history of urinary calculi is coded, as it is applicable to the current symptoms, but the brain tumor is not. There is no definite diagnosis, so the symptoms of dysuria and abdominal pain are coded.

18. Name some unspecified malignant neoplasms of the histiocytic tissue.

ANSWER: Follicular dendritic cell sarcoma, interdigitating dendritic cell sarcoma, langerhans cell sarcoma, and malignant neoplasm of the bone marrow NOS.

19. What is the correct code(s) for a patient admitted to the hospital for fracture of his tibia when he fell from a roof? Patient has been diagnosed with AIDS but is asymptomatic at this time.

ANSWER: 823.80 and 042; Although the statement is made that the AIDS was asymptomatic, the fact that the patient has been diagnosed with AIDS indicates that he has been symptomatic in the past and is, therefore, coded with AIDS. The AIDS is listed second because the fracture is not caused by the AIDS.

20. What is the correct code(s) for a patient who is seen in the hospital for shortness of breath due to carcinoma, but leaves AMA shortly after admission?

ANSWER: 199.1 and 786.05; No specifications were provided regarding the site for the neoplasm, so unspecified is selected from the neoplasm table. The shortness of breath also is coded.

ACTIVITY 6.8

1. What is the correct code(s) for a patient who is seen in the emergency room for acute pneumonia due to SARS-associated coronavirus?

ANSWER: 480.3; This is viral pneumonia and the type is specified as the coronavirus.

2. What would be the code if the diagnosis in the previous question was chronic?

ANSWER: This condition is not defined as chronic.

3. Is it true that the only categories for thyroiditis are chronic and acute? Why or why not?

ANSWER: No, there is a category for subacute.

4. What is the difference between acute and chronic diabetes?

ANSWER: Diabetes is not differentiated as acute and chronic.

5. What is the correct code(s) for a patient admitted to the hospital for chronic ischemic heart disease?

ANSWER: 414.9; The description for this code states that it is NOS for ischemic heart disease.

6. Is code 416.8 correct for chronic pulmonary heart disease? Why or why not?

ANSWER: No, 416.9 should be used because no information is provided, so the disease is chronic, but also unspecified.

7. What is the correct code(s) for a patient who is seen in outpatient for chemotherapy with diagnosis of pineoblastoma?

ANSWER: V58.1 and 194.4; Remember that the chemotherapy is listed first, because this is the purpose of the visit.

8. Is code 590.80 the correct code for acute pyonephrosis? Why or why not?

ANSWER: No, code 590.10 is the correct code for the acute condition.

9. What is the correct code(s) for acute necrotizing ulcerative stomatitis?

ANSWER: 101; This is known as Vincent's angina.

10. Is code 743.00 the correct code for chronic anophthalmos? Why or why not?

ANSWER: No, anophthalmos is a congenital anomaly, not a chronic condition.

11. What is the correct code(s) for acute and chronic mastoiditis?

ANSWER: 383.00 and 383.1; These are listed in the tabular index as separate, so both must be coded.

12. What is the correct code(s) for a patient who is seen in outpatient for radiation treatment for malignant melanoma?

ANSWER: V58.0 and 172.9; Remember that the radiation therapy is coded first, because this is the reason for the service.

13. What is the correct code(s) for chronic appendicitis?

ANSWER: 542; Note that the description in the tabular index is listed as *other appendicitis,* but *chronic* is listed in the descriptors listed below the subcategory heading.

14. What is the correct code(s) for acute and chronic gastroenteritis?

ANSWER: 558.9; Note that there is no description in the tabular index of chronic or acute; however, within the alphabetical index it does direct the coder to code either/or, acute, or chronic condition as 558.9.

15. What is the correct code(s) for a patient who received chemotherapy for Schwannoma and returns 3 months later, for follow-up?

ANSWER: V67.2; A therapy code is not listed because the patient received follow-up exam services, not chemotherapy.

16. What is the correct code(s) for chronic venous insufficiency?

ANSWER: 459.81; Chronic venous insufficiency is listed under subcategory described as venous insufficiency, unspecified.

17. What is the correct code(s) for recurrent cyclitis?

ANSWER: 364.02; Although recurrent would seem to indicate a chronic condition, this is not so as indicated by the availability of an acute code, which is specified as recurrent for cyclitis.

18. What is the correct code(s) for a patient admitted to the hospital with chronic meningitis due to Lyme's disease?

ANSWER: 088.81 and 320.7; Meningitis is an infection as indicated here, so it is an acute condition, not chronic. Both the organism and meningitis must be coded.

19. What is the correct code(s) for a patient admitted to the Emergency Room for chronic nephritis and systemic lupus erythematosus?

ANSWER: 710.0 and 582.81; Note that the nephritis is a manifestation of the lupus, so the coding book directs you to code both.

20. What is the correct code(s) for patient seen as an outpatient for chronic bronchitis with acute exacerbation and obstruction?

ANSWER: 491.21; Note that the acute exacerbation would seem to indicate acute bronchitis, but this is not true. A patient can have a condition that has existed for some time and have episodes where it gets worse, which would be acute exacerbation of a chronic condition.

ACTIVITY 6.9

1. What code would have been the original code used for a third-degree burn of the left arm that occurred 6 months ago?

ANSWER: 943.30; The question is asking for the original code when the burn first occurred, which is the code in parenthesis in the late section.

2. What is the correct code(s) for a patient who is exhibiting facial droop due to a CVA 5 months ago?

ANSWER: 438.83; Note that only one code is necessary, because it includes both the current condition and the fact that it is a late effect.

3. What is the correct code(s) for a patient who is experiencing seizures due to viral encephalitis 6 months ago?

ANSWER: 780.39 and 139.0; Both the seizures and late effect of viral encephalitis are coded.

4. What is the correct code(s) for a patient with sequelae from an injury to the brachial plexus nerve?

ANSWER: 907.3; Note that there is specificity of codes within this late section, so use the codes in the parenthesis to determine if you are picking the right late effect code. Remember, the codes in parenthesis are the codes that were used to bill for the original injury. There is no other code except the late code because sequelae are nondefinitive and, therefore, cannot be coded. In other words, we do not know what sequelae are present.

5. What is the correct code(s) for repair of scar tissue on the face due to a third-degree burns?

ANSWER: 709.2 and 906.5; Scar tissue indicates that the burn is an old injury and today we are billing for repair of the scar tissue, not the burn.

6. What is the correct code(s) for malunion of fracture of the ulna?

ANSWER: 733.81 and 905.2; The malunion indicates that the fracture is a late effect and so *fracture* is located within the *late effect* category in the alphabetical index. Malunion also is coded.

7. What original codes would have been used to code for a current condition that is now described as a late effect of a sprain of the supraspinatus tendon?

ANSWER: 840.6; In the parenthesis listed within the alphabetical index under *late effect(s) of,* past conditions are selected from codes for sprain and strain.

8. What is the correct code(s) for a patient who is diagnosed with dysphasia due to a CVA three months ago?

ANSWER: 438.12; Note that two codes are not required because the one code includes the condition and the fact that it is a late effect of CVA.

9. What is the correct code(s) for the late effect of surgical amputation?

ANSWER: 977.60; Note that this late effect is described as postoperative and does not have any conditional codes listed in parenthesis, because a previous condition did not exist, as this is a result of surgery.

10. What other term is used in the coding book to describe late effects of encephalitis?

ANSWER: Myelitis, which can be found in the alphabetical index under *late effects of*.

11. What is the correct code(s) for a patient who is experiencing paresthesia in the toes after suffering a fracture of the fibula and tibia four months ago?

ANSWER: 782.0 and 905.3; The paresthesia is a late effect, so both the current condition and paresthesia should be coded.

12. What is the correct code(s) for a patient who presents today with an infection after a cesarean delivery 2 months ago?

ANSWER: 998.59 and 677; The code 677 is for late effect of complications of pregnancy. Be careful, because the infection code is specific to an infection caused by a surgical procedure.

13. What original codes would have been used to code for a current condition that is now described as a late effect of colitis due to *Entamoeba histolytica*?

ANSWER: 006.2; In the parenthesis listed within the alphabetical index under *late effect(s) of*, past conditions are selected from codes 001 through 136. *Entamoeba histolytica* is listed under amebiasis.

14. What is the correct code(s) for a patient who experienced a concussion with loss of consciousness for 45 minutes a month ago and is now experiencing severe headaches?

ANSWER: 784.0 and 907.0; Note that the late effect is for *intracranial injury*. We know this is the correct code, because the original condition coded with 850.12 is listed as covered within this category of late effect codes, which is denoted by the codes within the parenthesis.

15. What is the correct code(s) for a patient who is experiencing phantom limb pain?

ANSWER: 353.6; This code is not selected from the late codes. Although it would appear to be a late effect; it has its own code.

16. What is the correct code(s) for a patient who is experiencing effects from a skull and face fracture two months ago?

ANSWER: 905.0; The term *effects* does not provide any information concerning the present condition, so only the late effect code can be listed.

17. What original codes would have been used to code for a current condition that is now described as a late effect of a gunshot wound to the left hand?

ANSWER: 882.0; In the parenthesis listed within the alphabetical index under *late effect(s) of,* past conditions are selected from codes listed as *wound, open*.

18. What original codes would have been used to code for a current condition that is now described as a late effect from a laceration of the lung due to being crushed at a concert?

ANSWER: 926.19 and 861.22; In the parenthesis listed within the alphabetical index under *late effect(s) of,* past conditions are selected from codes for crushing. Note that the original condition required the use of codes for associated injuries, which would be the laceration of the lung.

19. What is the correct code(s) for a patient who has experienced hydrocephalus due to an intracranial abscess?

ANSWER: 331.4 and 326. Notice that the late effect code does not come from the 900 codes but is listed separately under the nervous system codes. In addition, there is a note specifying the coding of the condition in addition.

20. What is the current code(s) for a patient who has experienced sequelae due to an injury to the axillary nerve of the shoulder?

ANSWER: 907.4; Notice the specificity of this late effect code, which has numerous descriptions listed under *nerve*. Also, the term *sequelae* does not describe any condition that can be coded, so only the late effect is coded.

ACTIVITY 6.10

1. What is the correct code(s) for gastroenteritis due to Salmonella?

ANSWER: 003.0; This code includes both the organism and the condition, so only one code is necessary.

2. Does a patient have to have both heat stroke and sunstroke to qualify for code 992.0?

ANSWER: No, either one by itself or the two together would qualify for this code as indicated by *and*.

3. Does a patient have to have hyperosmolality and hypernatremia to qualify for code 276.0?

ANSWER: No, one or the other condition is necessary as required by the terms *and/or*.

4. What is the correct code(s) for a patient diagnosed with subacute sclerosis and pernicious anemia?

ANSWER: 281.0 and 336.2; Note that although *and* is used, the code for the sclerosis instructs you to also code the underlying condition, which indicates a causal relationship and functions more as a *with* statement.

5. What is the correct code(s) for a six-month pregnant patient diagnosed with UTI due to *E. coli*?

ANSWER: 646.63, 599.0 and 041.4; Note that although there only seems to be one existing condition, three codes are necessary, because coding directions instruct that the condition and organism must be coded in addition to the pregnancy condition.

6. What is the correct code(s) for a patient diagnosed with cardiopathy due to beriberi?

ANSWER: 265.0 and 425.7; Coding instructions direct that the underlying nutritional disease must be coded first, which indicates a causal relationship.

7. Must all digestive organs and spleen be diagnosed with neoplasm in order to use code 197.8? Why or why not?

ANSWER: No, this *and* is referring to any of these organs, not all.

8. What is the correct code(s) for a patient diagnosed with for eosinophilic meningitis?

ANSWER: 322.1; Note that although most meningitis codes require the coding the underlying disease, this particular meningitis does not.

9. What is the correct code(s) for a patient diagnosed with retinal microangiopathy?

ANSWER: 362.18; There are no other instructions for coding additional codes.

10. What is the correct code(s) for a diabetic patient with retinal microangiopathy?

ANSWER: 250.50 and 362.01; Note that this scenario differs from the previous question because diabetes is involved, which needs to be coded in addition since it is the cause.

11. What is the correct code(s) for a patient diagnosed with pericarditis who has TB and diaphoresis?

ANSWER: 017.9 and 420.0; This is a multiple coding scenario since the cause must be coded in addition to the condition as noted in the coding directions. The diaphoresis does not need to be coded, as it is a common symptom of TB, and there are no coding instructions requiring that it be coded.

12. What is the correct code(s) for a patient diagnosed with hypertensive malignant heart disease with congestive heart failure?

ANSWER: 402.01 and 428.0; Note that the one code includes the hypertension, heart failure, and benign heart disease, although there is a note stating that the type of heart failure also should be coded.

13. What is the correct code(s) for a patient diagnosed with malignant hypertensive heart disease with diastolic heart failure?

ANSWER: 402.01 and 428.30; Note that now two codes are required. The 402.01 still includes the heart failure, hypertension, and heart disease, but the type of heart failure is now known, so this must be coded in addition.

14. Do both lymphosarcoma and reticulosarcoma have to be coded together in order to use code 200? Why or why not?

ANSWER: No, although *and* is listed, these two conditions are not coded together as indicated by the further breakdown of the codes by fourth digits.

15. What is the correct code(s) for a patient diagnosed with septicemia due to Pseudomonas?

ANSWER: 038.43; This code includes both the condition and the organism, so only one code is necessary.

16. What is the correct code(s) for a patient diagnosed with gynecomastia and hemianopia with a pituitary tumor?

ANSWER: 237.0 and 611.1; The coding directions instruct that the functional activity also should be coded, which is the gynecomastia. Hemianopia does not need to be coded, as it is a common symptom of the tumor, and there are no coding instructions requiring it to be coded.

17. What is the correct code(s) for a patient diagnosed with galactosemia and cataracts?

ANSWER: 271.1 and 366.44; The coding directions instruct that the underlying condition must be listed first, so two codes are required.

18. What is the correct code(s) for a patient diagnosed with cirrhosis due to xanthomatosis?

ANSWER: 272.2; The alphabetical index instructs that the code includes both the cirrhosis and xanthomatosis, so only one code is required.

19. What is the correct code(s) for a patient diagnosed with hematoma due to open wound of the liver?

ANSWER: 864.11: Hematoma is associated with a wound, so only one code is required.

20. What is the correct code(s) for a patient diagnosed with vitamin B12 deficiency who is experiencing neuropathy and glossitis?

ANSWER: 266.2 and 357.4; In the alphabetical index, both codes are listed with the deficiency listed first. Glossitis does not need to be coded, as it is a common symptom of vitamin B12 deficiency.

ACTIVITY 6.11

1. What is the proper code(s) for a patient diagnosed with transient hypertension?

ANSWER: 796.2; Remember that transient means that the hypertension is temporary and, therefore, is not coded as hypertension.

2. What three categories are there for hypertension?

ANSWER: Benign, malignant, and unspecified.

3. What is the proper code(s) for renal failure and malignant hypertension?

ANSWER: 403.01; With renal failure, the hypertension is linked, whether or not it is indicated as such on the report, for it always is considered hypertensive renal failure. Notice the fifth digit of 1 because this is failure, not disease or involvement only.

4. Which of the following are not modifying terms for hypertension: vascular, idiopathic, secondary, etiological, diffuse, arterial, adaptive, subacute, primary, and paroxysmal?

ANSWER: Arterial, primary, idiopathic, vascular, and paroxysmal. These terms can be found with the term hypertension in the hypertension table in the beginning of the code book.

5. What is the correct code(s) for transient hypertension in a pregnant woman?

ANSWER: 642.3; Notice that this is different than when a woman is not pregnant, which would be 796.2.

6. What is the correct code(s) for secondary benign hypertension due to an aneurysm of the renal artery?

ANSWER: 405.11; This code can be found within the hypertension table under *secondary* and *renal*. It is benign, so it must be coded from this category.

7. What does secondary hypertension instruct you to code?

ANSWER: You will need to code the etiology of the hypertension, because secondary indicates that the hypertension has a cause.

8. What is the correct code(s) for hypertensive heart disease?

ANSWER: 402.90; The hypertension is described as contributing to the heart disease, so only one code is necessary, which contains both the heart disease and the hypertension. The type of hypertension is not specified, so unspecified is selected, rather than benign or malignant.

9. What is the correct code(s) for malignant hypertension due to polycythemia?

ANSWER: 405.09; Only one code is necessary because they are both included in the one code.

10. What is the proper code(s) for elevated blood pressure?

ANSWER: 796.2; The patient has elevated blood pressure, not hypertension.

11. What is the correct code(s) for pre-eclampsia with essential hypertension that is malignant?

ANSWER: There can be no malignant essential hypertension with pre-eclampsia as noted within the hypertension table, so you would need to check with the doctor as this scenario is not appropriate.

12. What is the correct code(s) for a patient diagnosed with diabetes with peripheral angiopathy and hypertension?

ANSWER: 250.70, 443.81, and 401.9; The diabetes is not specified, so a fifth digit of zero must be used. While the diabetes code does include the peripheral angiopathy, there is a note stating that the condition must be coded also, so code 443.81 must be used. The hypertension also must be coded, which is unspecified, so 401.9 must be used.

13. Is 401.9 the correct code for hypertension following surgery?

ANSWER: No, the correct code is 997.91.

14. What is wrong with the following scenario: patient with hemiparesis due to a CVA three months ago with hypertension, which is coded as 436, 402.90, and 342.9.

ANSWER: 438.20 is the correct code in addition to 401.9 for the hypertension. The code 438.20

includes the hemparesis in addition to the late effect of the CVA.

15. What does idiopathic mean?

ANSWER: Idiopathic means there is no known cause.

16. Who are you coding for if the code for chronic hypertension is coded as 760.0?

ANSWER: The fetus or newborn as listed within the hypertension table.

17. What is the correct code(s) for patient with edema, inflammation, and chronic hypertension due to deep vein thrombosis?

ANSWER: 459.12.

18. Which classification of hypertension did you select for the previous question (benign, malignant, or unspecified)?

ANSWER: The hypertension is coded as postphlebitic syndrome and does not have categories for benign and hypertension.

19. What is the code for malignant hypertension for a patient who has heart disease?

ANSWER: 429.9 and 401.0; This is not coded as hypertensive heart disease as it is not indicated that they are related, so both codes must be coded.

20. What are the diastolic and systolic pressures that usually are considered high enough to qualify as hypertension?

ANSWER: Diastolic 90 and systolic 140 mm of mercury.

ACTIVITY 6.12

1. If a malignant carcinoma has not invaded other adjacent structures, how would this be coded?

ANSWER: In situ, which is for neoplasms that have not spread to adjacent structures but are contained with a specific area.

2. Which of the following terms do not refer to neoplasms: gliomas, nodular, leukemias, hirsutism, sarcomas, astigmastism, seminoma, paronychium, and adenomas?

ANSWER: Sarcomas, leukemias, gliomas, adenomas, and seminoma are neoplasms. Nodular, hirsutism, and astigmasism are not neoplasms.

3. What is the correct code(s) for serous cystadenocarcinoma of the ovaries?

ANSWER: 183.0

4. How did you find the code for the previous question?

ANSWER: Checked under *cystadenocarcinoma* in the alphabetical index, which instructed you to check in the neoplasm table by anatomic site for a malignancy.

5. Provide descriptive terms for cystadenocarcinoma.

ANSWER: Serous, papillary, pseudomucinous, mucinous, and endometrioid, which are listed within the alphabetical index.

6. How would you code for cystadenocarcinoma that is benign?

ANSWER: Cystadenocarcinoma is always malignant as indicated within the alphabetical index.

7. What is the correct code(s) for a patient who is receiving chemotherapy for leiomyoma of the uterus?

ANSWER: V58.1 and 218.9; Note that the chemotherapy is listed first.

8. What did you look under in the Neoplasm Table to find the previous code?

ANSWER: You did not have to look under the Neoplasm Table, because it was listed directly under leiomyoma and uterus in the alphabetical index.

9. Is oat cell carcinoma benign, malignant, or *in situ*?

ANSWER: Malignant as stated in the alphabetic index.

10. What is the correct code(s) for oat cell carcinoma of the left lung?

ANSWER: 162.9 as listed within the neoplasm table.

11. What is the correct code(s) for Hodgkin's lymphoma of spleen and axillary lymph nodes.

ANSWER: 201.94 and 201.97; Note that although there is a fifth digit for multiple sites, this refers to lymph nodes only and not to the spleen, so two separate codes must be listed to indicate where the lymphoma is located.

12. Sarcomas can be found in which of the following anatomic sites: cardiac, muscle, glands, small intestine, bone marrow, uterus, bone, and fat?

ANSWER: Sarcomas are neoplasms of the connective tissue, which includes muscle, bone marrow, bone, and fat.

13. What is the correct code(s) for Schwannoma of the thigh?

ANSWER: 239.2; If you do not know what type of neoplasm Schwannoma is, you should look it up in the alphabetical index. In the index, Schwannoma is described as neoplasm of the connective tissue, either benign or malignant. We do not know whether it is benign or malignant, therefore we must code it as unspecified.

14. What is the correct code(s) for a patient who is diagnosed with bone metastasis. A year ago she had a mastectomy for carcinoma of the left breast and completed chemotherapy 5 months ago.

ANSWER: 198.5 and V10.3; Note that the breast cancer that was excised six months ago was the primary site, but since it no longer exists, it is coded as a history with the V code. The bone neoplasm is the secondary site and should be coded as such.

15. What is the correct code(s) for a patient receiving chemotherapy for rhabdomyosarcoma of the arm?

ANSWER: V58.1 and 171.2; Note that the chemotherapy is listed first. When you find rhabdomyosarcoma in the alphabetical index, it is described as neoplasm of the connective tissue and is always malignant.

16. What anatomic site does rhabdomyosarcoma refer to?

ANSWER: Striated muscle.

17. What is the correct code(s) for rhabdomyoma of the arm? Compare this to rhabdomyosarcoma with reference to coding.

ANSWER: 215.2; Rhabdomyoma is neoplasm of the connective tissue, but in contrast to rhabdomyosarcoma, it is benign as indicated by the suffix –oma.

18. What anatomic sites are included in leiomyosarcoma?

ANSWER: Smooth visceral muscle, which is indicated by –myo, which refers to the muscle and –leio, which refers to the smooth muscles.

19. What is the correct code(s) for a patient who had a mastectomy a year ago and then was treated with chemotherapy and returns today for a follow-up exam?

ANSWER: V67.2 and V10.3; This obviously is a follow-up visit since the excision and chemotherapy occurred a year before. The neoplasm was excised, so the patient only has a history of the breast cancer.

20. What is the correct code(s) for ovarian adenocarcinoma that invaded the gallbladder and peritoneum?

ANSWER: 183.0, 197.6, and 197.8; Note that the carcinoma has spread from the ovaries to the gallbladder and peritoneum which are, therefore, secondary sites with the ovaries the primary site.

ACTIVITY 6.13

1. What is the correct code(s) for a patient diagnosed with third-degree burns of the shoulder and chest areas?

ANSWER: 942.32 and 943.35; Notice that two codes are necessary so that both areas of the body are coded. These codes are used because the areas of the body are specified.

2. What is the correct code(s) for a patient diagnosed with ulcer of ankle resulting from postphlebitis that is inflamed?

ANSWER: 459.13 and 707.13; In addition, coding of the causal condition is required for postphlebitis.

3. What is the correct code(s) for a patient diagnosed with infection from a hand wound sustained while bicycling?

ANSWER: 882.0 and 958.3; Be sure to code for the infection in addition to the wound.

4. What is the correct code(s) for a patient diagnosed with infection of the hand from a wound that was repaired with sutures and debridement and was sustained while bicycling a month ago?

ANSWER: 958.3 and 906.1; Notice that a late effect is coded instead of a current wound in addition to the infection.

5. What is the correct code(s) for a patient diagnosed with infection resulting from a hand wound with injury of the tendons?

ANSWER: 882.2 and 958.3; Notice that the code changes from the previous examples, because there is involvement of the tendons, but the infection is still coded.

6. What is the correct code(s) for a patient diagnosed with infection resulting from a hand wound with injury to the tendons that occurred a month ago?

ANSWER: 958.3 and 905.8; Notice that the late effect includes the former diagnosis of the tendon injury as well as the wound.

7. What is the correct code(s) for a patient diagnosed with eight skin tags?

ANSWER: 701.9; Note that only one code is used for the diagnosis.

8. What is the correct code(s) for a patient diagnosed with endothelial sarcoma of the forearm?

ANSWER: 171.2; This type of sarcoma must be found under connective tissue, malignant, as the coding directions instruct.

9. What is the correct code(s) for a patient diagnosed with multiple lacerations of both knees and legs?

ANSWER: 959.7; No other diagnosis is necessary as this diagnosis includes all anatomic sites.

10. What is the correct code(s) for a patient diagnosed with lacerations of the left arm and fracture of the radius requiring sutures and application of cast?

ANSWER: 813.80 and 884.0; The fracture is the most severe injury, so it is coded first. It is coded as closed because the type (open or closed) is not specified. The lacerations are of the skin so are coded as wounds.

11. What is the correct code(s) for a patient diagnosed with an infection from diffuse cellulitis of the left cheek?

ANSWER: 682.0; Cellulitis of the external cheek area is coded, because infections with cellulitis are coded as cellulitis only. The term diffuse refers to external.

12. What is the correct code(s) for a patient diagnosed with infection and cellulitis at the colostomy site?

ANSWER: 569.61 and 682.2; The infection is coded as 569.61 and the cellulitis also is coded, which is not inclusive of the infection as the coding directions direct.

13. What is the correct code(s) for a patient diagnosed with burns over 43% his body with 17% described as third-degree?

ANSWER: 948.41; The fourth digit represents the total amount of body area burned for all types of burns. The fifth digit represents the total amount of third-degree burns only.

14. What is the correct code(s) for a patient diagnosed with cellulitis of the big toe of the right foot?

ANSWER: 681.10; The cellulitis is coded as 681.10 as directed in the alphabetical index.

15. What is the correct code(s) for a patient diagnosed with cellulitis of the big toe of the right foot, which is gangrenous?

ANSWER: 785.4; This code is used because the specified site is not listed in the alphabetical index, so the NOS code is used.

16. What is the correct code(s) for a diabetic patient diagnosed with cellulitis of the big toe of the right foot, which is gangrenous?

ANSWER: 250.70 and 785.4; The diabetes also must be coded in addition to the gangrene code.

17. What is the correct code(s) for a patient diagnosed with atherosclerosis who has cellulitis of the big toe of the right foot, which is gangrenous?

ANSWER: 440.24; Notice how this code differs from the previous questions in that all conditions are contained within this one code, including the atherosclerosis.

18. What is the correct code(s) for a patient diagnosed with scar tissue from burns to the left arm three months ago?

ANSWER: 709.2 and 906.7; Scar tissue is a late effect of the burn, so both the scar tissue and late effect codes must be listed.

19. What is the correct code(s) for a patient who was bitten in the nose during a fight?

ANSWER: 873.20; A bite is considered a wound for coding purposes.

20. Under what terms in the alphabetical index is the code found for the injury in the previous question?

ANSWER: Wound, open, by site.

ACTIVITY 6.14

1. What is the correct code(s) for a patient diagnosed with malunion of a traumatic fracture of the left arm?

ANSWER: 733.81 and 905.2; Upper extremity refers to the arms. The late effect is coded because the malunion indicated the fracture is a late effect.

2. According to the alphabetical index, what is the difference between arthritis and arthropathies for coding purposes?

ANSWER: They are coded from the same codes.

3. What is the correct code(s) for a patient diagnosed with fracture of the skull with subarachnoid hemorrhage.

ANSWER: 803.2; The code 803.2 includes both the hemorrhage and the fracture so only one code is necessary.

Be careful of indentations in the alphabetical index as this code is found under *skull* and then *with*.

4. What is the correct code(s) for a patient diagnosed with spontaneous fracture of the radius?

ANSWER: 733.12; This code is found under anatomic site for pathological.

5. Under what description in the alphabetical index is the code for spontaneous located?

ANSWER: Pathological.

6. What is the correct code(s) for a patient diagnosed with viral enteritis and arthritis?

ANSWER: 008.8 and 711.3; Although the enteritis is included in the code for the arthritis, an additional code is needed for the viral organism causing the enteritis.

7. What is the correct code(s) for a patient diagnosed with dislocation of the acetabulum:

ANSWER: 835.00; This code can be found in the alphabetical index under dislocation of pelvis or of hip.

8. What is the correct code(s) for a patient diagnosed with greenstick fracture of the lower epiphysis of the ulna?

ANSWER: 813.43; This is considered closed as indicated by the term greenstick.

9. What other anatomic sites are included in the code for the previous question?

ANSWER: Lower end of ulna, distal end, head, lower epiphysis, styloid process.

10. What is the correct code(s) for a patient diagnosed with infected fracture/dislocation of olecranon?

ANSWER: 813.11 and 958.3; The fracture/dislocation is coded as fracture only. It is coded as open, because infected indicates there is an opening. In addition, the infection should be coded.

11. What is the correct code(s) for a patient diagnosed with transient arthritis of hands and forearms?

ANSWER: 716.49; Notice the fifth digit is 9, which is for multiple sites, not unspecified as is often the description for fifth digits of 9.

12. What is the correct code(s) for a patient diagnosed with arthropathy with Behcet syndrome of the pelvis and thigh?

ANSWER: 136.1 and 711.25; Notice that in this case the arthropathy is located within multiple sites, but a fifth digit of 5 includes all of the areas, so no other coding is necessary.

13. Is code 828.0 the correct code for impacted fractures of ribs, sternum, and the left arm? Why or why not?

ANSWER: No, code 828.0 indicates that legs need to be included for this code to be used. Code 819.0 would be correct, as this includes the arm, sternum, and ribs.

14. What is the correct code(s) for a patient diagnosed with linear dislocation of the wrist?

ANSWER: 833.00; Linear indicates that this dislocation is closed, so the fourth digit is 0. The exact site of the wrist is not specified, so a fifth digit of 0 is used. This is not a fracture, only a dislocation, so no fracture is coded.

15. What is the correct code(s) for a patient diagnosed with arthritis in the bones in both hands and forearms?

ANSWER: 715.04 and 714.03; Arthritis in the bone is known as osteoarthritis, also which is known as osteoarthrosis. Both the hands and forearms must be coded as identified by the fifth digits, so there should be two codes.

16. What is the correct code(s) for a patient diagnosed with pathological fracture of the fibula?

ANSWER: 733.16; It is important to include pathological in the selection of the code.

17. What is the correct code(s) for a patient diagnosed with serum sickness and hypersensitive arthropathy?

ANSWER: 999.5 and 713.6; As the coding instructions direct, the serum sickness needs to be coded first.

18. What is the correct code(s) for a patient diagnosed with splenadenomegaly and leucopenia, including rheumatoid arthritis?

ANSWER: 714.1; This code includes all of the above mentioned diagnoses, so no other codes are necessary.

19. What is the eponym for the diagnosis in the previous question?

ANSWER: Felty's syndrome.

20. What is the correct code(s) for a patient diagnosed with a collapsed fracture?

ANSWER: 733.13; Notice that this fracture is described as a pathological fracture of the vertebra within the alphabetical index.

ACTIVITY 6.15

1. What is the correct code(s) for a patient diagnosed with CVA?

ANSWER: 434.91; CVAs now are referenced by the alphabetical index as classified under 434.91, not 436.

2. What is the correct code(s) for a patient diagnosed with hemiparesis due to a CVA 7 weeks ago?

ANSWER: 438.20; Notice that this code is for late effect of CVA.

3. Why or why not was the CVA coded in the previous question?

ANSWER: The CVA occurred 7 weeks ago and is not being currently treated, so it is not coded.

4. What is the correct code(s) for a patient being treated for apraxia resulting from a CVA a week ago?

ANSWER: 438.81; Notice that this code includes the late effect of the CVA.

5. Why or why not was the CVA coded in the previous question?

ANSWER: The CVA was not reported because it caused the late effect of apraxia, which is coded, not the CVA.

6. What is the correct code(s) for a patient diagnosed, upon admission to the hospital, with monoplegia of the left leg and then is treated for acute CVA?

ANSWER: 434.91 and 344.30; Notice that now we use two codes because the CVA is currently being treated, as well as the monoplegia.

7. Why or why not did you code the monoplegia in the previous question?

ANSWER: It was coded separately, because the CVA is acute and currently being treated.

8. Why or why not did you use a late effect code for the vertigo in the previous question?

ANSWER: A late effect code can be used because the coding book notes instruct to use these codes for effects any time after onset of CVA.

9. What is the correct code(s) for a patient diagnosed with facial droop due to cerebral thrombosis two weeks ago?

ANSWER: 438.83: The droop is coded as a late effect.

10. What is the correct code(s) for a patient diagnosed with transient neurological signs who is admitted to the hospital for care of transient ischemic attack?

ANSWER: 435.9; No other codes are coded because the description of the code explains that the transient signs are included.

11. What is the correct code(s) for a patient diagnosed with varicose veins of the rectum with bleeding?

ANSWER: 455.2; Notice that these varicose veins are described as included in this code according to the coding book description.

12. What term is the description in the previous questions commonly referred to as?

ANSWER: Hemorrhoids.

13. What is the correct code(s) for a patient diagnosed with phlebitis due to an elective abortion three weeks ago?

ANSWER: 639.8; This category includes complication of a previously performed abortion.

14. In the previous question, did you or did you not select a fifth digit? If yes, how did you select it?

ANSWER: No, there is no fifth digit required.

15. What is the correct code(s) for a patient diagnosed with malignant heart disease due to hypertension?

ANSWER: 402.00; The connection is described between the heart disease and the hypertension, so these are coded using one code.

16. What is the correct code(s) for a patient diagnosed with benign heart disease and hypertension?

ANSWER: 429.9 and 401.9; The connection is not made between the hypertension and the heart disease, so they are coded separately, since the coder cannot make the diagnosis that they are related nor can the hypertension be coded as benign.

17. What is the correct code(s) for a patient diagnosed with a long history of malignant hypertension, who is diagnosed with renal and heart disease with edema?

ANSWER: 403.00 and 428.1; Notice that we do code the renal disease with the hypertension, because it is known that they are related, but we cannot do this with the heart disease. The edema is not coded, as it is included in the heart disease code.

18. What is the correct code(s) for a patient diagnosed with transient hypertension?

ANSWER: 796.2; Transient hypertension is coded as elevated blood pressure in that transient indicates that it is not a permanent condition, so it is only considered to be elevated blood pressure and not hypertension.

19. What is the correct code(s) for a patient diagnosed with Wenckebach AV block?

ANSWER: 426.13; Notice that this is described as a Mobitz type I block.

20. What is the correct code(s) for a patient diagnosed with progressive muscular dystrophy that has resulted in heart disease?

ANSWER: 359.1 and 425.8; Notice that the etiology of the heart disease is listed as the dystrophy, so both codes must be listed as indicated within the alphabetical index.

ACTIVITY 6.16

1. What are the two ways hemolytic anemia can occur?

ANSWER: Hereditary or acquired.

2. What is the correct code(s) for a patient who is diagnosed with puerperal septicemia one week after delivery?

ANSWER: 670.04; The fifth digit indicates that the condition occurred after delivery.

3. What is the correct code(s) for a patient diagnosed with systemic sepsis due to anthrax?

ANSWER: 022.3; Notice that the code for the organism includes the sepsis.

4. What is the correct code(s) for newborn diagnosed with sepsis due to herpes?

ANSWER: 771.81 and 053.9; Pay attention to the notes concerning this condition, as there are many that seem to conflict with each other. Follow directions closely in such circumstances. The note for the newborn's sepsis directs us to code the organism, so that is what we do.

5. What is the correct code(s) for a patient diagnosed with acquired pancytopenia?

ANSWER: 284.8; The alphabetical index directs us to use code 284.8.

6. What is the description of the codes under which acquired pancytopenia is coded?

ANSWER: Aplastic anemia.

7. What is the correct code(s) for a patient who is diagnosed with anemia due to acute blood loss from injuries sustained during a car accident?

ANSWER: 285.1; The anemia is due to acute posthemorrhage.

8. How is the anemia in the previous question differentiated within the codes? Explain how this affected the choice of codes.

ANSWER: Chronic or acute; the coding book indicates that acute posthemorrhagic anemia should not be coded as chronic, but rather acute. The acute condition is indicated by the sudden onset caused by the accident.

9. Within the codes, is buccal sepsis specifically described as sepsis? Explain.

ANSWER: No, buccal sepsis is coded as cellulitis and abscess.

10. What is the correct code(s) for a patient diagnosed with sepsis of the urinary tract with septicemia due to Salmonella?

ANSWER: 003.1 and 995.91; The systemic infection also must be coded, which is septicemia due to Salmonella.

11. Was there another choice of codes for the previous question? If so, why did you pick the code you did?

ANSWER: Yes, the second code 599.0, however, this refers to the UTI, which was not described in the question.

12. What is the correct code(s) for a patient diagnosed with cryptogenic septicemia?

ANSWER: 038.9; Notice that this is for unspecified septicemia.

13. What is the correct code(s) for a patient diagnosed with bacteremia due to Strep B?

ANSWER: 790.7 and 041.02; Notice that bacteremia is not coded as septicemia although it, too, is an infection of the blood. Be sure to read the notes about coding the organism additionally.

14. What is the correct code(s) for a patient diagnosed with bacteremia with sepsis due to Clostridium?

ANSWER: 038.3; Notice that this is coded as septicemia since sepsis is described. If you check in the alphabetical index under bacteremia and sepsis, it will direct you to code it as septicemia.

15. What is the correct code(s) for a patient diagnosed with bacteremia due to RSV?

ANSWER: There is an error in this description because the condition is described as bacterial, but the organism described is a virus.

16. What is the correct code(s) for a patient diagnosed with pneumococcal septicemia with SIRS and hepatic failure?

ANSWER: 038.2, 995.92, and 570; The septicemia code instructs that the SIRS also must be coded. The SIRS code instructs that the organ failure must be coded as well.

17. What is the correct code(s) for a patient diagnosed with ESRD with anemia?

ANSWER: 285.21; Notice that only one code is needed since both conditions are included in the one code, and there are no notes specifying the need for additional codes.

18. What is the correct code(s) for a patient diagnosed with chronic hypertensive uremia?

ANSWER: 403.91; Notice that there is one code, which includes the hypertension and uremia.

19. What is the correct code(s) for newborn with anemia due to Rh incompatibility?

ANSWER: 773.0; Be sure to code for the newborn and not the mother.

20. What is the correct code(s) for a patient diagnosed with hyperchromic anemia due to acute blood loss?

ANSWER: 285.1; Be sure to code for the etiology of acute blood loss when selecting the code.

ACTIVITY 6.17

1. Asthma due to internal bodily conditions is known as what?

ANSWER: Intrinsic asthma.

2. If normal treatment does not alleviate asthma, this can be referred to as what?

ANSWER: Status asthmaticus.

3. What is the correct code(s) for a patient diagnosed with COPD and who is status asthmaticus?

ANSWER: 493.21; Notice that this one code includes everything. The fifth digit indicates the condition of status asthmaticus.

4. What is the correct code(s) for a patient diagnosed with COPD and pneumonia due to strep pneumoniae?

ANSWER: 481 and 496; COPD can be coded in addition to the pneumonia because the pneumonia codes are not excluded from use with the COPD codes.

5. What are the three types of pneumothoraces?

ANSWER: Spontaneous, iatrogenic, and traumatic.

6. What is the correct code(s) for a patient diagnosed with chronic sinusitis?

ANSWER: 473.9; Notice that this code is for chronic, not acute sinusitis with no other condition existing at the same time.

7. What is the correct code(s) for a patient diagnosed with acute and chronic sinusitis?

ANSWER: 461.9 and 473.9; Notice that two codes must be used to include both the acute and chronic, because one code does not exist that would include both conditions.

8. What is the correct code(s) for a patient diagnosed with upper respiratory infection?

ANSWER: 465.9; No sites are specified, so the unspecified code should be listed.

9. What is the correct code(s) for a patient diagnosed with lower respiratory infection?

ANSWER: 519.8; No codes are available that specifically code for lower respiratory infection.

10. What is the correct code(s) for a patient diagnosed with acute pulmonary edema?

ANSWER: 518.4; This edema is associated with lung problems and has its own code.

11. What is the correct code(s) for a patient diagnosed with edema due to congestive heart failure as associated with hypertension?

ANSWER: 402.91 and 428.0; The edema is included in the heart failure code, so it is not coded separately. The heart failure code instructs coders to code for the hypertensive heart condition additionally.

12. What is the correct code(s) for a patient diagnosed with TB associated with a pneumothorax?

ANSWER: 011.70; Notice that the one code includes both the TB and the pneumothorax.

13. What is the correct code(s) for a patient diagnosed with Rostan's asthma?

ANSWER: 428.1; Follow the directions precisely when coding for this type of asthma, as it is related to the cardiovascular system.

14. What exactly is the condition in the previous question described as?

ANSWER: Failure, ventricular, left.

15. What is the correct code(s) for a patient diagnosed with open sucking chest wound due to car accident?

ANSWER: 860.1; Although this is described as a sucking chest wound, it is traumatic, so it is coded from the 860 codes and not the 512 codes, as instructed by an exclusion notation.

16. What is the correct code(s) for a patient diagnosed with chronic tonsillitis and adenoiditis?

ANSWER: 474.02; Be sure to code both the tonsillitis and adenoiditis, which is contained within the one code.

17. What is the correct code(s) for a patient diagnosed with edema of the glottis?

ANSWER: 478.6; Although this is edema, it is of the larynx, so it is coded from the 478 codes.

18. What is the correct code(s) for a patient diagnosed with allergic bronchitis?

ANSWER: 493.90; Notice that this is listed as unspecified asthma.

19. What is the correct code(s) for a patient diagnosed with atelectasis?

ANSWER: 518.0; Atelectasis is collapse of the lungs.

20. What is the correct code(s) for a patient diagnosed with pleural effusion due to staph infection?

ANSWER: 511.1; Notice that although the organism is listed in the description, it is not coded separately because there is no note stating that the organism needs to be coded separately and also the staph is mentioned in the code's description.

ACTIVITY 6.18

1. If the urine of a patient is tasteless and watery, what type of diabetes do they have?

ANSWER: Diabetes insipidus.

2. What is the correct code(s) for a patient diagnosed with volvulus of the duodenum?

ANSWER: 537.3; No other information is provided, so volvulus is coded.

3. What is the correct code(s) for a patient diagnosed with hiatal volvulus gangrenous obstructive hernia?

ANSWER: 551.3; Notice that the gangrene must be coded, not the obstruction, because gangrene and obstruction are related.

4. What description did you look under in the alphabetical index for the previous question?

ANSWER: Hernia, by site, with gangrene.

5. What is the difference between CRF and ESRD?

ANSWER: ESRD is end stage renal disease and is the end stage of CRF, which is chronic renal failure.

6. What is the correct code(s) for a patient diagnosed with bronzed diabetes?

ANSWER: 275.0; Bronze diabetes is a disorder of metabolism with reference to iron, so it is coded from codes other than the 250 diabetes code.

7. A hernia that completely obstructs the bowels is known as what?

ANSWER: Incarcerated.

8. What is the correct code(s) for a patient diagnosed with retinal edema as a complication of insulin-dependent diabetes?

ANSWER: 250.51 and 362.01; There is a note in the coding book that directs the additional coding of the retinal edema condition, so two codes are required. The diabetes code is described as insulin-dependent, so a fifth digit of 1 must be used in addition to there being no statement of the diabetes being uncontrolled.

9. What is the correct code(s) for a patient diagnosed with sliding incarcerated hernia?

ANSWER: 550.10; Sliding hernia is coded as inguinal hernia if no other information is given. Incarcerated denotes obstruction, so this needs to be coded as obstructed. No information is given as to the hernia being bilateral, so an unspecified fifth digit of 0 must be used.

10. What is the correct code(s) for a patient diagnosed with acute renal failure?

ANSWER: 584.9; This code is for acute renal failure, not chronic.

11. What is the correct code(s) for a patient diagnosed with chronic uremia with neuropathy?

ANSWER: 585 and 357.4; Chronic uremia is classified under chronic renal failure. There is a note that requires an additional code be used to code for the condition of neuropathy.

12. What is the correct code(s) for a patient diagnosed with stones in the urinary bladder?

ANSWER: 594.1; Stones are known as calculus, so they are coded from the 594 codes.

13. What is the correct code(s) for a patient diagnosed with UTI due to *E. coli*?

ANSWER: 599.0 and 041.4; UTI is a urinary tract infection. The notes in the coding book indicate that the organism also must be coded.

14. What is the correct code(s) for a patient diagnosed with hypoglycemia due to diabetes?

ANSWER: 250.80; Notice that the hypoglycemia is coded form the diabetes codes, not the 251 codes for hypoglycemia. Although there is a note stating the manifestation also should be coded, the hypoglycemia is not coded additionally because the hypoglycemia code of 251 instructs not to use the 251 code when diabetes is the etiology.

15. What is the correct code(s) for a patient diagnosed with hormone-induced hypopituitarism?

ANSWER: 253.7; Hormone-induced indicates that this hypopituitarism is iatrogenic.

16. What is the correct code(s) for a patient diagnosed with diverticulosis with hemorrhage and diverticulitis of the ileum?

ANSWER: 562.03; There is a note indicating that diverticulosis is included in the code for diverticulitis.

17. What is the correct code(s) for a patient diagnosed with diverticulosis of the ileum with hemorrhage?

ANSWER: 562.02; No diverticulitis is indicated, so only the diverticulosis is coded with hemorrhage, which differs from the previous question.

18. What is the correct code(s) for a patient diagnosed with staghorn calculus?

ANSWER: 592.0; This calculus can be located in the alphabetical index under staghorn or kidney.

19. Where is a staghorn calculus located?

ANSWER: Kidneys.

20. What is the correct code(s) for a patient diagnosed with cholecystocolic fistula?

ANSWER: 575.5; This is a fistula of the gallbladder and can be described in many ways for indexing in the alphabetical index.

ACTIVITY 6.19

1. Within the codes, which of the following terms apply to hemiplegia: dominant, primary, suppurative, flaccid, and hysterical?

ANSWER: Dominant and flaccid.

2. If a patient presents with a long history of alcohol abuse, can they be diagnosed with acute episode of alcoholism? Explain.

ANSWER: Yes, because they can have an acute episode, where they ingest too much alcohol but still have the history of alcohol abuse. Chronic conditions do not necessarily prohibit an acute condition from occurring.

3. What is the correct code(s) for a patient diagnosed with acute iridocyclitis that is recurrent?

ANSWER: 364.02; This code indicates that the condition is recurrent, so the acute condition is coded as recurrent.

4. What is the correct code(s) for a patient diagnosed with continuous abuse of Dexedrine?

ANSWER: 305.71; Dexedrine is an amphetamine, so the patient is diagnosed with abuse of amphetamines. The patient has not been diagnosed with dependency, so nondependent abuse of drugs must be selected. The fifth digit indicates the continuous abuse.

5. What is the correct code(s) for a patient diagnosed with otitis externa due to impetigo?

ANSWER: 684 and 380.13; The underlying disease must be coded first as indicated in the notes.

6. What is the correct code(s) for a patient diagnosed with epileptic seizure?

ANSWER: 345.90; No other information is provided so this generalized code would be used.

7. What is the correct code(s) for a patient diagnosed with sarcoidosis with meningitis?

ANSWER: 135 and 321.4; The underlying disease also must be coded as indicated by the coding notes.

8. What is the correct code(s) for a patient diagnosed with drug-induced myelopathy?

ANSWER: 336.8; This code is described as *other* and is not specific to drug-induced only.

9. What is the correct code(s) for a patient diagnosed with myelopathy with cervical spondylosis?

ANSWER: 721.1; Although myelopathy is coded with 335 codes, when it is with spondylosis it is coded from the 721 codes.

10. What is the correct code(s) for a patient diagnosed with hemiplegia due to a CVA three weeks ago?

ANSWER: 438.20; Remember, hemiplegia as a late effect of CVA is coded from the 438 codes and not from regular hemiplegia codes.

11. What is the correct code(s) for a patient diagnosed with vision loss due to senile punctate cataracts?

ANSWER: 366.12: Senile refers to the aging process and how it can produce adverse effects. Punctate means small pinpoint punctures in the surface.

12. What is the correct code(s) for a patient diagnosed with visual impairment of the left eye at 20/250?

ANSWER: 369.71; 20/250 is considered severe impairment of the left eye as indicated in the eye chart within the coding book. The right eye is not specified, so 369.71 is the correct code.

13. What is the correct code(s) for a patient diagnosed with intractable clonic epilepsy?

ANSWER: 345.11; The fifth digit represents that the epilepsy is intractable and clonic is represented in the general code itself. Clonic refers to the alternating contractions and relaxation that can occur during an epileptic seizure.

14. What is the correct code(s) for a patient who is a complete quadriplegic at C5–C7?

ANSWER: 344.03; Quadriplegic codes are specified according to complete or incomplete and location of spinal injury.

15. What is the correct code(s) for a patient diagnosed with exophthalmos due to hyperthyroidism?

ANSWER: 242.90 and 376.21; The thyroid disorder must be coded first as directed within the notes.

16. What is the correct code(s) for a patient diagnosed with delirium tremens due to alcohol withdrawal?

ANSWER: 291.0; The regular code for alcohol withdrawal, 291.81, cannot be used because deliriums are specified, so code 291.0 must be used.

17. What is the correct code(s) for a patient diagnosed with serous otitis media?

ANSWER: 381.01; This condition does not describe suppurative, so nonsuppurative would be coded.

18. What is the correct code(s) for a left-handed patient diagnosed with spastic hemiparesis affecting the right side?

ANSWER: 342.12; This hemiparesis is spastic and affecting the nondominant side for this patient who is left-handed.

19. What is the correct code(s) for a patient diagnosed with migraines with characteristic auras?

ANSWER: 346.00; There is no mention of the migraine being intractactble, so it is coded as without specifications for the fifth digit.

20. What is the correct code(s) for a patient diagnosed with collapse of external ear canal due to inflammation?

ANSWER: 380.53; Collapse of the external ear canal is described as acquired stenosis of the canal.

ACTIVITY 6.20

1. What does G3, P2, Ab1 mean?

ANSWER: 3 pregnancies, 2 viable births, 1 abortion.

2. What is the correct code(s) for a patient who has retained a dead fetus at 23 weeks gestation?

ANSWER: 654.40; This is described as occurring at 23 weeks gestation, which is then termed intrauterine death of a fetus instead of a missed abortion (code 632).

3. What is the correct code(s) for a pregnant patient with malpositioning of the fetus, which was a successful version with delivery of a single liveborn?

ANSWER: 652.11 and V27.0; Successful version means the baby was successfully turned from a breech to head first birth.

4. Prolonged delivery means delivery after how many weeks of gestation?

ANSWER: 42 weeks.

5. What is the definition of a habitual aborter?

ANSWER: A patient who has had three or more consecutive pregnancies that abort spontaneously.

6. What is the correct code(s) for a newborn who is affected by the mother who is diagnosed with pre-eclampsia and edema?

ANSWER: 760.0; Do not use the 640 codes, because these are for the fetus who is the patient and is affected by the mother's condition.

7. What is the correct code(s) for a patient who delivered a single liveborn three days ago and returns today with deep vein thrombosis?

ANSWER: 671.44; The patient returned after being discharged for delivery, so only the postpartum condition is coded, not the delivery.

8. What is the correct code(s) for a 20-week pregnant patient who is diagnosed with mild hypertension with albuminuria?

ANSWER: 642.43; Although the physician does not state that the patient has pre-eclampsia, the alphabetical index within the hypertension table directs the use of the 642 code for pre-eclampsia for these symptoms.

9. What is the correct code(s) for a 24-week pregnant patient who is diagnosed with diabetes mellitus?

ANSWER: 648.03 and 250.00; The notes require that diabetes is coded in addition to the pregnancy code with diabetes. Notice that the pregnancy code is listed first.

10. What is the correct code(s) for a 22-week pregnant patient who is diagnosed with carcinoma of the fibular bone?

ANSWER: 170.7 and V22.2; Both the carcinoma and incidental pregnancy should be coded.

11. What is the correct code(s) for a 28-week pregnant patient who is seen today for abnormal fetal heart rate?

ANSWER: 659.73; The condition of the fetus is affecting the management of the mother's care. These codes can be found in the alphabetical index under *pregnancy,* then *management affected by.*

12. What is the correct code(s) for a 21-week pregnant patient who is diagnosed with rubella, which is suspected to have affected the fetus?

ANSWER: 655.33; 647.53 is a code for rubella in a pregnant mother, but since it is suspected that this condition has affected the fetus, code 655.33 is used instead as noted in the exclude note.

13. What is the correct code(s) for a patient who delivered a single liveborn with the use of forceps?

ANSWER: 669.51 and V27.0; The use of forceps indicates an abnormal vaginal delivery, so 650 code cannot be used.

14. What is the correct code(s) for a patient with corneal ectopic pregnancy?

ANSWER: 633.80; There are specific codes for ectopic pregnancies.

15. What is the correct code(s) for a 13-week pregnant patient who miscarried with shock?

ANSWER: 634.50; The fourth digit represents the complication of shock and the fifth digit refers to incomplete or complete abortion, which is coded as 0 because this is not specified.

16. What is the correct code(s) for a 36-week pregnant patient who delivered a single liveborn?

ANSWER: 644.21 and V27.0; Early onset of delivery is described as before 37 weeks, so this must be coded and includes the delivery. The normal delivery code, 650, cannot be used, which the coding book indicates.

17. What is the correct code(s) for a patient who delivered a single stillborn at 24 weeks gestation after rupture of amniotic sac 30 hours before?

ANSWER: 658.21, 656.41, and V27.1; The rupture of the amniotic sac at least 24 hours prior to delivery must be coded.

18. What is the correct code(s) for a 29-week pregnant patient who is diagnosed with rubella?

ANSWER: 647.53 and 056.8; The coding directions instruct that the condition of rubella must be coded additionally.

19. What is the correct code(s) for a patient who delivered a single liveborn at 32 weeks with a history of premature births and who sustained obstetrical laceration of the cervix?

ANSWER: 665.31, 644.21, and V23.41 and V27.0; The laceration is obstetrical trauma, so this must be coded to include the delivery. The outcome of delivery also must be coded. The history of premature delivery also should be coded, as this was a factor in the management of the mother's care.

20. What is the correct code(s) for a patient who was previously treated for tubal pregnancy and who returns today for oliguria?

ANSWER: 639.3; The tubal pregnancy currently is not being treated, although the oliguria resulted from it, so the pregnancy is not coded.

ACTIVITY 6.21

1. What is the correct code(s) for a patient who has rejected a kidney transplant?

ANSWER: 996.81; There is no other complication that can be coded.

2. What is the correct code(s) for a patient who is diagnosed with failure of an artificial skin graft?

ANSWER: 996.55; Be sure to not code this as autogenous graft, since the graft did not come from the patient.

3. What is the correct code(s) for a patient who is diagnosed with stertorous respirations following endotracheal surgery?

ANSWER: There is no code; Stertorous respirations are an expected complication of endotracheal surgery.

4. What is the correct code(s) for a patient who is diagnosed with shock occurring during delivery?

ANSWER: 669.11; This is not a postpartum condition, so the fifth digit should be coded as 1.

5. What is the correct code(s) for a patient who is diagnosed with an ulcer of the ankle due to a cast for a fractured tibia and fibula?

ANSWER: 707.06; This complication is not coded from the regular 996 codes for surgical complications.

6. What is the correct code(s) for a patient who is diagnosed with phlebitis following chemotherapeutic infusion?

ANSWER: 999.2; This code is particular for infusions, perfusions, and transfusions.

7. What is the correct code(s) for a patient who is diagnosed with phlebitis due to inserted arteriovenous fistula?

ANSWER: 996.62; Unlike the previous question, this is an implant of a device that has caused the phlebitis, which requires the use of this code as distinct from the code in the previous question.

8. What is the correct code(s) for a patient who is diagnosed with second-degree burns of the chest from phototherapy?

ANSWER: 990; This code is specifically for effects from radiation treatment.

9. What is the correct code(s) for a patient who is diagnosed with traumatic mydriasis after surgery for exotropia?

ANSWER: Nothing is coded as this is a condition that commonly occurs after eye surgery. Notice that if you check this condition in the coding book, it states that only persistent mydriasis is coded.

10. What is the correct code(s) for a patient who is diagnosed with continuing cardiac insufficiency after implantation of a pacemaker a year ago?

ANSWER: 429.4; This code is not from the regular surgical complication codes. Note that this code is used for long-term effects from cardiac implants, as effects within the postoperative (global) period are excluded from this code.

11. What is the correct code(s) for a patient who is diagnosed with phantom limb syndrome?

ANSWER: 353.6; Although phantom limb syndrome is a complication following the surgical procedure, it has its own specific code.

12. What is the correct code(s) for a patient who is diagnosed with malfunction of enterostomy?

ANSWER: 569.62; Malfunction of colostomy or enterostomy is coded as a mechanical complication.

13. What is the correct code(s) for a patient who is diagnosed with thrombus of graft?

ANSWER: 996.74; This is a code for the complication of a graft.

14. What is the correct code(s) for a patient who is diagnosed with apraxia following hospitalization for a CVA three weeks ago?

ANSWER: 438.81; Although apraxia is a result of the CVA a week before, there are codes for late effects of CVAs. Do not confuse this with effects resulting from medical procedures.

15. What is the correct code(s) for a patient who is diagnosed with bradycardia following angioplasty two days ago?

ANSWER: 997.1 and 427.89; Bradycardia is listed as a surgical complication that should be coded as noted within the coding book. The complication also must be coded.

16. What is the correct code(s) for a patient who is diagnosed with a nonhealing surgical wound:

ANSWER: 998.83; Nonhealing indicates that this condition is not a common side effect of the surgery, so it should be coded.

17. What is the correct code(s) for a patient who is diagnosed with accidental laceration of the left lung by a catheter following surgery?

ANSWER: 512.1; This occurred after the surgery so is not coded from 998.2.

18. What is the correct code(s) for a patient who is diagnosed with elephantiasis following mastectomy?

ANSWER: 457.0; Although this is a complication of a surgical procedure, it is coded with codes for noninfectious disorders of the lymphatic channels.

19. By what other medical terms is the condition in the previous question known as?

ANSWER: Postmastectomy lymphedema syndrome.

20. What is the correct code(s) for a patient who is diagnosed with meningitis due to vaccination?

ANSWER: 997.09 and 321.8; Both the complication code and the meningitis must be coded. Notice that in the 321.8 code for meningitis, you are directed to code the underlying disease, but there is not a disease, rather a complication that produced the meningitis.

ACTIVITY 6.22

1. What is the correct code(s) for a patient who is experiencing severe stomach pains after ingestion of one full bottle of Clonidine in a suicide attempt?

ANSWER: 972.6 and 789.00 and E950.4; The Clonidine was obviously taken against physician advice, so this is coded as a poisoning.

2. If a person takes another person's medication, is this considered a poisoning? Explain.

ANSWER: Yes, it is a poisoning because the person was taking another person's medication, which is against medical advice.

3. What is the correct code(s) for a 3-year-old child who ingested petroleum jelly?

ANSWER: 976.3; This code can be found in the Table of Drugs and Chemicals under Petroleum jelly/ointment.

4. What is the correct code(s) for a patient who was stung by a sea anemone?

ANSWER: 989.5; This code can be found in the Table of Drugs and Chemicals under Sea.

5. What is the correct code(s) for a patient who is seen today for gum swelling due to use of Coumadin?

ANSWER: 784.2 and E934.2; The gum swelling must be coded as well as the therapeutic use of the Coumadin as an E code.

6. What is the correct code(s) for a patient who was admitted after taking Medrol during a drinking binge?

ANSWER: 962.0 and 980.0; The Medrol is coded as a poisoning since the ingestion of alcohol is contradictory to physician orders. Medrol is known as methylprednisolone. The alcohol is coded as a poisoning also.

7. What is the correct code(s) for a patient who developed fever due to the use of Lidex?

ANSWER: 780.6 and E946.0; Lidex is a corticosteroid and so the fever due to the proper use of the steroid must be coded in addition to the E code for the Lidex.

8. What is the correct code(s) for a patient who developed numerous side effects from ingestion of Sinequan?

ANSWER: 969.0 and 995.2; Sinequan is known as doxepin and is coded as a poisoning. The side effects are not definitive, so these are coded as adverse effects.

9. What is the correct code(s) for a patient who experienced erythema while taking Dilantin and alcohol?

ANSWER: 966.1 and 693.0; The erythema is coded in addition to a poisoning code for the drug. This is a poisoning since the prescription was taken with alcohol, which is against physician advice.

10. What is the correct code(s) for a patient seen today for anaphylactic shock due to accidental ingestion of Solfoton?

ANSWER: 967.0, 995.0, and E851; Solfoton is a barbiturate and was taken by accident, so this is coded as a poisoning with an accompanying E code for accident. The anaphylactic shock also should be coded.

11. What is the correct code(s) for a patient who had an allergic reaction to the Rocky Mountain spotted fever vaccine?

ANSWER: 995.2 and E949.6; This code can be found in the Table of Drugs and Chemicals under Rocky Mountain spotted fever vaccine under therapeutic use, as the patient had an allergic reaction and no conditions of poisoning were noted. The use of the code for the allergic reaction is coded also.

12. What is the correct code(s) for a patient who was seen for side effects from thorazine, which was taken as directed?

ANSWER: 995.2 and E939.1; The thorazine was taken properly, so this is coded as therapeutic use with an E code. The side effects are coded as 995.2, since they are not specified.

13. What is the correct code(s) for a patient who was exposed to cupric sulfate fumes while at work and fainted?

ANSWER: 780.2 and 973.6; The fainting should be coded in addition to the exposure to the fumes. The codes can be located in the Table of Drugs and Chemicals.

14. What is the correct code(s) for a patient who is admitted for extreme fatigue who has been taking digitalis glycoside as prescribed.

ANSWER: 780.79 and E942.1; The digitalis glycoside is coded as therapeutic use, and the fatigue also must be coded.

15. What is the correct code(s) for a patient who has polyneuropathy due to proper use of Elavil?

ANSWER: 357.6 and E939.0; Elavil is amitriptyline, so the polyneuropathy is due to the correct use of the drug.

16. What is the correct code(s) for a patient who ingested over-the-counter antihistamines in addition to Coumadin without advising her physician?

ANSWER: 964.2 and 963.0; The Coumadin is coded as a poisoning since the over-the-counter was taken against physician advice as indicated by lack of physician notification.

17. What is the correct code(s) for a patient who took a friend's Ativan and developed dyskinesia?

ANSWER: 969.4 and 781.3; Ativan is known as lorazepam and should be coded as a poisoning, because the patient took the drug against physician advice since the prescription was not the patient's.

18. What is the correct code(s) for a patient who developed side effects after proper ingestion of Tylenol with codeine?

ANSWER: 995.2, E935.4, and E935.2; The side effects are listed as unspecified effects. The Tylenol and codeine were taken as prescribed, so they are coded from the therapeutic column of the Table of Drugs and Chemicals. Tylenol is known as acetaminophen.

19. What is the correct code(s) for a 2-year-old child who was forced to ingest cigarette lighter fluid by the stepfather and experienced burns of the throat?

ANSWER: 981, 947.2, and E962.1; Notice that this is coded as an assault. The poisoning code and E code for assault both must be listed for the lighter fluid.

20. What is the correct code(s) for a patient who experienced seizures after ingestion of ethyl alcohol?

ANSWER: 980.0 and 780.39; Notice that ingestion of ethyl alcohol is never considered to be therapeutic. Since it was taken against physician advice, it would be coded as a poisoning.

ACTIVITY 6.23

1. What is the correct code(s) for a patient who received tetanus antitoxin?

ANSWER: V07.2; This administration is coded as prophylactic immunotherapy.

2. What is the correct code(s) for a patient who was given a blood test for employment?

ANSWER: V70.4; This code is for medical/legal reasons, which would include employment checks.

3. What is the correct code(s) for a patient who had to be fitted with another cast three months after removal of a previous cast?

ANSWER: V53.7; Normally application of a cast would not be coded as a V code, because the initial fracture would have been coded instead. But if a patient has to have another cast applied by either the same physician or another physician, then this V code would be used.

4. What is the correct code(s) for a patient who is seen for severe headaches due to a brain shunt inserted one month before?

ANSWER: 784.0 and V45.2; Note that there is a code for headache due to loss of spinal fluid, which may be associated with this patient's diagnosis, but it was not stated as such by the physician, so the inclusion of the history of the shunt is important.

5. What is the correct code(s) for a patient who received vaccinations for tetanus and diphtheria?

ANSWER: V06.5; This code includes both tetanus and diphtheria, so only one code is necessary.

6. What is the correct code(s) for a pregnant patient whose fetus is diagnosed with spina bifida?

ANSWER: 655.03; Although there are pregnancy-related codes in the V code section, this is not coded from this section, but instead is a regular code.

7. What is the correct code(s) for the birth of a premature newborn weighing 1525 grams with forcep delivery?

ANSWER: V30.0 and V21.34; We are coding for the newborn whose birth weight is low, so this should be coded. The birth using V27 codes is not listed because we are coding for the newborn. The forcep delivery is not coded because the newborn was not affected by the use of the forceps.

8. What is the correct code(s) for a patient who is seen for fatigue with a family history of leukemia?

ANSWER: 780.79 and V16.6; The family history is coded as a V code because it may be related to the patient's condition.

9. What is the correct code(s) for a patient who sustained a concussion with loss of consciousness but who refused care due to religious beliefs?

ANSWER: 850.5 and V62.6; There is a code for a religiously discriminating act.

10. What is the correct code(s) for a patient who has developmental dyslexia?

ANSWER: 315.02; This is not coded from the V codes.

11. What is the correct code(s) for a patient who worried well?

ANSWER: V65.5; This is coded as a feared complaint in which no diagnosis was found but for which a patient was seen.

12. What is the correct code(s) for a patient who requires intermittent renal dialysis for renal disease?

ANSWER: V45.1 and 593.9; Be careful, for there are many codes that may be applicable, but this V code is specific to this description of service. Also, code the renal disease.

13. What is the correct code(s) for a homeless patient who was admitted for heat stroke?

ANSWER: 992.0 and V60.0; The V code is for the homelessness.

14. What is the correct code(s) for a patient who has episodic abuse of cocaine?

ANSWER: 305.62; This is not coded from the V codes for episodic abuse of drugs although it would seem to be a code associated with the Table of Drugs and Chemicals, but abuse of drugs has its own code that is not a V code.

15. What is the correct code(s) for a patient who was seen for a PAP smear?

ANSWER: V72.31; There is a specific code for the PAP smear as the reason for the visit.

16. What is the correct code(s) for a patient who is diagnosed with osteosarcoma of the fibula who had a mastectomy a year ago?

ANSWER: 170.7 and V10.3; The cancer is coded as malignant bone neoplasm. The mastectomy is coded as a history of breast cancer with a V code.

17. What is the correct code(s) for a patient who has claustrophobia?

ANSWER: 300.29; This is not coded from the V codes although it might seem to be a category that would be, instead it is coded from the regular codes as the alphabetical index indicates.

18. What is the correct code(s) for a patient who receives implantation of an artificial eye?

ANSWER: V43.0; There are V codes specific to organ or tissue replacements.

19. What is the correct code(s) for a single liveborn baby born in the hospital before the mother was admitted?

ANSWER: V30.1; Note that this is not the V27 codes for outcome of delivery, but is for the birth of the newborn.

20. What is the correct code(s) for a patient who is admitted today for sterilization who gave birth to triplets a year ago?

ANSWER: V25.2 and V61.5; The birth of triplets the year before should be coded with a V code known as multiparity.

Chapter 7: CURRENT PROCEDURAL TERMINOLOGY (CPT)

Objectives

(1) Understand the process of development and purpose of the CPT book.

(2) Navigate the structure of the book to find correct codes.

(3) Properly interpret and use the terminology in the CPT book.

(4) Distinguish between tabular and alphabetic indices and use them properly.

(5) Understand how to use appendices properly.

(6) Analyze details and codes skillfully.

(7) To be flexible when coding specialty areas and to be knowledgeable about the idiosyncrasies within specialties.

Key Terms

Alphabetic Index—Index in the CPT coding book that lists procedures and services in alphabetic order to assist in proper selection of codes.

Bundling—Bundling is the containment of all associated services, supplies, and materials within one main code that describes the major procedure/service provided so only one code is necessary for reimbursement purposes.

CPT—Current Procedural Terminology

Guidelines—The guidelines are information and notes located at the beginning of each section in the CPT book. These are extremely helpful and informative for selecting the proper codes. An important part of preparing for the national exams is familiarizing yourself with the guidelines.

Modifiers—Modifiers are two-digit alphanumeric codes that follow the CPT code and are separated from it with a hyphen. Modifiers are located in Appendix A of the CPT book or the HCPCS book. Modifiers indicate additional circumstances influencing the CPT or HCPCS code that may alter reimbursement; they help ensure accurate payment for services.

Tabular Index—The tabular index is a list of the actual codes in numeric order with notations to assist with correct coding.

CURRENT PROCEDURAL TERMINOLOGY (CPT)

INTRODUCTION

CPT means **Current Procedural Terminology**. CPTs are used by hospitals for outpatient billing. It is not used for physician billing by the hospitals since the hospitals are not billing for physicians. CPTs are not used for inpatient billing; Volume 3 of the ICD-9 coding book is used for coding procedures for inpatient hospital.

CPTs are five-digit numeric codes used to charge for services and procedures. Codes are categorized according to anatomic site, specific condition or service, and synonyms, eponyms, and abbreviations. There are six sections, which are: Evaluation and Management (E/Ms), Anesthesiology, Surgery, Radiology, Pathology/Laboratory and Medicine. The E/Ms are first, because they are used so often, even though this disrupts the numerical ordering of the codes. The codes are classified by anatomical site, procedures or services, condition, and other modes of description, such as abbreviations or synonyms.

Do not code more than the documentation you have, but do code to the greatest specificity that you can. Some codes are generalized, but others list specific procedures.

CPT codes are Level I HCPCS codes. HCPCS are Level II codes. Level III are local codes, are not on the exam, and have been eliminated under HIPAA. There are no HCPCS Level II codes on the abstracting portion of the CCS exam.

Every coding scenario must have a CPT/HCPCS code when you are coding for a physician, otherwise the physician does not get paid. CPT codes must be linked to at least one ICD-9 diagnostic code.

At the beginning of each section in the CPT coding book, there are **guidelines**. There also are guidelines and notes at the beginning of each subsection and within codes. Read them and understand them! They are crucial for the exams.

If a code contains all of the information regarding a service, then you do not need to code any additional information unless directed to do so by the notes associated with the code.

Remember to code for the services as documented in a written report—not documented, not done. Although there may be other sources of information, the physician's documentation is the final word in determining codes! For instance, labs and tests are not reported unless the physician has documented the services. On the exams, all that exists is what is in front of you!

Code to the proper level according to what services were provided. Selecting a code that is higher than the level of service documented is called *upcoding* and is wrong. Although this may become confusing, simplify it by focusing on what services actually were provided. Fear of upcoding may intimidate coders into coding less than is documented in the file, which is called *downcoding* and is also wrong.

LINKAGE

The purpose of the combined ICD-9, CPT, and HCPCS codes is to summarize a patient's care to someone who does not have access to the patient's file or time to review it. A prime purpose of coding is proper reimbursement and compliance with governmental and contractual regulations. A critical aspect of determining proper reimbursement is proof of medical necessity as demonstrated by the linkage of the codes. The diagnosis must prove the need for the service, which means the ICD-9 codes must be linked to the CPT/HCPCS codes in the order that they coexist together on the CMS-1500 billing form.

ACTIVITY 7.1

Select the correct CPT code according to its proper linkage to the ICD-9 code given.

1. 707.10
   A. 47562
   B. 88235
   C. 47700
   D. 11000

2. 443.9
   A. 35456
   B. 26910
   C. 66984
   D. 85547

3. 652.41
   A. 59100
   B. 59409
   C. 64680
   D. 29834

4. 820.8
   A. 27134
   B. 34900
   C. 51550
   D. 58920

5. 946.1
   A. 27635
   B. 11440
   C. 16000
   D. 16010

6. 427.0
   A. 11900
   B. 50200
   C. 33240
   D. 37650

7. 173.7
   A. 17261
   B. 37650
   C. 17000
   D. 11900

8. 706.1
   A. 41018
   B. 15780
   C. 15820
   D. 17260

9. 441.4
   A. 69950
   B. 47785
   C. 35082
   D. 33508

10. 711.9
    A. 36580
    B. 29871
    C. 64614
    D. 29835

11. 271.1 and 366.44
    A. 66920
    B. 26540
    C. 67005
    D. 11471

12. 174.9
    A. 31750
    B. 19367
    C. 53440
    D. 64831

13. 487.0
    A. 92997
    B. 33015
    C. 87400
    D. 78350

14. 410.00
    A. 62287
    B. 37195
    C. 52283
    D. 84484

15. 585
    A. 50080
    B. 90918
    C. 43460
    D. 26479

16. 594.1
    A. 65275
    B. 58960
    C. 36823
    D. 52353

17. 336.8
    A. 20500
    B. 22554
    C. 22556
    D. 33970

18. 456.0
    A. 52400
    B. 36470
    C. 43130
    D. 91030

19. 550.9
    A. 25360
    B. 49321
    C. 49560
    D. 49495

20. 733.82
    A. 64776
    B. 51820
    C. 25431
    D. 92568

Details

Because a CPT code can include a wide range of services or can vary slightly from another CPT code, it is critical that you read details, such as the notes. Before the exam, make sure you have read all of the notes in the CPT book, highlighting them if desired. Take the time to check before and after a selected code and to check any other possible code listed in the **alphabetic index** since there are slight details that may make one code better than another. Read the details, because some services are not coded within the same section as similar services, but may have their own codes or be in another section. One example is repair of lacerations, which could include repair of tendons and nerves and require the use of codes other than just repair of a wound.

Conventions

One standard notation is a circle with a 51 in it marked through with a slash. This indicates that a modifier -51 is not used with this code. This is helpful to have in your CPT book; however, you cannot use your books in the first part (multiple choice) of the AHIMA exam and **modifiers** are not required in the second part (abstracting) of the AHIMA exam.

A plus sign (+) indicates that a code is an add-on code and must be used in addition to the primary code, as noted in the CPT book. Be careful that you code these add-on codes in addition to their primary procedure code. Additional conventions in the book can include symbols indicating revised codes and new codes.

Triangles indicate a change in the description of a code. A bullet indicates a new procedure.

Indexes

Like the ICD-9 book, there are primarily two indexes—the tabular codes and the alphabetic index. As with the ICD-9 codes, do not code directly from the alphabetic index, but refer to the tabular list.

The alphabetic index is not as extensive as it is in the ICD-9 book. For instance, with CPTs, you may find that a specific service is not listed in the index or is described differently. At this point, you need to find familiar terminology and get yourself as close as you can to the right codes. In the index, you may even find several code areas listed. Until you are highly knowledgeable in the area, you should check out all of the suggested codes. Once you find an area of codes that appear correct, read the descriptions for each code until you find the right one. If the service is not specifically listed, you may have to select a less specific code. *Be careful!* Take a few seconds more to read before and after the code or to read the other codes listed in the index because you may think a code is correct, but CPT codes can be vague. They also can be very specific. Correct codes often are missed by one digit because of not investigating further.

ACTIVITY 7.2

1. What would be wrong if 99091 was reported twice for dates 1/2/04 and 1/8/04?

2. What is a trophoblastic tumor GTT?

3. What codes could be used to code for a pregnancy test?

4. What is the difference between the codes listed in the previous question?

5. If a coding scenario were 49905 and 44700, would it be correct? Why or why not?

6. Can submucous resection of turbinates be coded with 30520? Why or why not?

7. If a coding scenario was coded as 15220, 15221 × 6, and 15000, would this be correct? Why or why not?

8. Are codes for magnetic resonance angiography of the head distinguished as either with or without contrast? Why or why not?

9. If code 30465 is used to code for repair of left-sided nasal vestibular stenosis, is this correct? Why or why not?

10. How should a posterior capsulotomy be coded when there is extraction of the lens?

11. Evisceration is also known as what?

12. A Jones and Cantarow test is for what substance?

13. For a transurethral resection, can code 51530 be used? Why or why not?

14. If a bone graft is coded as 20930-62, is this correct? Why or why not?

15. If a coding scenario is coded as 62318 and 01996, is this correct? Why or why not?

16. What are the classifications for coding acupuncture?

17. ACTH means what?

18. If a coding scenario is coded as 45341 and 76975, is this correct? Why or why not?

19. Does selective vascular catheterization include introduction of the catheter? Explain.

20. If a coding scenario was coded as 31633 and 31629, would this be correct? Why or why not?

Bundling

Using one code for all services, supplies, and materials incurred during a procedure is called *bundling and the procedure is known as a global procedure.* For example, the procedure, approach, debridement, catheters, and follow-up care are all components of the main procedure and, therefore, would not be billed separately. This one code describes the major procedure/service provided so that only the one code is necessary for reimbursement purposes. These associated codes cannot be separately coded. Separating codes associated with a global code is called *unbundling.* Do not unbundle packaged codes.

In particular, do not unbundle Evaluation and Management codes for pre and postoperative care, which is part of the main procedural code. Absolutely do not do this on the exams! Time periods and services associated with a bundled code do vary.

Sometimes a service that is normally included within the global package may require more time and care than expected. There are codes and modifiers that specifically address the need for additional charges because of increased services or changes in such situations. This is particularly true in situations where patients have other conditions, such as diabetes or AIDS affecting their care and for whom healing may be longer and more complicated.

Semicolon

The semicolon is an important notation in the CPTs. *Be sure you understand it!*

When there is a semicolon in a coding description, it means that *everything* after the semicolon disappears when you go to an indented code listed after the original main code and is replaced by what is in the indented description. Everything before the semicolon is included in the indented codes. This means that *nothing* after the semicolon is included in the codes that follow within that category. If the services provided describe both the information after the semicolon within the primary code and the information contained within an indented code, then you will need to list both codes in order to include all of the information. The exception to this is when the descriptors, such as notes, instruct you to select only one of these codes. Read the *details*!

ACTIVITY 7.3

1. If a tendon transplantation of the wrist extensor is performed with grafts for each tendon, is it correct to code it as 25310 and 25312? Explain.

2. Would a test for total protein through collection of a urine sample be coded as both 81000 and 84160? Explain.

3. If codes 90920 and 90924 are coded together, is this correct? Explain.

4. If codes 69990 and 31536 were listed together, would this be correct? Explain.

5. If codes 82948 and 83026 are coded together, is this correct? Explain.

6. If a patient receives a limited lymphadenectomy with staging with prostatectomy, would it be correct to code it as 38562? Explain.

7. If coronary artery bypass grafts were performed utilizing four coronary veins and an arterial graft, would code 33513 and 33533 be correct? Explain.

8. Is it correct to use code 25315 times two to code for a flexor origin slide of the wrist and forearm? Explain.

9. Is it correct to code a 33 sq cm full-thickness graft of the chest as 15200? Explain.

10. Is it appropriate to code 3 antepartum visits as 59414, after which the patient discontinued visits with the physician? Explain.

11. If an open biopsy of internal mammary lymph nodes is performed with an axillary lymphadenectomy, can this be coded as 38530 and 38740? Explain.

12. If a complete cystectomy is performed with bilateral pelvic lymphadenectomy of the external iliac, hypogastric and obturator nodes, and an ureterosigmoidostomy, is it correct to code it as 51580 and 51585? Explain.

13. If a surgical hysteroscopy is performed with resection of the intrauterine septum with sampling of the endometrium, is it correct to code this as 58560? Explain.

14. Is it a correct coding scenario to code separately both a radical neck dissection and total thyroidectomy for malignancy? Explain.

15. Is it correct to code a layered closure of a 5.2 cm laceration of the cheek and a simple repair of a 4.8 cm laceration of the chin as 12053 and 12013? Explain.

16. If an x-ray of the pelvis is taken with three views, is it correct to code it as 72170 and 72190? Explain

17. If an ureteroneocystostomy and vesical neck revision were performed with anastomosis from a single ureter to the bladder, would it be correct to code it as 50780 and 51820? Explain.

18. If 18 skin tags were coded as 11200 and 11201, would it be correct? Explain.

19. If destruction of a sacral paravertebral facet joint nerve at two levels by neurolytic agent is coded as 64622 and 64623, is this correct? Explain.

20. If skin and subcutaneous tissue are excised for axillary hidradenitis, which required complex repair, is it correct to code this as 11450 and 11451? Explain.

CPT TERMINOLOGY

Be careful of the details. You must ensure that you have read all of the description, because not coding for each unit will cost you serious points. Some codes you think should be coded by *each* or *additional* units may not be coded this way, but others will be, so you must read the details.

EACH/SEPARATE/ADDITIONAL

When a code's description includes *each, separate,* or *additional,* you must charge for each unit of that code. *Add-on* codes are similar and are denoted by a plus sign (+) in the book. The CPT book provides a notation that an add-on code cannot be used without the prior, or main code, (for the first incident) and the add-on code is for the additional incident.

CROSS REFERENCES

Cross-references are terms that provide you with more information for correct coding. *See* and *See also* are such terms.

ANATOMICAL SITES

Beware of anatomical sites! Sometimes a body part may be listed separately, and sometimes it is included in another part of the body. Many sections are divided by anatomical sites. To ensure that you code the anatomic site with the proper code, check the alphabetical index under that anatomical site.

PROFESSIONAL VERSUS TECHNICAL

Professional component means supervision and interpretation. This means the physician reviewed the reports and interpreted them. Supervision and interpretation are services provided within certain codes, such as radiology codes.

Some testing codes include supervision and interpretation and some do not include this component. If a code is listed as supervision and interpretation, it means that the code does not include the technical component of testing. If you use a code that says supervision and interpretation, and the physician administered the tests, then the technical component has to be billed in addition. If the test code does not specify supervision and interpretation, it means that it includes the professional and technical components. In this case, if the physician provided only the professional component (supervision and interpretation), then you would have to use the 26 modifier to signify that the physician provided only the professional component. In addition, if only the technical component is performed in this scenario, then a TC modifier should be attached.

UNLISTED PROCEDURES

Documentation, including hospital reports and research articles, must be provided with unlisted procedures. Do not select these codes unless there absolutely are no codes for this service.

ACTIVITY 7.4

1. What is the correct code(s) for a patient who received a chest x-ray with four views without interpretation and supervision?

2. What is the correct code(s) for a patient who received a subsequent extended ophthalmoscopy with retinal drawing with interpretation?

3. Should modifier 26 be attached to the code in the previous question? Explain.

4. Under what terms can you find Burrow's Operation?

5. What is the correct code(s) for a patient who receives an open transluminal balloon angioplasty of two vessels in the brachiocephalic branches?

6. What is the correct code(s) for a patient who had three lacerations on his left arm and one on his neck? Two of the arm wounds were 2.2 and 5.2 cm and required closure of the superficial fascia. The other one on the arm was 0.8 cm and was more severe with avulsions and required retention sutures. The one on the neck was 2.2 cm and required repair, including the subcutaneous tissues.

7. Eutelegenesis is known as what?

8. What is the correct code(s) for a patient who received an MRI for excision of intracranial lesion with contrast, which was read and interpreted by the physician?

9. Is modifier 26 necessary for the previous coding scenario? Explain.

10. What is the correct code(s) for a patient who had three calluses removed?

11. What is the correct code(s) for a patient who receives closed treatment of a distal radial fracture and ulnar styloid?

12. What is the correct code(s) for a patient from whom an intramuscular biopsy of the elbow is obtained?

13. What is the correct code(s) for a patient who receives a capsulotomy of two metatarsophalangeal joints with a tenorrhaphy?

14. P&P is known as what in the CPT book?

15. What is the correct code(s) for a patient who receives a tenotomy of the Achilles tendon?

16. What is the correct code(s) for a patient who received an MRI for excision of intracranial lesion with contrast?

17. Is a modifier 26 necessary for the previous coding scenario? Explain.

18. What is the correct code(s) for a patient who received an excision of a bone cyst from the head of the radius?

19. What is the correct code(s) for a patient who had biopsies of three retroperitoneal lymph nodes, which include a diagnostic laparascope?

20. What is the correct code(s) for a patient whose physician reviews and interprets the results of her Bucky study from a chest x-ray?

ANESTHESIA

The AMA has developed codes for anesthesia services, but the American Society of Anesthesiologists (ASA) developed their own codes. Within the CPT codes, anesthesia generally is classified by anatomic site and is general in its descriptions. There are some codes, however, that specify certain procedures. The codes are global and include the additional usual services associated with administering anesthetic, such as monitoring services, administration of fluids and/or blood, and related evaluation and management (E/M) services. The codes also include supplies and materials normally used for these services.

The anesthesia codes are used when the physician provides only the anesthesia, and not the procedure. If the physician performs the surgery, then the procedure code is listed in addition to modifier 47.

With services, there may be certain types of anesthesia normally associated with them, such as a local for suturing. However, in certain circumstances, a different form of anesthesia may need to be administered, such as a general anesthesia when suturing a difficult patient. This is denoted by the use of modifier 23.

Anesthesia codes include additional codes and modifiers for physical status modifiers and qualifying circumstances. The physical status modifiers describe a patient's condition at the time anesthesia is administered: P1 (normal), P2 (mild systemic disease), P3 (severe systemic), P4 (severe systemic that is threat to life), P5 (not expected to live without surgery), and P5 (declared brain-dead). Qualifying circumstances should be listed in addition to the anesthesia code (add-on codes) and include 99100 for extreme age (under 1 and over 70), 99116 (total body hypothermia), 99135 (controlled hypotension), and 99140 (emergency conditions which must be specified). Do not forget these!

ACTIVITY 7.5

Select the proper anesthesia codes.

1. What is the correct code(s) for a 19-year-old patient who had anesthesia for a tenodesis of the bicep?

2. What is the correct code(s) for this 67-year-old patient who was admitted to the Emergency Room with critical injuries sustained in a car accident that included respiratory collapse of the left lung requiring ventilation?

3. What is the correct code(s) for a 48-year-old patient who received anesthesia during a coronary angiography?

4. What is the correct code(s) for a 43-year-old patient who received anesthesia for a cranioplasty?

5. What is the correct code(s) for this 6-month-old baby who received anesthesia for a chest tube?

6. What is the correct code(s) for a patient who received anesthesia for Harrington rod surgery?

7. What is the correct code(s) for a patient who received anesthesia for Torek procedure?

8. What is the correct code(s) for a healthy 28-year-old patient who had a total knee arthroplasty?

9. What is the correct code(s) for a 23-year-old patient who received daily management of her epidural while in the hospital?

10. What is the correct code(s) for a 57-year-old patient who receives anesthesia for a revision of her dialysis shunt?

11. What is the correct code(s) for a 16-year-old patient who received anesthesia for a double osteotomy of the tarsals?

12. What is the correct code(s) for a 62-year-old patient who received anesthesia for replacement of an aortic valve?

13. What is the correct code(s) for a 45-year-old patient who received anesthesia for a hysterectomy?

14. What is the correct code(s) for a 29-year-old patient who received anesthesia for a lumbar sympathectomy?

15. What is the correct code(s) for a 13-month-old baby who received anesthesia for removal of a toy from the larynx?

16. What is the correct code(s) for a baby who is 40 weeks gestationally and who received anesthesia for an incisional hernia repair?

17. What is the correct code(s) for a 73-year-old patient who received anesthesia for an osteotomy of the humerus?

18. What is the correct code(s) for a 24-year-old patient who received anesthesia for a cesarean hysterectomy following labor?

19. What is the correct code(s) for a 45-year-old patient who received anesthesia during a procedure for advancement of the flexor tendon in no man's land of the right hand?

20. What is the correct code(s) for an 18-year-old patient who received anesthesia for a bronchoscopy and who has mild systemic disease?

SURGERY

The surgical section is broken down by anatomic sites, which are further subdivided by procedure, classifiable to incisions, excisions, removal/introduction, repair, and other. Most surgery is coded as a package (global), i.e., all normally associated services are included in the one code for the surgical procedure. Usually with physician billing, preoperative and postoperative E/Ms are included. Within the global package for a procedure, however, this differs with the hospital, because hospitals do not provide professional long-term care as physicians do for patients. Supplies and materials that are a normal expected part patient care are included as part of the global. If additional care or supplies are required beyond what is normally expected, then the additional services can be coded also.

In contrast, some procedures are coded singularly, are not part of a package deal, and may be termed separate procedures within the coding book.

Code 99070 is the CPT code for supplies and materials; however, HCPCS can be used to specify the materials and supplies used and actually is preferable to code 99070.

SEPARATE PROCEDURES

As explained earlier, some codes in the CPT book are described as *separate*. This means that the procedure is normally associated with other procedures (globals) and should not be billed separately from the global code. However, on occasion the separate codes may be performed by themselves or with an unrelated code and may, therefore, be coded. Preoperative and postoperative care is not included in this code and can be coded separately, as well as supplies and materials provided by the physician. Remember the definition of separate procedure, because when you are coding a smaller procedure or service with a larger procedure, it could be included in the code for the larger procedure.

SIGNIFICANT PROCEDURES

A surgical procedure that has risks or requires special training is a significant procedure, which is used to classify procedures for ordering and classification status.

CLASSIFICATION

Surgical codes are categorized by anatomical site and are divided then into various categories such as Incision, Destruction, Excision, Removal, Repair, and Grafts, which have a consistent order throughout each anatomic site section. If you are unable to find a code in the alphabetic index, go to the anatomic site and look for the procedure. The procedures are listed in a procedural order.

SCOPES

Many procedures use scopes, although there are codes for the same procedures without scopes. You must remember to check for a scope when coding for procedures. It also is critical to remember that if the scope is diagnostic but proceeds into a surgical procedure, the scope is coded as surgical and not diagnostic. The diagnostic scope code is included within the surgical scope code.

Scopes include microscopes. Look for words, such as micro, that indicate a microscope was used. Also, look for these descriptors within the codes, because if they are described in the code, then the use of the microscope does not need to be coded. If not included in the code, then you must code for the use of a microscope with code 69990. This code lists various procedures that should not be coded in addition to 69990 because the use of the microscope is bundled into the listed procedures. For example in a direct laryngoscope, a scope is inserted into the mouth and pharynx but in an indirect laryngoscope, mirrors are used to visualize the larynx and a microscope is not used.

BIOPSIES

Similar to scopes, code biopsies when they are performed, unless they are included within a code. Biopsies are coded when a specimen is obtained by a needle aspiration, brushing, cell washings, a punch, or incision. Sometimes, specific biopsies will have their own codes, so be sure that you do your research. For example, bone marrow biopsy with fine needle aspiration is coded with 38220 instead of using another fine needle aspiration code. Biopsies with excisions are not coded separately but are a part of the excisional code.

Be sure to code for evaluation of the aspirate if performed, using code 88172 or 88173.

Be careful since biopsy codes can include multiple biopsies. List the biopsy code only once for biopsies within the same area, even if there is more than one biopsy performed. However, if a biopsy is performed in addition to an excision, but the sites are different, then both are coded.

Incisional biopsy refers to a small portion of the specimen being removed. Excisional biopsy refers to the entire specimen, such as a lesion, being removed, which is an excision. In some cases, a wire is used to mark the site for a specimen removal, which can be coded additionally; however, beware of codes that include the use of the wire for marking, e.g., 19125.

STENTS

Beware of stents. Stents are placed to provide access, usually long-term. If they are used but not included in the code, then they must be coded.

ACTIVITY 7.6

1. Which of the following codes are not separate procedures: surgical endoscopy, open tenotomy of the toe, arterial catheterization for transfusion, injection of ganglion cysts, drainage of external ear canal, and extensive biopsy of vaginal mucosa?

2. What is the correct code(s) for a patient who received a replacement of an inflatable bladder neck sphincter with pump, reservoir, and cuff, including removal?

3. What is the correct code(s) for a patient who received a lung needle biopsy?

4. What is the correct code(s) for a patient who received a biopsy by indirect laryngoscopy with removal of a lesion?

5. For the code(s) chosen in the previous question, are any procedures separate? Explain.

6. For the code(s) listed in the two previous questions, are any of them diagnostic scopes? Explain.

7. What is the correct code(s) for a patient in whom a cystourethroscopy was performed and a Gibbons stent was placed within the ureters?

8. What is the correct code(s) for a patient who received debridement of infected tissue with replacement of an inflatable bladder neck sphincter, which included the pump, reservoir, and cuff, including removal?

9. What is the correct code(s) for a patient who received a polypectomy with a diagnostic endoscopy and sinusotomy?

10. What is the correct code(s) for a patient who received a biopsy of a lesion in the pharynx?

11. What is the correct code(s) for a patient who received an upper GI endoscopy of the esophagus, stomach, and duodenum by brushing, including placement of a stent?

12. Is code 11011 considered a separate procedure? Explain.

13. What is the correct code(s) for a patient who received a fine needle lung biopsy?

14. What is the correct code(s) for a patient from whom a biopsy was obtained when a malignant 0.8 cm lesion was removed from the back?

15. What is the correct code(s) for a patient who received a percutaneous biopsy of the breast for needle core with imaging?

16. What is the correct code(s) for a patient who received a biopsy of a lesion of the pharynx that was excised?

17. What is the correct code(s) for a patient who received a biopsy of the cervix?

18. What is the correct code(s) for a patient who received a transcatheter placement of a stent within two coronary arteries after PTCA was performed?

19. What is the correct code(s) for a patient who received endovascular placement of a device for occlusion of the iliac artery in addition to placement of femoral-femoral prosthetic graft with exposure through a groin incision on the left side for repair of an endovascular aortic aneurysm?

20. Which of the codes in the previous question are separate procedures? Explain.

EXCISIONS

Excisions are denoted by the suffix *–ectomy*, whereas incisions are denoted by the suffix *–tomy*. Incisions usually are not coded as they are considered a part of most procedures. Excisions vary from one anatomic site to another and by extent. A simple excision involves superficial subcutaneous tissue and a radical excision involves deeper subcutaneous tissues. For a partial excision, the code states how much is excised.

SURGICAL DESTRUCTION

Surgical destruction is different from excision in that the destruction destroys the specimen, but is only removed with excision. Destruction can be part of a procedure code and, therefore, would not be coded separately. If a service exceeds normal expectations of what the service should be or no other procedure is performed that includes the destruction, then surgical destruction is coded. Destruction can occur through methods such as cryosurgery, laser, chemical, or electrocautery. The add-on codes, i.e., 17003 for destruction, are for destruction of additional lesions. The first surgical destruction of a lesion uses code 17000. Code 17003 is for additional lesions. It includes up to the fourteenth lesion and states that each lesion is to be coded. For example eight lesions would be coded as 17000 and 170003 × 7. If more complex repair is required than a simple closure, e.g., grafts, the additional repair also should be coded.

INTEGUMENTARY

Integumentary includes the epidermis, dermis, subcutaneous, nails, hair, sebaceous, and sweat glands. For skin, do not use modifiers to designate the part of the body because skin is not considered a body part in the sense that a toe or finger is; do not use LT for left or F7 for the digit.

Incision and Drainage

Abscesses can occur in many forms such as cysts and carbuncles. A common treatment for abscesses is incision and drainage (I&D), simple or complicated, single or multiple. Terms indicating complicated I&Ds include ligation or packing. Removal of foreign body is coded with its own codes, not from the I&D codes.

Lesions

Lesions can be removed by paring, excision, shaving or destruction. If the shaving involves the full-thickness of the skin, it is considered an excision. Means of destruction include cryosurgery, laser, chemical or electrosurgery. Destruction of lesions and appropriate coding will depend on where they are located. Location of lesions can be marked with needle localization in which a wire is inserted into the site to indicate the location of the lesion. Needle localization is coded. Radiological codes also must be coded with needle localization codes to identify the actual placement. Once located, a lesion can be marked by a radiological marker, which is usually included in the code.

Lesions are coded individually and in centimeters (cm). If the lesions are listed in inches, you will have to convert in the report (1 inch equals 2.54 cm). The lesion size is used to determine the proper code. The size of the lesion is defined as the size of the lesion, plus a narrow margin that may require excision. Report the largest size (e.g., a lesion that is 2.5 × 1 × 1.5 is coded as 2.5 cm). The size of the lesion should be obtained from the physician's notes, not from the pathology report.

Note: If a physician elects to excise two or more lesions together that are closely situated, then you will code only the excision of one lesion, because there is only one excision and one specimen. Additionally, if a physician has to re-excise a lesion site that was already excised, the new larger area would be coded with the excision codes for lesions. Do not code the previous size of the excised lesion.

If there are several lesions, code the most complex lesion first. Be sure to read the code's description, as some codes will include the first lesion; other codes will include multiple lesions listed either separately or as a group.

It is important to determine the morphology of the lesion(s), which is either benign, malignant, or uncertain. If a physician returns to excise more tissue surrounding a previously excised malignant lesion; then this code also is categorized as malignant.

If closure of lesions is anything other than a simple closure, code for the more complex closure (e.g., layered or graft) also. There is no closure required for shaving, but it does require a different code, 11300.

An excision is a full-thickness removal, including simple, nonlayered closure, and requires a separate code. A complex closure also requires an additional code and includes closure of the dermis with sutures in one of the deeper layers, which would require an additional code for the suturing. The only exception to coding for the graft in addition to the excision of the lesion is with adjacent flaps. With an adjacent flap, such as Z-plasty or W-plasty, the excision for the graft removes the lesion and, therefore, the two procedures are combined together in one procedure, the adjacent flap.

Moh's surgery is a staged procedure, performed by one physician, and used to destroy certain types of skin cancers. The first stage is destruction of the lesion and the second stage involves the microscopic examination and mapping of the lesions. For instance, one layer may be removed and examined to determine if cancer cells are present. This removal of layers continues until no cancer cells are found, which can occur on the same day or different days, and are known as stages. Within each stage performed, several specimens may be taken, which are a basis for the codes. Both functions—the removal and the pathological examination—must be performed by the same physician to bill for this type of surgery. If biopsy and pathology are performed at the same time as the Moh's, then they should be coded also. If a pathologist examines the specimens, then the pathologist will bill for the examination of the specimen and the physician can bill for the excision of the lesion only. If extensive repair is necessary, other than normal closure, use additional codes for these, such as grafting codes. If more than five specimens are examined, code 17310 is used in addition to the regular codes (177304–17306) for each additional specimen.

Wound Repair

Treatment of lacerations is coded by *adding* their lengths, in cm, together *if* they are of the same repair type and anatomic group. Normal debridement, simple exploration and ligation of blood vessels, nerves, or tendons is considered part of the repair procedure, so do not code these in addition to the repair code. Types of repair are classified as simple, intermediate, or complex.

Simple repair requires superficial repair that can include suturing the top level of the skin. Simple wound repair with adhesive only is coded for Medicare with G0168, but wound repair codes are used for other types of simple repair, such as sutures and staples. Intermediate repair requires layered closure of the top layer of skin and the underlying fascia and includes debridement. Complex repair includes wound closure and debridement. The most complicated repair is listed first.

Although debridement is included in these repair codes, if it is more extensive than normal, it can be coded in addition to the repair codes. Debridement also may be coded if there is no repair, only debridement.

If other anatomic sites are repaired (such as blood vessels, tendons, and nerves), you must code the repair of these sites separately from their own specific sections, such as nervous or cardiovascular. Typical wound closures include sutures, staples, and adhesives. If closure is more complicated, code such closures as extra with their own codes. Applying a bandage is bundled in the surgical code and is not coded separately. Only those supplies and materials that exceed normally expected supplies/materials can be coded as additional.

Debridement

Certainly any service provided by the physician should be compensated; however, as discussed earlier, global codes include costs for providing all normally associated services and procedures, such as debridement, whether for injuries or preparation for surgery. There are instances when services such as debridement are rendered that exceed what would normally be associated with a global procedure. This is true especially with lacerations, since they may be heavily contaminated and require extensive cleaning. If a debridement service is unusually extensive, then it must be well documented and can be coded additionally using codes 11000–11044. Notice that debridement codes are differentiated by whether or not an injury includes a fracture. Debridement can be accomplished either surgically or nonsurgically.

Grafts/Flaps

There are two types of grafts—adjacent and free. With adjacent grafts, a portion of the graft purposely remains attached, not created by accident. The graft is then manipulated to cover the adjacent recipient site. These types of grafts would include a Z-plasty, double pedicle flap, and rotational flaps. The types are listed under the code in the CPT book.

Free grafts are completely detached from the body and transferred to an area not adjacent to the graft. They are classified by split-thickness or full-thickness, which is determined by the depth of the graft. Split-thickness contains the epidermis and possibly part of the dermis; however, they result in more contracture and do not have the characteristics of normal skin. Full-thickness contains the epidermis and all of the dermis. Full-thickness grafts are desirable, because they are less likely to result in contracture, have a better appearance, and are stronger. However, full-thickness grafts remove all of the dermis from the donor site, which means the donor site must be closed and will develop scar tissue.

A pinch graft requires first elevating the skin with a needle and then slicing off a piece of skin. The donor site is the healthy tissue from where the graft is taken. The recipient site is where the graft is going to be affixed. The codes specify whether they are determined by recipient or donor site. The codes are determined by the size of the recipient site per 100 square (sq) cm for adults and percentage of body for children under the age of 10.

The grafts are measured by sq cm. The primary code will contain the first 20 sq cm. Additional add-on codes are available for when more area is involved. Even though the add-on codes are in increments of 20 sq cm, they can be used for additional area involved, even if that additional area is not the full size of the area listed in the code. For example, if there is an additional area of 2 sq cm and the add-on code is for 20 sq cm, the add-on code can still be applied.

If another procedure is performed and requires a graft for repair, such as an excision of a lesion, which then requires a graft to close it, then the graft should be coded in addition to the other procedure.

If a free graft is used to replace a site where there was an excision of a lesion, the site already is prepared and you need only to code for the graft and the excision of the lesion. However, if the graft is for a site that is not prepared such as an area with scarring, then the site must be prepared. Use codes 15000 and 15001 in this case.

If a graft other than adjacent tissue transfer is used to repair a defect, such as a pressure ulcer, then code the graft also. Normally, the size coded for the graft is the size of the recipient site. However, if a flap is formed for later use, then the size of the donor site is the size coded for the graft.

It is important to know the difference between autologous, xenograft, and allograft. Autologous means the graft came from the patient's own body. Allografts are taken from the same species, and xenografts are taken from a foreign source, such as an animal or man-made, such as Gortex.

Flaps are an interesting version of a graft and can appear more radical. With flaps, a portion of tissue remains attached with blood supply maintained. Flaps can be taken from nonadjacent sites and extended to another site, such as the thumb attached to the belly. Flaps can be direct, intermediate, muscle, myocutaneous, fasciocutaneous, tube, or delayed.

Burns

Services and procedures for burns are classified by percentage of body burned (using the rule of nines), type of burn, initial or subsequent treatment, and presence or absence of anesthesia. Treatments can include escharatomy (removal of dead skin) and grafting. Codes 15000 and 15001 are used to report preparation of burn sites for grafting. Rule of nines provide percentages for each anatomic site so that size of burn, i.e., small, medium, or large, can be determined for the code, which varies between adults and small children. Dressings of man-made material (alloplastic) can be used for covering the burns.

ACTIVITY 7.7

1. What is the correct code(s) for a patient who received repair, which included the subcutaneous tissue and superficial fascia with ligation of vessels for two 3.8 cm lacerations on both forearms?

2. What is the correct code(s) for a 3-year-old patient who suffered third-degree burns on both arms and who received dressing and debridement with anesthesia for one week and then without anesthesia for eight more days?

3. What is the correct code(s) for a patient who had a malignant lesion that measured 3.1 cm removed by surgical curettement from the upper back?

4. What is the correct code(s) for a patient who has percutaneous biopsy of the breast with placement of a clip for visualization?

5. What is the correct code(s) for a patient who had a radical mastectomy, including the axillary lymph nodes and pectoral muscles?

6. What is the correct code(s) for a patient who receives a TRAM flap for breast reconstruction with microvascular anastomosis for single pedicle with closure?

7. What is the correct code(s) for a patient who returns for additional excision of a 1.8-cm malignant lesion that was removed a week ago with an area measuring 2.2 cm today?

8. What is the correct code(s) for a patient who had a biopsy for possible skin cancer combined with initial Moh's surgery, which was evaluated in pathology and confirmed as malignant?

9. What is the correct code(s) for a patient who had two malignant lesions from the upper back excised that measured 1.0 and 0.8 cm, respectively? The physician was able to cut them out as one, including the border.

10. What is the correct code(s) for a patient who sustained a laceration of the forehead that required debridement?

11. What is the correct code(s) for a patient who has a malignant 0.8 cm lesion in the breast removed, which involves placement of wire for visualization?

12. What is the correct code(s) for a patient who received repair for a 2.2 cm laceration on the left hand that required suturing of the ulnar nerve and suturing of fascia? Debridement required removal of many small and medium pieces of glass.

13. What is the correct code(s) for a patient who received a biopsy of a lesion on the right hand with biopsy and excision of a malignant 1.2 cm lesion on the right forearm?

14. What is the correct code(s) for a patient who had a malignant lesion from the upper back excised that

measured 8.0 cm, which included a border? A split-thickness graft was performed to cover the area that included part of the epidermis.

15. What is the correct code(s) for a patient who had two malignant lesions removed, 0.7-cm lesion of the left forearm and 0.4-cm lesion of the neck?

16. What is the correct code(s) for a patient who returns for a second time for Moh's surgery, which includes seven skin neoplasms?

17. What is the correct code(s) for a 1-year-old patient who had sustained third-degree burns over her right arm eight months ago and is now receiving a 35 sq cm full-thickness graft from her leg to be applied to the scar tissue?

18. What is the correct code(s) for a patient who received injections for four lesions for chemotherapy?

19. What is the correct code(s) for a patient who had an excision of an ischial pressure ulcer with split-thickness myocutaneous flap?

20. What is the correct code(s) for a patient who had a malignant lesion from the upper back excised that measured 8.0 sq cm, which included a border? A Z-plasty was performed to cover the area.

MUSCULOSKELETAL

This section is classified by anatomic site, then by procedure. To begin coding, you must determine the specific anatomic site. In addition, you may need to determine if it was traumatic, included grafting or fixation, performed on soft tissue or bone, and/or at a single or multiple sites.

Wound Repair

Although typical wound repair is included in the integumentary section, wounds involving repair of other anatomic sites must be coded from their own areas. For instance, nerves are coded within the nervous system and, for the purpose of this section, wounds that involve the muscles and bones are coded from the musculoskeletal section.

Fractures/Dislocations

It is important to distinguish between coding for procedures for fractures/dislocations as opposed to coding for the diagnosis. A closed or open fracture diagnostic code is not the same as a closed or open procedural code. When coding for a procedure caused by a fracture/dislocation, you must determine if the procedure was open, closed or percutaneous fixation, what anatomic site was involved, if manipulation or traction was provided, and the reasons for the treatment. If it is not known whether treatment was open or closed, then code it as closed. If a procedure begins as closed, but is converted to open, then it should be coded as open only.

As assumed from its name, an open procedure indicates an incision was made with exposure or visualization of the bone. Closed treatment refers to repair such as traction, reduction, or manipulation in which the body is not opened. Percutaneous fixation is neither open nor closed and means that implants, such as pins, are inserted into the bone but the bone is not visualized. Usually visualization is achieved through imagery, such as x-rays.

Traction can be performed on the skin or on the bone. Skin traction usually involves the use of adhesive or elastic wraps, which can pull and tighten the skin. Buck's traction and Boot traction are two types of skin traction. Skeletal traction can involve pins, screws, or wires, such as Kirschner wires, which are inserted into the bone. Common names for this type of traction are Dunlop's or Russell's. With external fixation, implants are used, such as pins or screws which are attached to an externally placed device. Manipulation attempts to realign the bone properly through manual force and usually is included within the fracture care code.

Open procedures can include fixations, such as percutaneous and external. Fixation uses codes 20690 or 20692 in addition to the codes for the treatment of the fracture. Percutaneous fixation involves a stabilizer, such as pins. Internal fixation involves the placement of pins, wires, or screws through an incision to stabilize the fracture site. A common internal fixation procedure is known as ORIF, open reduction internal fixation. Sometimes this hardware may be left in the body permanently. If removed, then it should be coded using 20670 or 20680. With external fixation, the hardware is placed within the body but the fixation device is placed on the outside of the body. Removal of this type of hardware is part of the global package for fixation and should not be unbundled.

With fractures/dislocations, the first cast, splint, or strapping is included in the procedure code, i.e. open, closed, or percutaneous fixation. The materials and supplies also are included. Although there is a code for application of casts and strapping, this code is not used for initial treatment or first application of the fracture/dislocation but can be used when an additional cast or strapping is necessary or when another physician provides the service later.

Musculoskeletal procedures performed on the spine are based on the area of the spine that is treated and can include arthrodesis, fracture care, and grafts. Arthrodesis is fixation through fusion. These codes usually are indicated by spaces as denoted by the disks above and below, for instance T1–T2, which counts as one; T1–T3 would count as two. There are add-on codes for more than one segment. Injuries above C5 and at T1 and below usually result in quadriplegia. Remember that there are 7 cervical vertebrae, 12 thoracic, 5 lumbar, 5 sacrum, and 4 fused pieces for the coccyx, which is known as the tailbone. A vertebral segment is composed of a cerebral bone with articular processes and laminae. The vertebral interspace is the non-bony part lying between two vertebrae, which is composed of the intervertebral disk, nucleus pulposus, annulus fibrosus, and two cartilaginous endplates. Know your terminology!

Be careful to code for instrumentation when required and for fractures and arthrodesis, if they are not included in the initial code for the fracture or arthrodesis.

Introduction and Removal

Introduction and removal procedures include removal of foreign bodies, as well as insertion of fixation devices, such as pins and wires. Remember that fixations and instrumentations may need to be coded separately from other major procedures, so be sure to read the coding notation.

Injections

Injections are common within the musculoskeletal areas. Trigger injections are the injecting of anti-inflammatory agents into an area of the body such as tendons or ligaments. Other injections can occur in parts of the body such as joints. The substance injected is coded in addition to the actual injection code using the HCPCS codes. This also includes centesis in which fluid is aspirated from the joint by means of an injection.

Bone Grafts

As with other grafts, bone grafts can be autografts, allografts, alloplastic, or even composite and are based on the size of the donor area. Composite is a combination of the first three previously mentioned grafts. Spinal grafts have their own specified codes (20930–29038).

Foot Surgery

Foot surgery is performed for various reasons. One reason is to correct deformities such as hallux valgus and to repair bunions. Bunions are a prominence of the first metatarsophalangeal joint with inflammation that causes a displacement of the big toe laterally or as a valgus. Know your terminology!

Foot surgeries include the Silver procedure, which is the removal of the medial eminence that lies on the outside of the largest joint of the big toe. Keller procedure is the removal of the medial eminence and a portion of the proximal phalanx, which lies near the medial eminence. Sometimes a wire, such as a Kirschner, may be implanted for stability. The Keller-Mayo procedure is the same as the Keller procedure but includes an implant at the resection site and usually includes a total double stem implant in the toe.

A Joplin procedure includes the removal of the medial eminence, fusion between the two bones at the largest joint, and the transfer of the tendon to the head of the largest joint, instead of extending to the end of the toe. A Mitchell Chevron (or Austin) procedure includes removal of the medial eminence with an osteotomy within the neck of the bone at that location that is then repositioned and wired. A Lapidus-Type procedure involves the removal of the medial eminence, tenotomy at the large toe joint, and an arthodesis where the big toe and second toe join. A Phalanx osteotomy (or Akin procedure) is the removal of the medial eminence and a portion of the proximal phalanx, similar to the Keller, but a Kirschner wire is implanted for stability. Some CPT books provide demonstrations of these procedures.

ACTIVITY 7.8

1. What is the correct code(s) for a patient in the Emergency Room who has removal of a cranial halo that is complicated by infection, for which there is I&D?

2. What is the correct code(s) for a patient who received an osteotomy with removal of a wedge from the proximal phalanx of the big toe with insertion of Kirschner wire for correction of a bunion?

3. What is the correct code(s) for a patient who had a lesion of the tendon sheath removed from the leg?

4. What is the correct code(s) for a patient who had closed treatment for a comminuted fracture of the shaft of the tibia and fibula without manipulation?

5. Arthroplasty of the acetabulum and proximal femoral with a prosthesis is known as what?

6. What is the correct code(s) for a patient who received an osteotomy of the proximal tibia and fibula for *genu varus*?

7. What is the correct code(s) for a patient who received a fasciotomy with debridement of muscle and decompression?

8. What is the correct code(s) for a patient who received a partial synovectomy of the wrist?

9. What is a Monteggia fracture?

10. What is the correct code(s) for a patient who received a replacement for total shoulder arthroplasty?

11. What is the correct code(s) for a patient who received treatment for fracture/dislocation of shaft of the tibia and fibula?

12. What is the correct code(s) for a patient who had a portion of the mandible bone removed due to carcinoma? A bone graft was also performed to repair the defect.

13. What is the correct code(s) for a patient who received a partial claviculectomy and arthroscopy?

14. What is the correct code(s) for a patient who had a posterior approach for open treatment and reduction of fractures of C3–5 with wiring of the spinous processes?

15. What is the correct code(s) for a patient who received open fracture care for the shaft of the tibia and fibula with the insertion of screws?

16. What is the correct code(s) for a patient who received percutaneous repair of a ruptured Achilles tendon, which was repaired with a graft?

17. What is the correct code(s) for a patient who received an arthodesis of two carpometacarpal joints of the thumb and second digit?

18. What is the correct code(s) for a patient who received a cranioplasty involving the forehead with an allograft?

19. Which of the following are not part of the vertebral interspace: annulus fibrosus, endplates, laminae, bursa, nucleus pulposus, and carinatum?

20. What is the correct code(s) for a patient who received open procedure with internal fixation for dislocation of joint of the tibia and fibula?

CARDIOVASCULAR

Cardiovascular codes are categorized first by anatomic site, then procedure, and further divided by arteries and veins. Cardiovascular codes may come from the 30000s but also from 90000s, the Medicine section. You must know your medical terminology well when coding for cardiovascular, including the structure of the heart and venous and arterial systems.

As indicated within the code, a stress test includes an electrocardiogram (ECG). Do not code the ECG separately. Drugs used during the test, such as Technetium, should be coded using HCPCS codes.

When coding for cardiovascular procedures, beware of including references to the use of a heart-lung machine.

Valve procedures are categorized by type of valve—mitral, aortic, tricuspid, or pulmonary.

Bypass Grafts

When the flow of blood is restricted by blockage, such as plaque in the vessels, a bypass graft procedure can be used to replace the diseased portion of the vessel with a healthy vessel by attaching the healthy vessel above and below the diseased portion. Often, healthy veins or arteries are harvested from the saphenous vein in the leg or the internal mammary artery. A coronary artery bypass graft (CABG) is performed specifically coronary arteries, but other vessels can be treated with grafts also. Endartectomies and angioplasties are an integral part of the procedure and should not be coded separately.

To code, you must know how many vessels are being replaced, the approach, what vessel is used for replacement, what type of material is used, and whether they are venous, arterial, or a combination. The approaches are either the aorta or internal mammary artery. For the aortic approach, the number of coronary arteries involved must be known to code correctly. For the mammary approach, the use of either one or both mammary arteries is used to determine the code.

CABG usually come from the patient's saphenous leg vein or internal mammary artery. Harvesting from the donor site is included in the CABG code, so do not bill separately.

Coronary bypass grafting for the heart can be venous, arterial, or arterial-venous combined. When coding for venous, use the codes for venous grafting, which include 33510–33516. When coding for arterial only, use the codes for arterial grafting, which are 33533–33545. When coding for a combined arterial-venous graft, you must provide the code for combined arterial-venous grafting, 33517–33639, and a code for arterial grafting.

As noted in the CPT book, some grafts for coronary bypass include the procurement of the graft, but some do not. For example, procurement of the saphenous vein graft is included for an arterial-venous graft but harvesting of an upper extremity artery is not and should be coded additionally. Review your CPT book for further information.

When bypasses must be redone more than a month after initial surgery code 33530 is used to indicate this increase in the level of service. This is an add-on code and must be coded in addition to the procedure. Be sure to read the notes indicating which procedures this code can accompany.

Pacemakers

Pacemakers and cardioverter-defibrillators are devices that include a pulse generator, which is inserted into a surgically constructed pocket, and electrodes that are attached to the heart. The electrodes then may be inserted through a vein (transvenous) or placed on the surface of the heart (epicardial). Thoracotomies must be included in the code for epicedial placement of electrodes.

Pacemakers are electrical stimulations that produce a regular rhythm. Occasionally, a replacement or reinsertion must be performed. Pacemakers can be either single or dual with either one or two electrodes placed into the heart chamber.

When coding for a pacemaker, you must know if it is an initial placement, replacement, or a repair. You also need to know if the electrode lead is inserted into the atrium, ventricle, or both.

Cardioverter-defibrillators deliver electrical shocks to the heart to restore proper rhythm. To code for a cardioverter-defibrillator, you should know if it is a replacement, repair, or removal of components, what approach was used, and if it is a replacement or revision. If a repair or replacement, then codes for both the removal of the previous pacemaker and insertion of the new one must be used.

Cardiac Catheters

Cardiac catheterizations determine the severity of heart disease and possible treatment. There are two groups of cardiac-related codes regarding catheters that may be confusing. The 36000 codes are injection procedures for treatment or diagnostics. Cardiac catheterization codes are for left and right heart catheterizations and use codes 93500 through 93581. In addition, radiology codes can be reported for angiographies or the technical portion of this type of cardiovascular service. These radiological codes may be provided by the physician or by a radiologist.

With catheter codes, you must know where the catheter was inserted, where it went, and where it ended. Some of these codes (36000s) are selected according to vascular families, which are divided into orders of first, second, third, and so forth. The first order is the main vessels from which the second and third order vessels branch, and so forth.

Code only for the greatest level of division within a family. For instance, if the catheter enters a third order vessel, then code only for a third order in addition to placement. If however, the physician goes into a different order, code for this different order additionally. For example, if the physician goes into a third order, then goes into another third order of the same first and second order family, then each third order should be coded. There are two codes, one for each third order, plus placement, but not first or second order.

Selective catheterization means the catheter is manipulated into areas other than where placed (see code 36012), which includes placement and introduction of contrast material, in addition to anesthesia and repositioning, so these are not coded additionally. Nonselective catheterization means the catheter remains in the vessel where originally placed without further manipulation. This is considered an introduction only of catheter/needle (36000, 36200). Remember, if a catheter is reinserted into a separately distinct access site, this is considered an additional code.

The materials and supplies (catheters, drugs, and contrast media) are sometimes not included in these codes and can be coded separately, the notes in the CPT book state. If cardiovascular radiology codes indicate that they include contrast material, do not code for injection and contrast material.

For left and right cardiac catheterization, you must use medicine codes (93501 through 93572). Introduction is included in the code. For a left heart catheterization, the catheter is inserted into the aorta and then into the left ventricle. For a right heart catheterization, the catheter is inserted into the superior or inferior vena cava and into the atrium. There are several types of catheterizations, including combination of left and right, so be careful coding the specific type performed. There are separate codes for the injection procedures (93539–93545), which are used to increase visualization of vessels. The injection procedures are coded in addition to the catheterization code (right or left) as noted in the guidelines prior to the codes. These codes are selected based on the injected vessels. In addition, supervision and interpretation codes (93555–93556) can be used. If the physician provides supervision and interpretation, code this also (e.g., 93555).

When using the 36000 codes, additional codes may be used for supervision and interpretation of the

VESSEL ORDERS EXAMPLES

First Order	Second Order	Third Order
Brachiocephalic	Right common carotid	Right internal or external carotid
Brachiocephalic	Right subclavian	Right vertebral or internal mammary
Left common carotid	Left internal or external carotid	Right or left anterior cerebral, maxillary, facial
Left subclavian/ axillary	Left vertebral	Right or left cerebral basilar

Table 7-1

catheterization readings. Removal of a catheter usually is not coded.

Other Cardiovascular Devices

Vascular access devices (VADs) are catheters placed in the central venous system to provide long-term access for drugs, nutrition, chemotherapy, and laboratory tests. Total parenteral nutrition is possible through VADs, which are surgically implanted. Access can be gained centrally through the jugular, subclavian, or femoral vein. They can be accessed through an exposed catheter, subcutaneous port, or pump. The repair, replacement, or removal of a VAD also should be coded.

Angioplasties widen constricted vessels by inserting a catheter with a balloon on the end through the vessel, which then can be expanded when necessary. They can be performed percutaneously (PTCA) or through an open chest.

Fistulas

A fistula is an abnormal connection that can occur between arteries and veins (arteriovenous). These can be the result of trauma or congenital or surgically created, such as in hemodialysis. If fistulas need repair, they are coded with codes 35180 through 35286.

Thrombectomies/Embolectomies

A thrombus is a blood clot that occludes a vessel and an embolus is a mass of undissolved matter (thrombus that has become free) in the blood. When a thrombus is dislodged, it becomes an embolus. These are removed by either a thrombectomy or embolectomy, which includes removing debris through a balloon or catheter or through an incision in the vessel. If a thrombectomy or embolectomy is performed during any other procedure, such as a graft, then they are not coded.

ACTIVITY 7.9

1. What is the correct code(s) for a patient who received a limited study with duplex scan of veins of the legs and toes?

2. What is the correct code(s) for a patient who received a thrombectomy of a nonautogenous hemodialysis graft for a revision of an open fistula?

3. What is the correct code(s) for a patient who received a venous bypass graft of two coronary arteries from two segments in the arm?

4. What is the correct code(s) for a patient who received implantation of a dual-chamber cardioverter/defibrillator with epicardial electrodes?

5. What is the correct code(s) for a patient who had a PTCA with coronary atherectomy?

6. What is the correct code(s) for a patient who has an arteriovenous fistula inserted for hemodialysis using Gortex?

7. What is the correct code(s) for a 26-year-old patient who had a vascular access device inserted through the subclavian vein?

8. What is the correct code(s) for a patient who received an open balloon angioplasty of the renal artery?

9. What is the correct code(s) for a patient who received a patch graft for occlusive disease of the iliac artery?

10. What is the correct code(s) for a patient who received an aortofemoral bypass graft with Gortex?

11. What is the correct code(s) for a patient who received an arterial-venous bypass graft for two coronary arteries?

12. What is the correct code(s) for a patient who received a combined left and right heart catheterization including selective right and left ventricular angiography?

13. What is the correct code(s) for a patient who received a permanent pacemaker with a transvenous electrode placed into the right atrium, one into the right ventricle, and one into the left ventricle?

14. What is the correct code(s) for a patient who received an insertion of a catheter, by neck incision, for embolectomy of the subclavian artery?

15. What is the correct code(s) for a patient who received a patch graft for an aneurysm and occlusive disease of the iliac artery?

16. What is the correct code(s) for a patient who was provided an arteriovenous cannula for access during hemodialysis?

17. What is the correct code(s) for a patient who received a transluminal percutaenous angioplasty of the renal artery?

18. What is the correct code(s) for a patient who receives an upgrade of his pacemaker to a dual-chamber and new pulse generator?

19. What is the correct code(s) for a patient who received a bypass graft from the internal mammary artery with angioplasty?

20. What is the correct code(s) for a patient who received a single injection of sclerosing agent for telangiectasia of the legs?

DIGESTIVE

Coding for the digestive system is difficult because there are so many parts of the body involved. It is particularly important with this type of coding that you know your terminology well, because codes are classified based on various combining forms. The codes begin in descending order from the lips to the anus.

With digestive codes be sure that the various parts are included, e.g., with scopes. Be careful to not code biopsies as separate codes if the biopsy is already included in the code's description. Read the descriptions carefully.

Anastomosis

Sometimes a section of an anatomic site may be removed and reconnected, such as with the intestine. Sometimes an opening to the outside of the body may be created, such as a colostomy, which may be temporary or permanent. Stomies are performed when sections of the intestines are removed and a stomy is created, either on a temporary or permanent basis, to treat conditions such as colitis. If the stomy is temporary, there must be a final repair. Be careful to determine if the repair is included in the primary surgical code.

Endoscopies

Endoscopic codes are described by where they go, how far down or up. The codes must include all sites visited. Remember, do not code for a diagnostic scope if a surgical scope is performed since the surgical includes the diagnostic. If the endoscopic procedure is changed to an open procedure, then code the open procedure, not the endoscope. In such cases be sure to include V64.41, V64.42, and V64.43 codes for endoscopic procedure changed to open procedure. A scope of the upper GI system is known as an EGD (esophagogastroduodenoscopy).

Ureter procedures, such as cystourethroscopy, may require a stent. A temporary stent is included in the procedural code; a permanent stent needs to be coded in addition to the procedure code by using code 52332. If tubes are placed, changing the tubes, even within the global period of the insertion procedure, can be coded separately, as well as fluoroscopic guidance, as indicated by the codes.

Colonoscopies may be performed through a colostomy.

Dilation of the digestive system can include balloon, Bougie, Hurst, and Maloney dilators. Dilators are used to enlarge an opening. Hurst and Maloney dilators contain mercury to pull the dilator down. Be careful, however, for some dilations, in addition to other services such as calibration and meatotomy, are included in the major procedure as stated in the coding notes in that section (52000).

Lesions in the digestive system usually are removed with a scope. Procedures to remove lesions include a snare (loop slipped over lesion and used to cut it), electrocautery (use of electricity), hot or cold biopsy (forceps with or without electrical current), ablation (control of hemorrhaging), and bipolar cautery (electrodes).

A percutaneous endoscopic gastrostomy (PEG) tube placed in the stomach area provides nutrition to patients. Certain procedures are included in this code such as placement, insufflation, and passage of the tube. Remember, the endoscopic procedure also is included as indicated by the name of the procedure. An endoscopic retrograde cholangiopancreatography (ERCP) uses dye in the stomach area that is then visualized through the use of a scope.

Hernia Repair

Reducible hernias can be repaired by manipulation, but nonreducible hernias, such as incarcerated hernias, cannot be repaired. Mesh is often used to repair a hernia by holding it in place. An additional code is used to code for mesh when the hernia is ventral or incisional.

ACTIVITY 7.10

1. What is the correct code(s) for a patient who received a partial bowel resection with colostomy and creation of mucofistula?

2. What is the correct code(s) for a patient who received a temporary stent during a retrograde nephrostomy with wire with cystourethroscopy?

3. What is the correct code(s) for a patient who received an EGD with control of bleeding and dilation of the esophagus with a balloon?

4. What is the correct code(s) for a patient who received a PEG tube with insufflation?

5. What is the correct code(s) for a 33-year-old patient who received repair of an incisional incarcerated hernia with mesh?

6. What is the correct code(s) for a patient who received a diagnostic laparascopy and partial colectomy with anastomosis that was converted to open procedure due to complications?

7. What is the correct code(s) for a patient who received a cholecystotomy for removal of calculus with choledochostomy?

8. What is the correct code(s) for a patient who received repair of bilateral reducible recurrent femoral hernias?

9. What is the correct code(s) for a patient who received dilation of renal stricture with balloon with cystourethroscopy?

10. What is the correct code(s) for a 56-year-old patient who received repair of an incarcerated inguinal hernia with enterectomy?

11. What is the correct code(s) for a 44-year-old patient who received manipulation for a repair of an incarcerated inguinal hernia?

12. What is the correct code(s) for a patient who received sphincterotomy and ERCP with stent into the pancreatic duct?

13. What is the correct code(s) for a patient who received dilation of her esophagus with a balloon 25 mm for achalasia?

14. What is the correct code(s) for a patient who received flexible sigmoidoscopy for removal of lesions by cautery with endoscopic ultrasound?

15. What is the correct code(s) for a patient who received placement of a permanent stent during a retrograde nephrostomy with wire with cystourethroscopy?

16. What is the correct code(s) for a patient who was admitted for exploratory laparascopy and then partial colectomy with colostomy and closure of distal segment?

17. What is the correct code(s) for a patient who received percutaneous nephrostomy?

18. What is the correct code(s) for an 11-year-old patient who received a tonsillectomy and adenoidectomy?

19. What is the correct code(s) for a 23-year-old patient who had a reducible umbilical hernia with mesh repaired?

20. What is the correct code(s) for a patient who received an injection for percutaneous transhepatic cholangiography with professional services provided?

NERVOUS SYSTEM

Terminology is very important when coding the nervous system. You should know the following anatomical terms: lamina, disks, vertebral body, intervertebral, facet, and foramen.

The nervous system codes contain procedures and services for the brain, spinal cord, nerves, peripheral and central nervous system. Codes for services with the pituitary and pineal glands also are included in this section of codes and not the section of codes for the endocrine systems. If musculoskeletal services are provided, such as grafting or fracture care, use musculoskeletal codes.

To locate procedures on nerves, the specific nerve can be referenced in the alphabetical index. Sympathetic nerves stimulate the body when stressed, which includes increased heartbeat, blood pressure, and secretion of epinephrine. Parasympathetic nerves act as a counter effect to the sympathetic nerves; they slow down the heart rate and decrease the blood pressure.

The first group of codes is for procedures requiring holes or openings, such as twist or burr, to insert monitoring devices, place tubing, relieve pressure, allow drainage, or inject contrast material. Shunts also are used for drainage and are coded from codes 62180–62258.

Procedures for removal of lesions in the skull base (61580–61619) may require three separate codes—for the approach, the definitive procedure, and repair/reconstruction. The first two procedures, the approach and the definitive procedure, must be coded, which is the actual procedure performed. The repair/reconstruction is coded only if the services are extensive, such as flaps or grafts.

Procedures on the spine and spinal cord are classified according to condition, approach, unilateral or bilateral, and multiple procedures.

If nerves are repaired, they are coded from this section, and not integumentary, according to the specified nerve. Procedures performed on nerves include decompression, suture, repair, injection, and destruction. Decompression releases the nerves from tissue, such as scar tissue from previous surgery.

ACTIVITY 7.11

1. What is the correct code(s) for a patient who received a nerve block of the vagus nerve?

2. What is the correct code(s) for a patient who had a 3.2 cm laceration of the neck repaired that required retention sutures with split-thickness graft due to the extensive damage of fascia, muscle, and dermis?

3. What is the correct code(s) for a patient who had placement of a pump with programming?

4. What is the correct code(s) for a patient who had an intradural resection of the neoplasm of the cavernous sinus with dural repair with graft and infratemporal approach with ligation of the carotid artery with graft?

5. What is the correct code(s) for a patient who had a subarachnoid/subdural-jugular shunt placed for cerebrospinal fluid?

6. What is the correct code(s) for a patient who had neurostimulator electrodes implanted in the autonomic nerve and returns today for analysis?

7. What is the correct code(s) for a patient who received a craniectomy and creation of burr holes through the posterior fossa?

8. What is the correct code(s) for a patient who had a decompression of sciatic nerve with transposition?

9. What is the correct code(s) for a patient who received arthrodesis of T3–5?

10. What is the correct code(s) for a patient who had repair of a deep 2.2 cm laceration of the right hand that required layered closure and microscopic suturing of two digital nerves?

11. What is the correct code(s) for a patient who had a catheter inserted with daily management for continuous infusion of an anesthetic agent into the femoral nerve?

12. What is the correct code(s) for a patient who had a transthoracic vagotomy?

13. What is the correct code(s) for a patient who received a resection of neoplasm of the midline skull base with infratemporal approach requiring a myocutaneous flap?

14. What is the correct code(s) for a patient who received an extracranial reconstruction due to a benign tumor?

15. What is the correct code(s) for a patient who received burr holes for a ventricular catheter?

16. What is the correct code(s) for a patient who received an excision of an interdigital neuroma?

17. What is the eponym for the previous procedure?

18. What is the correct code(s) for a patient who had excision of neuromas from both digital nerves of two toes of the foot?

19. What is the correct code(s) for a patient who had a laparoscopic vagotomy?

20. What is the correct code(s) for a patient who received a laminectomy with drainage of a cyst to the pleural space?

OBSTETRICS/GYNECOLOGY

Normal obstetric care includes antepartum, delivery, and postpartum care. Antepartum care includes monthly visits up to 28 weeks gestation, biweekly visits up to 36 weeks gestation, and weekly visits until delivery. Delivery services include normal delivery or cesarean, episiotomy, and forceps. Postpartum care includes care up to six weeks after delivery. The code for normal care for pregnancy with vaginal delivery is 59400 and with cesarean it is 59510. If the patient had a previous cesarean section and then delivers vaginally (VBAC), then codes 59610 through 59614 are used.

Breech birth corrected by the physician's turning of the baby is known as *external cephalic presentation* and is coded in addition to the delivery code.

Rh incompatibility occurs when the mother is Rh-negative and the baby is Rh-positive. Particularly after the firstborn child, this difference can result in death of future siblings since the mother's immune system will produce antibodies to attack the Rh of the baby, which can result in erythroblastosis fetalis in which hemolysis occurs (the breakdown of the fetus' red blood cells). This is treated by administering RhoGAM, which is coded using an HCPCS code.

Usually, dilation and curettage (D&C) is performed with other procedures, such as abortions, and should not be coded additionally. However, on occasion a D&C may be performed alone and should be coded.

Hysterectomies

There are many variables with hysterectomies. A hysterectomy may be performed vaginally or through an incision. A hysterectomy may include various anatomical parts. A total abdominal hysterectomy with bilateral salpingo-oophorectomy (TAHBSO) includes the excision of the uterus, both fallopian tubes, and both ovaries. In contrast, a hysterectomy includes the excision of only the uterus. A total abdominal hysterectomy includes the excision of the uterus via an abdominal approach, whereas a vaginal hysterectomy is the excision of the uterus via a vaginal approach. A salpingectomy includes the excision of only the fallopian tubes, and oophorectomy is the excision of the ovaries. Terminology is critical!

With a vulvectomy, you must understand the terminology to code correctly. Removal of 80 percent of the vulvar area is considered complete; removal of less than 80 percent of the vulvar area is considered partial. Removal of no more than superficial subcutaneous tissue is considered simple and removal of deep subcutaneous tissue is considered radical.

OCULAR/AUDITORY

Ophthalmology is a specialized field with specific codes. Many of the codes are bilateral. Modifiers need to be used where necessary if the procedure does not agree with the code. For example, the code is for a bilateral procedure but the procedure is performed unilaterally, so a 52 modifier is needed to indicate this reduction of services. Codes may be taken from the surgical section or from the medicine section.

Be careful not to unbundle codes. Use proper codes that contain the entire anatomic site, i.e., upper and lower lids of the eye.

While glasses are considered part of the general ophthalmological exam, contact lenses are not, unless specified.

Strabismus is a common surgery for the eyes. Strabismus occurs when there is an imbalance in the eye muscles, which results in improper alignment of the eyes as sometimes demonstrated by the fovea of each eye not focusing in the same direction. Codes are selected based on the muscles involved in the procedure. Horizontal muscles are described as lateral and medial. The vertical muscles are described as superior and inferior. Procedures include a resection in which the muscles are cut or recession in which the muscle(s) are moved.

Cataract extractions can be extracapsular (ECCE) or intracapsular (ICCE). Extracapsular extraction is the removal of the lens without the removal of the posterior capsule. Intracapsular extraction is the removal of the entire lens including the capsule, so the lens is not removed from the capsule. ECCE is performed by phacofragementation or by ultrasound. The lens is broken into fragments and then removed. If an intraocular lens prosthesis is inserted, it is coded with 66985.

Evisceration is the removal of the intraocular contents but not the scleral shell. Exenteration is the removal of the eye.

If a physician attempts to remove a foreign body from an eye but is not able to complete it for whatever reason, then either the procedure should not be reported or modifier 52 should be used to indicate the reduction in service.

Some procedures on the ears include other services such as suctioning of fluid and/or incision of the eardrum. These procedures include myringotomy and typanostomy. With a tympanostomy, a ventilating tube is placed in the ear to allow the release of fluids.

Some services on the ears and eyes, such as special services and comprehensive evaluations, are included in the Medicine section of the CPT coding book. Within the Medicine section, there are many services included in the evaluations. Do not unbundle any service and bill it separately. So, be sure to read the notes in the CPT book, which describe many of the services that are included.

RADIOLOGY

Hospital coding for radiological services usually is performed by coders who specialize in these codes.

TYPES OF TESTS

There are four categories of radiological services: diagnostic imaging (x-rays), diagnostic ultrasound (the use of high-frequency sound waves to create an image), radiation oncology (radiation to treat tumors), and nuclear medicine (radioactive materials *in vivo*).

The first category, diagnostic imaging, includes CAT/CT (computed axial tomography), MRI (magnetic resonance imaging), mammography, and angiography. CAT/CT uses ionizing radiation to create slices of images and MRI uses magnetism to produce more comprehensive images, which are highly detailed. The second category, ultrasound, includes A-mode, M-mode, B-scan, and Real-Time. A-mode is a one-dimensional scan; B-mode is a two-dimensional scan; M-mode is a one-dimensional scan that monitors motion; and Real-Time is a two-dimensional scan that monitors motion and time. Ultrasounds use sound waves to create images. Types of ultrasound tests include sonograms and echograms. The third category, radiation oncology, is measured in rads and based on the number of areas, ports, and blocks. Brachytherapy is the implantation of radioactive material directly into an anatomic site with the use of ribbons and sources. A source is a container that holds the radioactive material, which is implanted into the body. A ribbon is seeds placed on tapes that are inserted into the body. The fourth category, nuclear, includes the injection of radioactive materials during either diagnostic or therapeutic testing, such as a stress test. Different anatomic sites will absorb different types of isotopes, which can then be photographed.

Interventional radiology is the combination of a surgical procedure (e.g., biopsy) and a radiological service.

PROFESSIONAL SERVICES

Although a physician may not provide the radiological service, he may read and interpret it (known as the professional component), which would be coded either with a modifier or a specified code if billing for the physician. If using the CMS-1450 (UB-92) form with hospital coding, you do not need the modifier to indicate the professional or technical component of the code because the revenue code will indicate this. Some codes, however, will state that it includes *supervision and interpretation*. A code that includes supervision and interpretation cannot be used to code for the technical parts of tests.

CONTRAST MATERIAL

Some tests can be performed with or without contrast material and are designated as such in the codes. Contrast material provides greater definition of anatomical sites and functions that are radiopaque, meaning they do not allow light through where the contrast material is absorbed. The contrast material is visible on the test. If the code indicates possible inclusion of contrast material, do not code additionally for contrast materials and the injection. The administration of contrast materials either orally or rectally (not injected into vessels) is coded as without contrast. Digital subtraction, which is used to examine blood vessels, is an example of a test using contrast material. This test produces images before and after injecting contrast material into vessels, which will indicate blood flow inconsistencies in these vessels. In the small bowel, the progress of the contrast material can be seen at regular intervals as it moves through the small intestines.

Typical contrast materials are barium or iodine-based and enter the body by ingestion, injection, or through an enema. Barium sulfate is used for tests on the digestive system. Tests include barium enema and upper GI series. Iodinated contrast materials, such as iothalamate or Isovue, can cause serious effects, so caution is exercised when these types of materials are administered.

POSITIONING

Patients can be placed in several different positions, including posteroanterior (PA), anteroposterior (AP), lateral (LAT), oblique, or decubitus. The various views used for testing include stereo, apical, right anterior oblique, left anterior oblique, left posterior oblique, right posterior oblique, odontoid, and swimmer.

RADIATION ONCOLOGY

Radiation oncology includes codes for management, planning, devices, and delivery of radiation, including hyperthermia and brachytherapy. There are a wide variety of services offered within these codes, some of which are bundled. Services include tumor localization, time/dosage determination, determination of ports, and selection of devices. Planning and simulation are defined as simple, intermediate, or complex as determined by number of ports, blocks, or treatment areas.

Simulation can be defined as three-dimensional. It uses three-dimensional methods to reconstruct the tumor's image using methods such as MRIs or CTs. Treatment is designated as consisting of five treatments; however, there is a code for less than five if needed.

PATHOLOGY/LABORATORY

The Pathology/Laboratory section includes codes for ordering tests, procuring and preparing samples, actual test codes, and codes for interpreting results. Remember that the laboratory performing the test can code for the services, which is common in hospitals. Hospital coding for pathology/laboratory services usually is performed by coders who specialize in these codes.

Tests can be qualitative or quantitative. Quantitative tests, or assays, determine the amount of a specific material, known as an analyte. They are classified by generic name, not brand. Qualitative tests determine the identity of the material (analyte) and are coded by number of analytes tested since the analyte is not known. Tests are considered quantitative unless stated otherwise.

Evocative or suppression tests measure the effects of certain drugs or chemicals on medical treatment, such as levels of thyroxine for hypothyroidism.

Surgical pathology is the accession, examination, and reporting of specimens. It is differentiated by various levels based on the anatomic site and if microscopic services are provided. The first code is strictly for gross examination. The remainder of the codes include gross and microscopic examination. Each specimen is coded, even if from the same anatomic site.

Surgical pathology should not be confused with anatomic pathology or cytopathology. Surgical pathology is a field within anatomic pathology, which is concerned with specimens from living patients to determine the diagnosis and anatomic pathology. Cytopathology focuses on the examination and testing of cells. Special stains can be coded separately, e.g., Gram and crystal violet.

Some tests are very specific, like blood work and urinalysis. Codes for blood tests are based on the method used for blood-drawing and the test performed. Blood counts can be manual or automated and are coded according to the method of obtaining a sample, manual or automated, and the factor being tested.

PANELS

One troublesome area in laboratory coding is the constantly changing lab panels. These are services that are performed together frequently, so they are included in one code. They can change frequently. With panels, all of the listed tests must have been performed to select the code; otherwise, each test must be listed separately. If all of the tests in a panel are performed, but additional tests are done, then the panel plus the additional tests would be coded.

MEDICINE

The Medicine section contains codes for a wide variety of services and procedures. Familiarize yourself with them, but you do not have to memorize them all! Some services, particularly within the Medicine section, will not be coded by the hospital. These are services that typically are performed by the physician and for which the hospital typically does not bill.

IMMUNIZATIONS/INFUSIONS/INJECTIONS

Code for materials as well as administration, if the hospital provided these services. HCPCS codes may be required if not listed in the CPT book.

IMMUNIZATIONS

Some services will not be coded by the hospital; these are services typically performed by the physician. Immunization is one such service. If the hospital does administer an immunization, code for the administration of the vaccine/toxoid and the material.

There are several codes for administration of vaccines/toxoid. Code 90471 is for the first administration by routes such as percutaneous, intradermal, subcutaneous, or intramuscular. There is an add-on code for additional vaccines administered by these methods. The add-on code is for each administration, so multiples can be used. If, for example, three vaccines are administered, the codes would be 90471 and 90472 × 2.

There is a separate code, 90473, for oral administration. Its add-on code is 90474.

Some vaccine/toxoid codes contain only one vaccine or toxoid; others contain several. Read carefully!

INFUSIONS

A therapeutic infusion is the introduction of liquid into the body over a long period of time, such as several hours. Medicine codes are for administering material. Use HCPCS codes for coding material. This does not include chemotherapy, which is coded with 96400 through 96549 codes. A physician must be present when chemotherapy is administered.

INJECTIONS

Injections require needles, which make them distinct from infusions. The codes are categorized by type of administration, i.e., subcutaneous (SQ), intramuscular (IM) or intra-arterial or intravenous (IV). These medicine codes are for administering material. The material should be coded using HCPCS codes.

CHEMOTHERAPY

Chemotherapy infusion codes are coded by time and by means of administration. There are add-on codes for additional hours, which are counted separately, i.e., an infusion that lasts 3 hours would use codes 96422 and 96423 × 2. Materials are coded using HCPCS codes.

Chemotherapy codes are separate and distinct from office visits. If catheters are required, use the codes from the cardiovascular section to code them. Each type of chemotherapy administration should be coded separately.

DIALYSIS

Dialysis cleans waste from the blood mechanically when the body is no longer able to do so. Dialysis can be temporary or permanent. With hemodialysis, waste is removed directly from the blood. In peritoneal dialysis, waste is removed indirectly through an infusion using fluid (dialysate). The dialysate is flushed into the peritoneum through a catheter, not through the blood, and is then removed with the body's waste products. The catheter must be coded separately. Ultrafiltration removes excess fluid from the blood using a dialysis machine but without the dialysate solution. All services relating to dialysis, and done on the same day, are included in the code.

The codes are broken down into charges per month or charges per day and by a patient's age. Per day codes are used only when the patient's care is disrupted and there is no full month service. Dialysis codes include patient education for self-administration of dialysis.

ACTIVITY 7.12

1. What is the correct code(s) for a patient who had an MRI with oral contrast?

2. What is the correct code(s) for a patient who had repair of a perforation of the cornea due to glass in the eye?

3. What is the correct code(s) for a patient who received debridement with a scissors and tweezers from a wound with topical application?

4. What is the correct code(s) for a patient who had a mastoidectomy with tympanoplasty?

5. What is the correct code(s) for a patient who received KUB of upper GI with films?

6. What is the correct code(s) for a patient who received extracapsular removal of her cataract in the left eye with phacoemulsification and insertion of lens?

7. What is the correct code(s) for a patient who received chemotherapeutic infusion for 4 hours?

8. What is the correct code(s) for a patient who receives an infusion with saline with Doppler and sonohysterography?

9. What is the correct code(s) for a patient who received dialysis in the hospital for ESRD?

10. What is the correct code(s) for a patient who received an EEG for cerebral seizure with 16 channels for 48 hours?

11. What is the correct code(s) for a patient who was given a venipuncture to test lipids?

12. What is the correct code(s) for a patient who is tested for quantity of cocaine?

13. What is the difference between hemodialysis and peritoneal dialysis?

14. What is the correct code(s) for a patient who received an extracapsular removal of a cataract from his left eye with capsulorrhexis and returns today for insertion of lens?

15. What is the correct code(s) for a 10-year-old patient who received an IM injection of tetanus, diphtheria and oral polio?

16. What is the correct code(s) for a patient who received closure of a congenital ventricular septal defect with a right heart catheterization with percutaneous transcatheter and implant?

17. What is the correct code(s) for a patient who received an esophageal acid reflux test?

18. What is the correct code(s) for a patient who received a two-dimensional ultrasound of the chest?

19. What is the correct code(s) for a patient who was tested for opioid use?

20. What is the correct CPT code(s) for a patient who received an IM injection of penicillin?

APPENDICES

The CPT code book has several appendices. Appendix A contains the modifiers; Appendix B provides a summary of additions and deletions; Appendix C provides clinical examples; Appendix D provides a list of add-on codes; and Appendix E provides a list of modifier 51 exempt codes.

MODIFIERS

CPT MODIFIERS are two-digit numeric codes that follow a CPT code and are separated from the code by a hyphen. There are additional modifiers listed in the HCPCS that contain alphanumeric or alphabetic letters, i.e., E1 for upper left eyelid and T1 for left foot, second digit.

Modifiers indicate additional circumstances relating to the CPT code that may alter reimbursement. Modifiers, therefore, help ensure accurate and full payment for services. Some modifiers are used with all CPT codes, but others can be used only with certain ones. (Each modifier will be discussed later in this book.) Some modifiers can be applied only to certain codes, such as only to E/M or surgery. Multiple modifiers can be attached to a single code, e.g., 73580-59-50.

Remember that modifiers are used in the outpatient hospital exam when CPT codes are applied, e.g., with use of Ambulatory Payment Classifications (APCs). Use of modifiers with outpatient hospital billing differs from modifier use with physician billing, because physicians usually provide services on a long-term basis whereas outpatient hospital services often occur in a single day. Each time a patient seeks care at a hospital outpatient facility, it is considered another distinct service. Global packaging, therefore, functions differently for hospital services than for physician services. Also, there usually is not one physician providing all hospital services, but rather hospital billing and coding is for an the entire hospital facility—Emergency Room, hospital-based Ambulatory Surgery Center (ASC) etc.

Modifiers are listed in Appendix A and in some information sections of major headings, such as Surgery, in the code book.

On the CCS and CPC-H exams, modifiers may appear in any section, except for the case studies on the CCS exam. Modifiers can be confusing, but you do not need to know every situation in which they might be used. Focus on what you can absorb and understand. Remember, nobody knows everything and you do not have to be perfect to pass the exam. Try to understand coding and how modifiers work so that you can make educated guesses.

Bilateral

If a code is described as bilateral and the procedure is bilateral, you do not need a modifier. If a code does not say bilateral, then unilateral is assumed. If a unilateral code is used for a bilateral procedure, then you must use modifier 50. Using the modifier 50 is preferable when coding bilateral procedures or services, because it is more descriptive and familiar than billing twice or placing a 2 in the unit box on the billing form.

If a procedure is described as bilateral, but only a unilateral procedure is done, then use modifier 52, which indicates reduction of service.

However, be forewarned—procedures done on both sides of the body are not always considered bilateral, but rather multiple procedures. For instance, a heart catheter is not listed as bilateral even if it goes through both sides of the body, because the code used is for course of the catheter, not whether it is done on opposite sides or not. Another example is lacerations repaired on both sides of the face. This also is not considered a bilateral procedure.

SPECIFIC MODIFIERS

Other modifiers include:

Modifier 25—for significant, separately identifiable E/M services on the same day of another service. This modifier is important because normally only one E/M can be charged per day. If, however, a patient visits again later in the same day, for a different diagnosis, additional E/M services should be billed and paid. The diagnosis should verify this.

Modifier 26 is not used with hospital coding, as the revenue code will indicate if a professional service was performed, even with outpatient services.

Modifier 27 is for multiple hospital outpatient E/M encounters in a single day, such as a patient being seen in an outpatient clinic and later the Emergency Room.

Modifier 52 is used for reduced services based upon a physician's discretion, such as a procedure listed in the codes as bilateral, but service is only unilateral.

Modifiers 54, 55, and 56 are not used by hospital facilities for outpatient services as the service occurs on the same day.

Modifier 58 is used for staged or related procedures that were either planned or more extensive than the original. This modifier must be attached even to a code that states it is staged. This modifier is added to a procedure each time it is performed.

Modifier 59 is used to bill for a distinct and different procedure or service performed on the same day, which is not a part of a bundled package of another procedure or service. The procedure or service may be distinct because of a different anatomical site, different session, or encounter. Linking CPT codes with proper ICD-9 codes is critical for outpatient coding since the diagnoses can indicate the differences. Using this modifier also ensures that full payment is received for the procedure or service and not reduced payment.

For radiological services, if a test is repeated on the same day to determine treatment progress, then modifier 91 can be used to indicate repeating the test; however, modifier 91 cannot be used when repeats are conducted because of faulty equipment or confirmation of the original test.

Modifier 53 is not used for APCs. Instead, modifier 73 is used for discontinued procedures in outpatient hospital/ASC facilities before anesthesia is administered and modifier 74 is for discontinued procedures after anesthesia is administered. Use a V code to indicate the reason why the procedure was terminated. There is 100 percent reimbursement for procedures that are discontinued after anesthesia has been administered and only 50 percent reimbursement before anesthesia is administered.

Modifier 76 is used when the physician is the same for a repeat procedure and modifier 77 is used when the physician is not the same for a repeat procedure, which occurs on the same day. Without this modifier, payment will be denied because the code will be considered a duplicate bill.

Similar to modifiers 76 and 77, modifier 78 is used when an additional procedure is related to a prior procedure on the same day or within a global period as another provided procedure.

Modifier 79 indicates that the same physician performed a distinct and different procedure on a patient during a global period as indicated by the link to a different diagnosis.

Modifier 91 refers to a repeat test done on the same day on the same patient. This modifier can be used only when the test is run again to confirm results. It cannot be used to confirm the results of a prior test.

When the patient is a Medicare patient, there may be reasons to bill Medicare even when you know Medicare will not pay for the services. To bill Medicare for charges that are not allowable, there are two modifiers—GY and GZ—which can be used. GY is for services or procedures billed to Medicare, which are not payable, because they are either excluded or not a benefit such as cosmetic surgery. GZ is for services or procedures that Medicare is expected to deny.

ACTIVITY 7.13

1. What is the correct code(s) for a patient who returned three days later for a lithotripsy by the same physician?

2. What is the correct code(s) for a patient who had a chest tube inserted a week ago and returns today for a rotation flap on his left arm, which is 8 sq cm, by the same physician?

3. What is the correct code(s) for a patient who had a Z-plasty with fasciectomy performed on the right palm with release of the second and third digits?

4. What is the correct code(s) for a patient who had two malignant lesions removed by cryosurgery in the morning from the chest, measuring 0.4 and 0.8 cm respectively, who returns later that day to have another lesion removed from the chest, which was also found to be malignant, size 1.5 cm?

5. What is the correct code(s) for a patient who was seen by a different physician who provided closed treatment with manipulation of a distal radius fracture after having been treated by another physician last week for closed treatment with manipulation of an ulnar fracture?

6. What is the correct code(s) for a patient who was returned to the operating room for enterolysis by the same physician in the same visit?

7. What is the correct code(s) for a patient who received a preoperative and postoperative glucagon tolerance test?

8. What is the correct code(s) for a patient who received a total abdominal hysterectomy a week after having her tubes and ovaries removed?

9. What is the correct code(s) for a patient who had repositioning of a permanent dual-chamber pacemaker with transvenous electrodes that was implanted one week ago?

10. What is the correct code(s) for a patient who had preoperative and postoperative chest x-rays performed with two views?

11. What is the correct code(s) for a patient who had a cerebrovascular shunt removed without replacement three days ago and returns today for a nephrectomy with partial ureterectomy by same physician?

12. What is the correct code(s) for a patient who received arthroplasties on both wrists with internal fixation?

13. What is the correct code(s) for a patient who had a laparoscopic of the abdomen with collection of a specimen by brushing two weeks ago and who returns today for a direct thrombectomy of the iliac vein?

14. What is the correct code(s) for a patient who received an arthrotomy of both elbows?

15. What is the correct code(s) for a patient who had replacement of a temporary single chamber pacemaker with a dual-chamber permanent pacemaker two weeks later?

16. What is the correct code(s) for a patient who was taken to the operating room for repair of an inguinal hernia, which was cancelled before anesthesia was administered due to patient noncompliance?

17. What is the correct code(s) for a patient who had repair of the fascia and dermis for 2.1-cm and 2.2-cm lacerations on his arms?

18. What is the correct code(s) for a patient who received an excision of a tumor from the right knee?

19. What is the correct code(s) for a patient who received a biopsy of the axillary lymph nodes, which were excised?

20. What is the correct code(s) for a patient who was admitted to ASC for osteotomy of the calcaneus, which was discontinued because of the deteriorating condition of the patient once anesthesia had been administered?

SUMMARY

All reimbursement forms must have at least one procedural/service code listed to ensure proper payment. These are CPT/HCPCS codes for outpatient hospital services. Proper use of CPT/HCPCS codes ensures that services are properly billed. The CPT book has a definite structure that allows the coder to navigate through the book and locate the correct codes. The book's structure is based primarily on anatomic sites, which are divided by procedures and services. For example, in the Surgery section, anatomic sites are listed and then divided into procedures and services, such as incisions, excisions, and grafts.

Guidelines are included at the beginning of each section of the CPT book, in addition to the notes, which are critical for proper coding, located throughout the coding sections. CPT code books have an alphabetic index that helps locate procedures. The index includes terms for anatomic sites, procedural names, eponyms, and abbreviations. The actual code selection always must be selected from the tabular list of codes to ensure that the correct code is selected. Proper use of modifiers, which can be selected from either the CPT or HCPCS book, is critical for reimbursement.

ANSWERS

ACTIVITY 7.1

Select the correct CPT code according to its proper linkage to the ICD-9 code given.

1. 707.10
   - A. 47562
   - B. 88235
   - C. 47700
   - **D. 11000**

   Debridement, 11000, can be performed for ulcers.

2. 443.9
   - **A. 35456**
   - B. 26910
   - C. 66984
   - D. 85547

   An angioplasty is performed for artherosclerosis.

3. 652.41
   - A. 59100
   - **B. 59409**
   - C. 64680
   - D. 29834

   The only possible procedural code for the pregnant woman is delivery.

4. 820.8
   - **A. 27134**
   - B. 34900
   - C. 51550
   - D. 58920

   The revision of the hip replacement (arthroplasty) would be a procedure related to a hip fracture.

5. 946.1
   - A. 27635
   - B. 11440
   - C. 16000
   - **D. 16010**

   Codes A–C relate to burns.

6. 427.0
   - A. 11900
   - B. 50200
   - **C. 33240**
   - D. 37650

   Heart rhythm problems can be treated with cardioverter-defibrillators.

7. 173.7
   - **A. 17261**
   - B. 37650
   - C. 17000
   - D. 11900

   This is a malignant lesion, so the removal code must reflect this.

8. 706.1
   - A. 41018
   - **B. 15780**
   - C. 15820
   - D. 17260

   Dermabrasion can be performed for acne scarring.

9. 441.4
   - A. 69950
   - B. 47785
   - **C. 35082**
   - D. 33508

   Aneurysm repair can be linked to a diagnosis of aneurysm.

10. 711.96
    - A. 36580
    - **B. 29871**
    - C. 64614
    - D. 29835

    The procedure is for treatment of an infected knee.

11. 271.1 and 366.44
    - **A. 66920**
    - B. 26540
    - C. 67005
    - D. 11471

    The cataracts are removed, which the diagnosis reflects.

12. 174.9
    A. 31750
    B. 19367
    C. 53440
    D. 64831

    A TRAM procedure can be performed when a neoplasm has been removed from the breast.

13. 487.0
    A. 92997
    B. 33015
    C. 87400
    D. 78350

    Code 87400 is for influenza, so this would be the correct code.

14. 410.00
    A. 62287
    B. 37195
    C. 52283
    D. 84484

    Testing for Troponin, a serum enzyme is used to test for MI.

15. 585
    A. 50080
    B. 90918
    C. 43460
    D. 26479

    Dialysis is provided for ESRD.

16. 594.1
    A. 65275
    B. 58960
    C. 36823
    D. 52353

    Calculus can be removed with ESWL, extracorporeal shock wave lithotripsy.

17. 336.8
    A. 20500
    B. 22554
    C. 22556
    D. 33970

    Code 22554 is for fusion of cervical vertebrae, which is consistent with the diagnosis.

18. 456.0
    A. 52400
    B. 36470
    C. 43130
    D. 91030

    To stop the bleeding, a sclerosing agent is injected into the vein.

19. 550.9
    A. 25360
    B. 49321
    C. 49560
    D. 49495

    Both the diagnostic and procedure code refer to the inguinal hernia.

20. 733.82
    A. 64776
    B. 51820
    C. 25431
    D. 92568

    The nonunion, which can result from a fracture, can be repaired.

ACTIVITY 7.2

1. What would be wrong if 99091 was reported twice for dates 1/2/04 and 1/8/04?

ANSWER: Coding notes instruct that code 99091 can be used only once in a 30-day time period.

2. What is a trophoblastic tumor GTT?

ANSWER: Hydatidiform mole as listed in the alphabetical index.

3. What codes could be used to code for a pregnancy test?

ANSWER: 84702–84703 and 81025 as listed in the alphabetical index.

4. What is the difference between the codes listed in the previous question?

ANSWER: Code 84702 is for gonadotropin quantitative, while 84703 is for the same substance but a qualitative test. Code 81025 is a urine pregnancy test.

5. If a coding scenario was 49905 and 44700, would it be correct? Why or why not?

ANSWER: No. Code 49905 is an add-on code and must be coded in addition to the code for the primary procedure. Notes indicate that code 49905 cannot be used in conjunction with 44700.

6. Can submucous resection of turbinates be coded with 30520? Why or why not?

ANSWER: No, it should be coded with 30140 as indicated in the coding notes.

7. If a coding scenario was coded as 15220, 15221 × 6, and 15000, would this be correct? Why or why not?

ANSWER: No, the amount of area prepared exceeds 100 sq cm, so 15001 also would need to be coded.

8. Are codes for magnetic resonance angiography of the head distinguished as either with or without contrast? Why or why not?

ANSWER: No. Although code 70546 is for coding with contrast materials only and code 70547 is for coding without contrast materials, code 70548 includes without contrast followed by contrast materials.

9. If code 30465 is used to code for repair of left-sided nasal vestibular stenosis, is this correct? Why or why not?

ANSWER: No, 30465 is for bilateral services, so a modifier 52 needs to be attached to use this code for a unilateral procedure.

10. How should a posterior capsulotomy be coded when there is extraction of the lens?

ANSWER: This pertains to cataract removal and the capsulotomy is included in the extraction code of 66830.

11. Evisceration also is known as what?

ANSWER: Exenteration as noted within the alphabetical index.

12. A Jones and Cantarow test is for what substance?

ANSWER: Blood urea nitrogen as indicated in the alphabetical index.

13. For a transurethral resection, can code 51530 be used? Why or why not?

ANSWER: No, coding notes indicate that transurethral resection should be coded with codes 52234–52240 or 52305.

14. If a bone graft is coded as 20930-62, is this correct? Why or why not?

ANSWER: No, coding notes indicate that for these codes, modifier 62 cannot be used.

15. If a coding scenario is coded as 62318 and 01996, is this correct? Why or why not?

ANSWER: Yes, because the coding notes indicate that these codes can be used together.

16. What are the types of acupuncture that can be coded?

ANSWER: With or without electrical stimulation as listed in the alphabetical index.

17. ACTH means what?

ANSWER: Adrenocorticotropic hormone, which can be found in the alphabetical index.

18. If a coding scenario is coded as 45341 and 76975, is this correct? Why or why not?

ANSWER: No, the coding notes indicate that these codes cannot be coded together.

19. Does selective vascular catheterization include introduction of the catheter? Explain.

ANSWER: Yes, the coding notes indicate that the introduction is included.

20. If a coding scenario was coded as 31633 and 31629, would this be correct?

ANSWER: Yes, coding notes indicate that these codes can be coded together.

ACTIVITY 7.3

1. If a tendon transplantation of the wrist extensor is performed with grafts for each tendon, is it correct to code it as 25310 and 25312? Explain.

ANSWER: No, *each tendon*, which is the additional statement that is listed after the semicolon in the code 25310, also is included in the statement for code 25312.

2. Would a test for total protein through collection of a urine sample be coded as both 81000 and 84160? Explain.

ANSWER: No, only the one code 81000 would be necessary as indicated in the coding notes.

3. If codes 90920 and 90924 are coded together, is this correct? Explain.

ANSWER: Yes, because the patient may have received ESRD services for more than 30 days, which allowed code 90920, with the patient continuing to have several more days of ESRD services but not more than 30; code 90924 would account for each extra day.

4. If codes 69990 and 31536 are listed together, would this be correct? Explain.

ANSWER: No, the use of the microscope cannot be listed in addition to code 31536 as noted in the coding book.

5. If codes 82948 and 83026 are coded together, is this correct? Explain.

ANSWER: Yes, these lab tests can be coded separately since there is no panel that includes both.

6. If a patient receives a limited lymphadenectomy with staging with prostatectomy, would it be correct to code it as 38562? Explain.

ANSWER: No, coding notes indicate that with a prostectomy, code 55812 or 55842 should be used instead.

7. If coronary artery bypass grafts were performed utilizing four coronary veins and an arterial graft, would codes 33513 and 33533 be correct? Explain.

ANSWER: No, there is a code (33521) for combined venous and arterial grafts that should be used instead.

8. Is it correct to use code 25315 times two to code for a flexor origin slide of the wrist and forearm? Explain.

ANSWER: No, the code 25315 includes both the wrist and/or forearm, so the code needs only to be listed once.

9. Is it correct to code a 33 sq cm full-thickness graft of the chest as 15200? Explain.

ANSWER: No, code 15200 is for 20 sq cm or less, so code 15201 should be coded in addition for the remaining 13 sq cm that is not included in the 15200 code.

10. Is it appropriate to code 3 antepartum visits as 59425, after which the patient discontinued visits with the physician? Explain.

ANSWER: No, code 59425 is coded only for 4 to 6 visits. Three visits should be coded with E/M codes as described in the notes.

11. If an open biopsy of internal mammary lymph nodes is performed with an axillary lymphadenectomy, can this be coded as 38530 and 38740? Explain.

ANSWER: No, coding directions state that codes 38530 and 38740 cannot be coded together.

12. If a complete cystectomy is performed with bilateral pelvic lymphadenectomy of the external iliac, hypogastric and obturator nodes, and ureterosigmoidostomy, is it correct to code it as 51580 and 51585? Explain.

ANSWER: No, code 51585 includes everything, so code 51580 should not be used.

13. If a surgical hysteroscopy is performed with resection of the intrauterine septum with sampling of the endometrium, is it correct to code this as 58560? Explain.

ANSWER: No, with code 58560 the sampling is not included.

14. Is it a correct coding scenario to code separately both a radical neck dissection and total thyroidectomy for malignancy? Explain.

ANSWER: No, the radical neck dissection is included in the total thyroidectomy, 60254.

15. Is it correct to code a layered closure of a 5.2 cm laceration of the cheek and a simple repair of a 4.8 cm laceration of the chin as 12053 and 12013? Explain.

ANSWER: Yes, these codes are correct as the types of repairs are different and, therefore, should be coded separately.

16. If an x-ray of the pelvis is taken with three views, is it correct to code it as 72170 and 72190? Explain.

ANSWER: No, code 72190 includes the three views. Code 72170 is not necessary.

17. If an ureteroneocystostomy and vesical neck revision were performed with anastomosis from a single ureter to the bladder, would it be correct to code it as 50780 and 51820? Explain.

ANSWER: No, only code 51820 is needed as it includes everything.

18. If 18 skin tags were coded as 11200 and 11201, would it be correct? Explain.

ANSWER: Yes, it would be correct as code 11200 includes the first 15 and code 11201 includes the next ten.

19. If destruction of a sacral paravertebral facet joint nerve at two levels by neurolytic agent is coded as 64622 and 64623, is this correct? Explain.

ANSWER: Yes, this is correct as the additional level is coded with the add-on code of 64623.

20. If skin and subcutaneous tissue are excised for axillary hidradenitis, which required complex repair, is it correct to code this as 11450 and 11451? Explain.

ANSWER: No, only code 11451 is necessary, because it is for the complex repair alone.

ACTIVITY 7.4

1. What is the correct code(s) for a patient who received a chest x-ray with four views without interpretation and supervision?

ANSWER: 71030-TC; The professional component was not provided as indicated by no interpretation and supervision having been provided, so a modifier of TC must be applied.

2. What is the correct code(s) for a patient who received a subsequent extended ophthalmoscopy with retinal drawing with interpretation?

ANSWER: 92226; Note that the professional component is included in this code as indicated by the terms *interpretation and report*.

3. Should modifier 26 be attached to the code in the previous question? Explain.

ANSWER: No, the code includes the professional component as denoted by the terms *interpretation and report*, which remained with the code even though it was indented.

4. Under what terms can you find Burrow's Operation?

ANSWER: Skin, adjacent tissue transfer; this is located in the alphabetical index, which directs you to *see* under these terms.

5. What is the correct code(s) for a patient who receives an open transluminal balloon angioplasty of two vessels in the brachiocephalic branches?

ANSWER: 35458 × 2; There is an indication that this code is for each vessel, so it must be multiplied by two.

6. What is the correct code(s) for a patient who had three lacerations on his left arm and one on his neck? Two of the arm wounds were 2.2 and 5.2 cm and required closure of the superficial fascia. The other one on the arm was 0.8 cm and was more severe with avulsions and required retention sutures. The one on the neck was 2.2 cm and required repair, including the subcutaneous tissues.

ANSWER: 12034 and 12001; Note that the more severe laceration is added into the intermediate repair, because it was not long enough to be coded as a complex repair, which is indicated in the coding notes. The lacerations are added together by anatomic sites and type of repair and are not coded individually.

7. Eutelegenesis is known as what?

ANSWER: Artificial insemination; There is a note indicating that eutelegeneis can be found by looking under *see, artificial insemination*.

8. What is the correct code(s) for a patient who received an MRI for stereotactic excision of intracranial lesion with contrast, which was read and interpreted by the physician?

ANSWER: 61751 and 70552; The code 61751 indicates that if supervision and interpretation were performed, then code 70552 also should be coded. The MRI was used to provide guidance for the excision.

9. Is modifier 26 necessary for the previous coding scenario? Explain.

ANSWER: No, the code 70552 includes supervision and interpretation as indicated by the physician's reading and interpretation.

10. What is the correct code(s) for a patient who had three calluses removed?

ANSWER: 11056; Note that the code 11056 is for the removal of 2 to 4 lesions (or calluses). Code 11055 is for a single lesion, but the codes are not for separate and individual lesions but for a specified number of lesions.

11. What is the correct code(s) for a patient who receives closed treatment of a distal radial fracture and ulnar styloid?

ANSWER: 25600; Without manipulation is assumed for this code since it is not indicated. Notice that this code is specific to the distal end of the radius and includes the ulnar styloid.

12. What is the correct code(s) for a patient from whom an intramuscular biopsy of the elbow is obtained?

ANSWER: 24066; Notice that this code is for the upper arm and elbow area.

13. What is the correct code(s) for a patient who receives a capsulotomy of two metatarsophalangeal joints with a tenorrhaphy?

ANSWER: 28270 × 2; There are two joints involved and the code indicates that it is applied for each joint, so there has to be a times two indicated.

14. P&P is known as what in the CPT book?

ANSWER: Proconvertin; The alphabetical indicates that you must *see proconvertin* to find the correct code for P&P.

15. What is the correct code(s) for a patient who receives a percutaneous tenotomy of the Achilles tendon?

ANSWER: 27605; Local anesthesia is assumed because there are no notes concerning the type of anesthesia, so code 27605 is used. Notice that this code includes the leg and ankle.

16. What is the correct code(s) for a patient who received an MRI for excision of intracranial lesion with contrast?

ANSWER: 61751; Unlike one of the previous questions, there is no indication of supervision and interpretation by the physician, so no other code is required.

17. Is modifier 26 necessary for the previous coding scenario? Explain.

ANSWER: No, unlike one of the previous questions, there is no professional component indicated.

18. What is the correct code(s) for a patient who received an excision of a bone cyst from the head of the radius?

ANSWER: 24120; Note that the head of the radius has a code distinct from the radius itself for this procedure.

19. What is the correct code(s) for a patient who had biopsies of three retroperitoneal lymph nodes, which include a diagnostic laparascope?

ANSWER: 38570; The scope was used to biopsy the nodes, so it is included in code 38570. The code is for single or multiple biopsies, so no other codes are necessary and the code is listed only once since there is no indication that the biopsies should be listed separately.

20. What is the correct code(s) for a patient whose physician reviews and interprets the results of her Bucky study from a chest x-ray?

ANSWER: 71035-26; Note that modifier 26 must be attached since the physician did not perform the x-ray, but did interpret the results, which is the professional component only.

ACTIVITY 7.5

1. What is the correct code(s) for a 19-year-old patient who had anesthesia for a tenodesis of the bicep?

ANSWER: 01716; This anesthesia code is located under upper arm and elbow.

2. What is the correct code(s) for this 67-year-old patient who was admitted to the Emergency Room with critical injuries sustained in a car accident that included respiratory collapse of the left lung requiring ventilation?

ANSWER: 00541-P5; Note this code is from the intrathoracic section and includes the ventilation. A physical status modifier of P5 should be included to indicate that the patient's condition is life-threatening.

3. What is the correct code(s) for a 48-year-old patient who received anesthesia during a coronary angiography?

ANSWER: 01920; There is a specific code for the angiography, which is included under the catheterization code for anesthesia.

4. What is the correct code(s) for a 43-year-old patient who received anesthesia for a cranioplasty?

ANSWER: 00215. The anesthesia is coded for the procedure on the head.

5. What is the correct code(s) for this 6-month-old baby who received anesthesia for a chest tube?

ANSWER: 00520 and 99100; An intrathoracic anesthesia code is used since this involves the chest. The code 99100 should be included to indicate the patient is less than one year of age.

6. What is the correct code(s) for a patient who received anesthesia for Harrington rod surgery?

ANSWER: 00670; Harrington rods are used within the spine and are found under the spinal section of the anesthesia codes.

7. What is the correct code(s) for a patient who received anesthesia for Torek procedure?

ANSWER: 00930; This procedure is specifically listed in the alphabetical index although it is not listed in the code itself.

8. What is the correct code(s) for a healthy 28-year-old patient who had a total knee arthroplasty?

ANSWER: 01402-P1; This procedure is specifically listed in the coding section for knee and popliteal. The health status of the patient is noted, so it should be coded with a modifier P1.

9. What is the correct code(s) for a 23-year-old patient who received daily management of her epidural while in the hospital?

ANSWER: 01996; There is a code for the epidural management that is used for drug administration while the patient is in the hospital.

10. What is the correct code(s) for a 57-year-old patient who receives anesthesia for a revision of her dialysis shunt?

ANSWER: 01844; This code is listed under forearm, wrist, and hand, because the shunt is placed in this area.

11. What is the correct code(s) for a 16-year-old patient who received anesthesia for a double osteotomy of the tarsals?

ANSWER: 01480; Notice that this code is selected from the lower leg anesthesia codes and is considered an open procedure on the bones of the foot as indicated in the description of the procedure, which includes the term osteotomy.

12. What is the correct code(s) for a 62-year-old patient who received anesthesia for replacement of an aortic valve?

ANSWER: 00560; The procedure was performed on the heart as indicated by the aortic valve, so the code is selected from the intrathoracic section of the anesthesia codes.

13. What is the correct code(s) for a 45-year-old patient who received anesthesia for a hysterectomy?

ANSWER: 00840; It is not stated that this is radical, so the general code for not otherwise specified within the lower abdomen is coded.

14. What is the correct code(s) for a 29-year-old patient who received anesthesia for a lumbar sympathectomy?

ANSWER: 00622; Note that this code comes from the spine and spinal cord codes.

15. What is the correct code(s) for a 13-month-old baby who received anesthesia for removal of a toy from the larynx?

ANSWER: 00320; The code 00320 is determined by the patient's age, which is over 1 year of age.

16. What is the correct code(s) for a baby who is 40 weeks gestationally and who received anesthesia for an incisional hernia repair?

ANSWER: 00836; Gestational age refers to the age including intrauterine, so this baby falls into the category of less than 50 weeks but more than 37 weeks, so code 00836 is correct and not code 00834. The additional age code of 99100 is not used as code 00836 references the age in addition to a note indicating that the 99100 code cannot be used with 00836.

17. What is the correct code(s) for a 73-year-old patient who received anesthesia for an osteotomy of the humerus?

ANSWER: 01742 and 99100; The osteotomy code is selected from the shoulder and axilla section of the anesthesia codes. Note that an additional code is necessary, since the patient is over 70-years-old.

18. What is the correct code(s) for a 24-year-old patient who received anesthesia for a cesarean hysterectomy following labor?

ANSWER: 01967 and 01969; Note that anesthesia for just the hysterectomy cannot be coded because labor also occurred. There is a note that an add-on code is necessary.

19. What is the correct code(s) for a 45-year-old patient who received anesthesia during a procedure for advancement of the flexor tendon in the right hand?

ANSWER: 01810; There are specific anesthesia codes for tendons, muscles, etc.

20. What is the correct code(s) for an 18-year-old patient who received anesthesia for a bronchoscopy and who has mild systemic disease?

ANSWER: 00520-P2; This is included as a closed chest procedure. A physical status modifier of P2 should be attached to indicate the patient has mild systemic disease complicating the procedure.

ACTIVITY 7.6

1. Which of the following codes are not separate procedures: surgical endoscopy, open tenotomy of the toe, arterial catheterization for transfusion, injection of ganglion cysts, drainage of external ear canal, and extensive biopsy of vaginal mucosa?

ANSWER: Surgical endoscopy, injection of ganglion cysts, drainage of external ear canal, and extensive biopsy of vaginal mucosa.

2. What is the correct code(s) for a patient who received a replacement of an inflatable bladder neck sphincter with pump, reservoir, and cuff, including removal?

ANSWER: 53447; This code includes the replacement and removal and there are no other codes required.

3. What is the correct code(s) for a patient who received a lung needle biopsy?

ANSWER: 32405; Note that this is specifically for needle biopsy, not fine needle.

4. What is the correct code(s) for a patient who received a biopsy by indirect laryngoscopy with removal of a lesion?

ANSWER: 31512; This scope includes removal of the lesion, so it is not coded as biopsy but as removal.

5. For the code(s) chosen in the previous question, are any procedures separate? Explain.

ANSWER: No, the separate procedure notation was included with the term *diagnostic,* which was eliminated when the lesion removal code was selected.

6. For the code(s) listed in the two previous questions, are any of them diagnostic scopes? Explain.

ANSWER: No, because the term *diagnostic* was eliminated when the lesion removal code was selected.

7. What is the correct code(s) for a patient in whom a cystourethroscopy was performed and a Gibbons stent was placed within the ureters?

ANSWER: 52332; Although there is a code for insertion of an indwelling ureteral stent, a scope was used, so code 52332 must be listed.

8. What is the correct code(s) for a patient who received debridement of infected tissue with replacement of an inflatable bladder neck sphincter, which included the pump, reservoir, and cuff, including removal?

ANSWER: 53448; No debridement codes are required even though debridement was performed, because the debridement is included in the code for the replacement.

9. What is the correct code(s) for a patient who received a polypectomy with a diagnostic endoscopy and sinusotomy?

ANSWER: 31237; Only the polypectomy needs to be coded with a surgical scope. The diagnostic scope and sinusotomy are included.

10. What is the correct code(s) for a patient who received a biopsy of a lesion in the pharynx?

ANSWER: 42804; Only a biopsy is performed, so no other codes are required.

11. What is the correct code(s) for a patient who received an upper GI endoscopy of the esophagus, stomach, and duodenum by brushing?

ANSWER: 43235; This code includes the brushing.

12. Is code 11011 considered a separate procedure? Explain.

ANSWER: No, although there is the statement to list it separately, it is an add-on code and needs to be coded with its primary code, 11000.

13. What is the correct code(s) for a patient who received a fine needle lung biopsy?

ANSWER: 10021; Fine needle aspiration for biopsies are coded with their own specific codes of 10021 and 10022, not the code for needle biopsy.

14. What is the correct code(s) for a patient from whom a biopsy was obtained when a malignant 0.8 cm lesion was removed from the back?

ANSWER: 11601; The lesion was excised, so only this is coded.

15. What is the correct code(s) for a patient who received a percutaneous biopsy of the breast for needle core with imaging?

ANSWER: 19102; This is a separate procedure.

16. What is the correct code(s) for a patient who received a biopsy of a lesion of the pharynx that was excised?

ANSWER: 42808; The lesion was excised so, although it was described as a biopsy, it is coded as an excision.

17. What is the correct code(s) for a patient who received a biopsy of the cervix with the use of a scope?

ANSWER: 57455; Only a biopsy was performed, so it is coded, but note that the biopsy was performed with a scope.

18. What is the correct code(s) for a patient who received a transcatheter placement of a stent within two coronary arteries after PTCA was performed?

ANSWER: 92980, and 92981; These are coronary arteries, so the codes are selected based on this. PTCA is included in the stenting code. More than one artery was treated, so an additional code must be listed.

19. What is the correct code(s) for a patient who received endovascular placement of a device for occlusion of the iliac artery in addition to placement of femoral-femoral prosthetic graft with exposure through a groin incision on the left side for repair of an endovascular aortic aneurysm?

ANSWER: 34812, 34813, and 34808; Notice that there are coding notes directing the additional coding of the other codes.

20. Which of the codes in the previous question are separate procedures? Explain.

ANSWER: None, although the statement *list separately* is provided, it does not mean the same as *separate procedure.*

ACTIVITY 7.7

1. What is the correct code(s) for a patient who received repair, which included the subcutaneous tissue and superficial fascia with ligation of vessels for two 3.8 cm lacerations on both forearms?

ANSWER: 12034; No code is required for the ligation of the vessels because simple ligation is included in the repair codes. This is an intermediate repair as indicated by the repair of the fascia and subcutaneous tissue. Both the anatomic sites are included within the same code, so the sites are added together to determine the total size for selection of the proper code.

2. What is the correct code(s) for a 3-year-old patient who suffered third-degree burns on both arms and who received dressing and debridement with anesthesia for one week and then without anesthesia for eight more days?

ANSWER: 16015 × 7 and 16025 × 8; In the infant, both arms would constitute a medium size area (9% for each arm). The care was given with anesthesia for 7 days and then without anesthesia for 8 days, so the number of times must be indicated.

3. What is the correct code(s) for a patient who had a malignant lesion that measured 3.1 cm removed by surgical curettement from the upper back?

ANSWER: 17264; This lesion is malignant, which is reflected in the selection of the proper code. In addition, it was removed by surgical curettement, which is a form of destruction, not excision, so it is coded as destruction.

4. What is the correct code(s) for a patient who has percutaneous biopsy of the breast with placement of a clip for visualization?

ANSWER: 19102 and 19295; The image guided placement as indicated by the clip must also be coded in addition to the biopsy.

5. What is the correct code(s) for a patient who had a radical mastectomy, including the axillary lymph nodes and pectoral muscles?

ANSWER: 19200; There are many codes for mastectomies that include various anatomic sites, so be careful in selecting the proper one.

6. What is the correct code(s) for a patient who receives a TRAM flap for breast reconstruction with microvascular anastomosis for single pedicle with closure?

ANSWER: 19368; This code includes all of the diagnostic statement within code 19367, since the semicolon appears at the end of the statement in 19367.

7. What is the correct code(s) for a patient who returns for additional excision of a 1.8 cm malignant lesion that was removed a week ago with an area measuring 2.2 cm today?

ANSWER: 11603; The same codes are used for excision, but the size coded is the size excised in the subsequent visit of the day.

8. What is the correct code(s) for a patient who had a biopsy for possible skin cancer combined with initial Moh's surgery, which was evaluated in pathology and confirmed as malignant?

ANSWER: 17304, 11100, and 88331; This is the first stage of Moh's with a biopsy and subsequent pathological services provided, so these also must be coded. There is only one specimen.

9. What is the correct code(s) for a patient who had two malignant lesions from the upper back excised that measured 1.0 and 0.8 cm including margins, respectively? The physician was able to cut them out as one, including the border.

ANSWER: 11602; Only one excision needs to be coded since the physician cut them out as one, so the total measurement is 1.8 cm.

10. What is the correct code(s) for a patient who sustained a laceration of the forehead that required debridement?

ANSWER: 11040; The debridement was simple and the laceration required no repair.

11. What is the correct code(s) for a patient who has a malignant 0.8 cm lesion in the breast removed, which involves placement of wire for visualization?

ANSWER: 19120 and 12920; The lesion is marked by the wire, so this must be coded also.

12. What is the correct code(s) for a patient who received repair for a 2.2 cm laceration on the left hand that required suturing of the ulnar nerve and suturing of fascia? Debridement required removal of many small and medium pieces of glass.

ANSWER: 64836, 12041 and 11042; The ulnar nerve repair must be coded separately from the suturing of the fascia. Suturing of the fascia constitutes intermediate repair. Debridement is coded also, because it is noted that additional time was spent removing the glass. Notice that in selection of the debridement code, there is the distinction between with or without fracture.

13. What is the correct code(s) for a patient who received a biopsy of a lesion on the right hand with biopsy and excision of a malignant 1.2 cm lesion on the right forearm?

ANSWER: 11602 and 11100-59; The biopsy of the forearm is included in the excision code, but the biopsy of the hand is not and should be coded. The modifier is attached to indicate that the biopsy was not distinctly different from the other procedure.

14. What is the correct code(s) for a patient who had a malignant lesion from the upper back excised that measured 8.0 cm, which included a border? A split-thickness graft was performed to cover the area that included part of the epidermis.

ANSWER: 11606 and 15100; The lesion is differentiated as malignant in the code. The graft was a free graft, so it must be coded. The measurements are taken as is because a small border can be included. The graft is a split-thickness, since all layers of the dermis and epidermis were involved. The site is already prepared because the graft was done at the same time the lesion was excised, so no code is required for preparation of site.

15. What is the correct code(s) for a patient who had two malignant lesions removed, 0.7 cm lesion of the left forearm and 0.4 cm lesion of the neck?

ANSWER: 11600 and 11620; Each lesion must be coded separately as malignant.

16. What is the correct code(s) for a patient who returns for a second time for Moh's surgery, which includes seven skin neoplasms?

ANSWER: 17305 and 17310 × 2; The patient returns indicating that it is not the first stage of the Moh's surgery. There are seven neoplasms, so the first code, 17305, includes the first five specimens that are not specified as each or individual, so only the one code is necessary. However, with the second code, 17310, each additional is described, so the code must be listed for each additional specimen, so times two is listed with the code.

17. What is the correct code(s) for a 1-year-old patient who had sustained third-degree burns over her right arm eight months ago and is now receiving a 35 sq cm full-thickness graft from her leg to be applied to the scar tissue?

ANSWER: 15220, 15221, and 15000; The site must be prepared, because it has been three months and scar tissue has formed. The graft is a full-thickness and is from the leg, which is reflected in the codes. An add-on code of 15221 must be included because the graft is more than 20 sq cm, which is contained within the first code of 15220. The second code, 15221, can be used even though it is not the full additional amount of 20 more sq cm

18. What is the correct code(s) for a patient who received injections for four lesions for chemotherapy?

ANSWER: 96405; This code is specific for chemotherapy and is not coded from the other lesional injection codes of 11900. The code is for up to 7 lesions that are not to be listed separately, as there are no notes to indicate separate coding.

19. What is the correct code(s) for a patient who had an excision of an ischial pressure ulcer with split-thickness myocutaneous flap?

ANSWER: 15946 and 15100; The excision is coded, but so is the graft, with code 15100.

20. What is the correct code(s) for a patient who had a malignant lesion from the upper back excised that measured 8.0 sq cm, which included a border? A Z-plasty was performed to cover the area.

ANSWER: 14000; The lesion is excised with a Z-plasty, so there is no need to code for the excision. There is no need for preparation either since the area is already prepared because of the Z-plasty.

ACTIVITY 7.8

1. What is the correct code(s) for a patient in the Emergency Room who has removal of a cranial halo, which is complicated by infection, for which there is I&D?

ANSWER: 20665 and 10180; Obviously if the patient is having the halo removed under emergency circumstances, the physician is not the primary physician, so the code specifies that this is not the same physician who initially put the halo on. The infection is treated with I&D postoperatively as the infection resulted from the halo.

2. What is the correct code(s) for a patient who received an osteotomy with removal of a wedge from the proximal phalanx of the big toe with insertion of Kirschner wire for correction of a bunion?

ANSWER: 28298; This procedure is known as an Akin procedure and involves the implantation of a Kirschner wire.

3. What is the correct code(s) for a patient who had a lesion of the tendon sheath removed from the leg?

ANSWER: 27630; Remember that codes that apply in the integumentary system also can apply in other sections, such as excisions and biopsies.

4. What is the correct code(s) for a patient who had closed treatment for a comminuted fracture of the shaft of the tibia and fibula without manipulation?

ANSWER: 27750; Limited information is given for the treatment, so it is assumed to be a closed treatment, so code 27750 is used. The type of fracture, whether open or closed, is irrelevant to the type of care the fracture receives, so comminuted is not considered when coding for the procedure.

5. Arthroplasty of the acetabulum and proximal femoral with a prosthesis is known as what?

ANSWER: Total hip replacement, code 27130.

6. What is the correct code(s) for a patient who received an osteotomy of the proximal tibia and fibula for *genu varus*?

ANSWER: 27455; Although there is another code, 27709, for osteotomy of fibula and tibia, for the specific purposes of correcting bow-leggedness (*genu varus*), the code 27455 is used.

7. What is the correct code(s) for a patient who received a fasciotomy with debridement of muscle and decompression?

ANSWER: 27892; There is another code, 27600, for fasciotomy but it does not include debridement.

8. What is the correct code(s) for a patient who received a partial arthroscopic synovectomy of the wrist?

ANSWER: 29844; Synovectomies can be either partial or complete.

9. What is a Monteggia fracture?

ANSWER: It is a fracture of the proximal end of the ulna with dislocation of the radial head, as described in codes 24620–24635.

10. What is the correct code(s) for a patient who received a replacement for total shoulder arthroplasty?

ANSWER: 23472; There are several different codes to select from, but this is for a replacement, so be sure to code it as 23472.

11. What is the correct code(s) for a patient who received treatment for fracture/dislocation of shaft of the tibia and fibula?

ANSWER: 27750; This fracture/dislocation is coded as a fracture that is closed, because no information is provided as to whether it is open or closed treatment or if manipulation was provided.

12. What is the correct code(s) for a patient who had a portion of the mandible bone removed due to carcinoma? A bone graft was performed to repair the defect.

ANSWER: 21044 and 21215; The bone is removed because of the malignancy, so the code for the excision of the mandible must indicate the malignancy. The bone graft must be coded with 21215.

13. What is the correct code(s) for a patient who received a partial claviculectomy and arthroscopy?

ANSWER: 29824; A scope was used, so code 23120 cannot be used as it does not include the scope. Remember that some procedures can be performed with or without scopes and must be coded likewise.

14. What is the correct code(s) for a patient who had a posterior approach for open treatment and reduction of fractures of C3–5 with wiring of the spinous processes?

ANSWER: 22326, 22328, and 22841; There are two cervical vertebrae involved, so add-on code 22328 must be listed. There is instrumentation involved in the fixation, so a code must be listed for that also.

15. What is the correct code(s) for a patient who received open fracture care for the shaft of the tibia and fibula with the insertion of screws?

ANSWER: 27758; This scenario is different from scenarios in previous questions, because this care is open and involved screws.

16. What is the correct code(s) for a patient who received percutaneous repair of a ruptured Achilles tendon, which was repaired with a graft?

ANSWER: 27652; The graft is included in this code.

17. What is the correct code(s) for a patient who received an arthodesis of two carpometacarpal joints of the thumb and second digit?

ANSWER: 26841 and 26843; Two codes are required, because the thumb has its own code, which are distinct from the remaining digits.

18. What is the correct code(s) for a patient who received a cranioplasty, with an allograft, which involved the forehead?

ANSWER: 21179; Cranioplasty is coded as reconstruction and includes the graft.

19. Which of the following are not part of the vertebral interspace: annulus fibrosus, endplates, laminae, bursa, nucleus pulposus, and carinatum?

ANSWER: Laminae, bursa, and carinatum; What constitutes the vertebral interspace is listed in the notes in some coding books.

20. What is the correct code(s) for a patient who received open procedure with internal fixation for dislocation of joint of the tibia and fibula?

ANSWER: 27832; This is a dislocation only and the care is open, so code 27832 applies.

ACTIVITY 7.9

1. What is the correct code(s) for a patient who received a limited study with duplex scan of veins of the legs and toes?

ANSWER: 93926; This code includes both the toes and legs, which are considered lower extremities.

2. What is the correct code(s) for a patient who received a thrombectomy of a nonautogenous hemodialysis graft for a revision of an open fistula?

ANSWER: 36833; Because this graft is for hemodialysis, it is coded separately from other thrombectomies, which are coded as 35875.

3. What is the correct code(s) for a patient who received a venous bypass graft of two coronary arteries from two segments in the arm?

ANSWER: 33511 and 35500; There are two coronary venous grafts, so code 33511 should be used in addition to 35500 for harvesting of the veins from the upper arm.

4. What is the correct code(s) for a patient who received implantation of a dual-chamber cardioverter/defibrillator with epicardial electrodes?

ANSWER: 33245; A thoracotomy is required for the placement of the electrodes on the epicardial surface of the heart, so this must be included in the code.

5. What is the correct code(s) for a patient who had a PTCA with coronary atherectomy?

ANSWER: 92995; There is no mention of additional vessels treated, so only one code is necessary.

6. What is the correct code(s) for a patient who has an arteriovenous fistula inserted for hemodialysis using Gortex?

ANSWER: 36830; Gortex is a nonautologous substance and repair is not direct, so code 36830 must be selected.

7. What is the correct code(s) for a 26-year-old patient who had a VAD inserted through the subclavian vein?

ANSWER: 36556; Subclavian indicates that the VAD was inserted centrally.

8. What is the correct code(s) for a patient who received an open balloon angioplasty of the renal artery?

ANSWER: 35450 and 75966; The angiography also must be coded as directed.

9. What is the correct code(s) for a patient who received a patch graft for occlusive disease of the iliac artery?

ANSWER: 35221; This repair does not include any diagnosis other than occlusive disease, so no additional code is applicable from the 35000 series.

10. What is the correct code(s) for a patient who received an aortofemoral bypass graft with Gortex?

ANSWER: 35647; Gortex is nonautologous and must be reflected in the code.

11. What is the correct code(s) for a patient who received an arterial-venous bypass graft for two coronary arteries?

ANSWER: 33518 and 33534; With an arterial-venous bypass graft, two codes are needed, one for the combination and one for the arterial as directed within the coding guidelines.

12. What is the correct code(s) for a patient who received a combined left and right heart catheterization including selective right and left ventricular angiography?

ANSWER: 93528, 93542, 93543, and 93555; There are many types of catheterizations that can occur, so be careful in selecting the proper one. The injection procedures for the selective angiography also must be coded.

13. What is the correct code(s) for a patient who received a permanent pacemaker with a transvenous

electrode placed—one into the right atrium, one into the right ventricle, and one into the left ventricle?

ANSWER: 33208 and 33225; Another electrode is inserted into the left ventricle, so this must be coded as 33225 in addition to the atrial and ventricular electrodes.

14. What is the correct code(s) for a patient who received an insertion of a catheter by neck incision for embolectomy of the subclavian artery?

ANSWER: 34001; The catheter is included in this code. These codes are specified by approach, which is a neck incision in this case.

15. What is the correct code(s) for a patient who received a patch graft for an aneurysm and occlusive disease of the iliac artery?

ANSWER: 35132; The occlusive disease and aneurysm are included in the code for patch graft.

16. What is the correct code(s) for a patient who was provided an arteriovenous cannula for access during hemodialysis?

ANSWER: 36810; This insertion is specifically for providing access for hemodialysis.

17. What is the correct code(s) for a patient who received a transluminal percutaenous angioplasty of the renal artery?

ANSWER: 35471, 93508, and 75722; Coding guidelines direct the coding of placement and supervision/interpretation of the angiography.

18. What is the correct code(s) for a patient who receives an upgrade of his pacemaker to a dual-chamber and new pulse generator?

ANSWER: 33214; The upgrade includes replacement of the various parts of the pacemaker, including the pulse generator.

19. What is the correct code(s) for a patient who received a bypass graft from the internal mammary artery with angioplasty?

ANSWER: 33533; Internal mammary artery is coded as arterial bypass graft. The angioplasty is an integral part of the procedure and is not coded.

20. What is the correct code(s) for a patient who received a single injection of sclerosing agent for telangiectasia of the legs?

ANSWER: 36468; Telangiectasia is known as spider veins and the sclerosing agent is used to treat it.

ACTIVITY 7.10

1. What is the correct code(s) for a patient who received a partial bowel resection with colostomy and creation of mucofistula?

ANSWER: 44144; The resection is coded as colectomy and includes the colostomy and fistula.

2. What is the correct code(s) for a patient who received a temporary stent during a retrograde nephrostomy with wire with cystourethroscopy?

ANSWER: 52334; The stent is temporary and is not coded additionally. A guide wire is used, which is included in the code.

3. What is the correct code(s) for a patient who received an EGD with control of bleeding and dilation of the esophagus with a balloon?

ANSWER: 43249 and 43255; Both codes must be included to code both the dilation and control of bleeding.

4. What is the correct code(s) for a patient who received a PEG tube with insufflation?

ANSWER: 43750; The insufflation is included in the insertion of the PEG tube, which is known as percutaneous endoscopic gastrostomy and can be found in the alphabetical index under gastrostomy tube.

5. What is the correct code(s) for a 33-year-old patient who received repair of an incisional incarcerated hernia with mesh?

ANSWER: 49561 and 49568; The mesh can be coded for the incisional hernia.

6. What is the correct code(s) for a patient who received a diagnostic laparascopy and partial colectomy with anastomosis that was converted to open procedure due to complications?

ANSWER: 44140 and V64.4; The procedure began with a diagnostic scope, but changed to surgical scope, but then changed to open procedure, so only the open procedure is coded.

7. What is the correct code(s) for a patient who received a cholecystotomy for removal of calculus with choledochostomy?

ANSWER: 47420; The choledochostomy and cholecystotomy are included in one code.

8. What is the correct code(s) for a patient who received repair of bilateral reducible recurrent femoral hernias?

ANSWER: 49555-50; The modifier 50 must be attached as this procedure is performed on both sides.

9. What is the correct code(s) for a patient who received dilation of renal stricture with balloon with cystourethroscopy?

ANSWER: 52343; This includes the balloon for opening the stricture and the use of a scope.

10. What is the correct code(s) for a 56-year-old patient who received repair of an incarcerated inguinal hernia with enterectomy?

ANSWER: 49507 and 44120; The enterectomy also must be coded since it is a repair of the intestine after being damaged by the incarceration.

11. What is the correct code(s) for a 44-year-old patient who received manipulation for a repair of an incarcerated inguinal hernia?

ANSWER: Incarcerated hernias are nonreducible and cannot be repaired with manipulation.

12. What is the correct code(s) for a patient who received sphincterotomy and ERCP with stent into the pancreatic duct?

ANSWER: 43268 and 43262; The spincterotomy must be coded additionally as noted in the coding guidelines. The stent is included in the code for the ERCP, endoscopic retrograde insertion of tube.

13. What is the correct code(s) for a patient who received dilation of her esophagus with a balloon 25 mm for achalasia?

ANSWER: 43220; There is a code for dilation over 30 mm, which is 43458, but this is not correct as indicated in the coding guidelines.

14. What is the correct code(s) for a patient who received flexible sigmoidoscopy for removal of lesions by cautery with endoscopic ultrasound?

ANSWER: 45333 and 45341; Two codes must be used to include the ultrasound and the removal of lesions.

15. What is the correct code(s) for a patient who received placement of a permanent stent during a retrograde nephrostomy with wire with cystourethroscopy?

ANSWER: 52334 and 52332; The permanent stent is coded in addition to the code for the nephrostomy. Notice that a scope is used.

16. What is the correct code(s) for a patient who was admitted for exploratory laparascopy and then partial colectomy with colostomy and closure of distal segment?

ANSWER: 44206; The exploratory scope changed to a surgical scope, so it is coded as such, with a colostomy and closure also included.

17. What is the correct code(s) for a patient who received percutaneous nephrostomy?

ANSWER: 50040; Only the nephrostomy is coded with no scope, stent, or wire.

18. What is the correct code(s) for an 11-year-old patient who received a tonsillectomy and adenoidectomy?

ANSWER: 42820; The one code includes both the tonsillectomy and adenoidectomy.

19. What is the correct code(s) for a 23-year-old patient who had a reducible umbilical hernia with mesh repaired?

ANSWER: 49585; No mesh is coded, as this is not an incisional or ventral hernia, so the mesh is included.

20. What is the correct code(s) for a patient who received an injection for percutaneous transhepatic cholangiography with professional services provided?

ANSWER: 47500 and 74320; The professional services are coded as 74320 and must be coded separately.

ACTIVITY 7.11

1. What is the correct code(s) for a patient who received a nerve block of the vagus nerve?

ANSWER: 64408; Nerve blocks are classified as introduction/injection of anesthetic agent.

2. What is the correct code(s) for a patient who had a 3.2 cm laceration of the neck repaired that required retention sutures with split-thickness graft due to the extensive damage of fascia, muscle, and dermis?

ANSWER: 15120 and 13132; There is no mention of nerve damage, so there are no codes included for repair of nerves.

3. What is the correct code(s) for a patient who had placement of a pump with programming?

ANSWER: 62362; This pump is programmable, so the code must reflect this.

4. What is the correct code(s) for a patient who had an intradural resection of the neoplasm of the cavernous sinus with dural repair with graft and infratemporal approach with ligation of the carotid artery with graft?

ANSWER: 61608, 61591, and 61609; Both the approach and the procedure must be coded. The ligation, which includes the graft, also must be coded.

5. What is the correct code(s) for a patient who had a subarachnoid/subdural-jugular shunt placed for cerebrospinal fluid?

ANSWER: 62190; Be careful in selecting the proper anatomic sites for this shunt.

6. What is the correct code(s) for a patient who had neurostimulator electrodes implanted in the autonomic nerve and returns today for analysis?

ANSWER: 95971; The patient receives services for the analysis, not the implantation, so code 95971 is used.

7. What is the correct code(s) for a patient who received a craniectomy and creation of burr holes through the posterior fossa?

ANSWER: 61305; This code includes crainectomy and burr holes, so no other code is necessary.

8. What is the correct code(s) for a patient who had a decompression of sciatic nerve with transposition?

ANSWER: 64712; Freeing of the sciatic nerve is known as *neuroplasty*. The transposition is included in the code.

9. What is the correct code(s) for a patient who received arthrodesis of T3–5?

ANSWER: 22556 and 22585; Two interspaces are fused, so two codes are required. Notice that although this is related to the spine, it is coded from the musculoskeletal section, not the nervous section.

10. What is the correct code(s) for a patient who had repair of a deep 2.2 cm laceration of the right hand that required layered closure and microscopic suturing of two digital nerves?

ANSWER: 64831, 64832, 13120, and 69990; Nerves are repaired in addition to the laceration repair, so both nerve repairs must be coded. A microscope is used, so it also must be coded.

11. What is the correct code(s) for a patient who had a catheter inserted with daily management for continuous infusion of an anesthetic agent into the femoral nerve?

ANSWER: 64448; The daily management, infusion, and placement of the catheter all are included in this code.

12. What is the correct code(s) for a patient who had a transthoracic vagotomy?

ANSWER: 64752; Vagotomy is coded in the section of transaction or avulsion.

13. What is the correct code(s) for a patient who received a resection of neoplasm of the midline skull base with infratemporal approach requiring a myocutaneous flap?

ANSWER: 61607, 61590, and 15732; The approach, procedure, and closure need to be coded with the closure being the flap.

14. What is the correct code(s) for a patient who received an extracranial reconstruction due to a benign tumor?

ANSWER: 21181; Reconstruction is involved, so this is not coded as excision with code 61563.

15. What is the correct code(s) for a patient who received burr holes for a ventricular catheter?

ANSWER: 61210; The burr holes are coded in addition to the catheter placement in this one code.

16. What is the correct code(s) for a patient who received an excision of an interdigital neuroma?

ANSWER: 28080; This is coded as an excision of a neuroma.

17. What is the eponym for the previous procedure?

ANSWER: Morton's neurectomy.

18. What is the correct code(s) for a patient who had excision of neuromas from both digital nerves of two toes of the foot?

ANSWER: 64776 and 64778; The terms, *One or both* as listed in code 64776, refers to the nerves, whereas the terms *each additional digit* refers to a separate toe.

19. What is the correct code(s) for a patient who had a laparoscopic vagotomy?

ANSWER: 43651; Remember that when a scope is involved, the procedural code that includes the scope is correct.

20. What is the correct code(s) for a patient who received a laminectomy with drainage of a cyst to the pleural space?

ANSWER: 63173; This laminectomy includes drainage and does not need to be coded separately.

ACTIVITY 7.12

1. What is the correct code(s) for a patient who had an MRI with oral contrast?

ANSWER: 74181; Oral contrast is coded as without contrast.

2. What is the correct code(s) for a patient who had repair of a perforation of the cornea due to glass in the eye?

ANSWER: 65275; The perforation is included in the code, which is limited to the cornea.

3. What is the correct code(s) for a patient who received debridement with a scissors and tweezers from a wound with topical application?

ANSWER: 97601; This code is not used in conjunction with the wound repair in the integumentary section, because it involves selective services, like the scissors and tweezers.

4. What is the correct code(s) for a patient who had a mastoidectomy with tympanoplasty?

ANSWER: 69635; The tympanoplasty was planned, so code 69604 cannot be used as it reflects the unplanned procedure following the planned mastoidectomy.

5. What is the correct code(s) for a patient who received KUB of upper GI with films?

ANSWER: 74241; The KUB is a test for the kidneys, ureter, and bladder.

6. What is the correct code(s) for a patient who received extracapsular removal of her cataract in the left eye with phacoemulsification and insertion of lens?

ANSWER: 66984-LT: This code includes the insertion of the lens and the removal of the cataract, which is extracapsular.

7. What is the correct code(s) for a patient who received chemotherapeutic infusion for 4 hours?

ANSWER: 96422 and 96423 × 3; Chemotherapy has its own infusion codes. The second code, 96423, is for each additional hour after the first hour, so it must be multiplied by three for the three extra hours of infusion.

8. What is the correct code(s) for a patient who receives an infusion with saline with Doppler and sonohysterography?

ANSWER: 76831 and 58340; The introduction of the saline also must be coded.

9. What is the correct code(s) for a patient who received dialysis in the hospital for ESRD?

ANSWER: 90935; This code is to be used when the dialysis is given in the hospital, as noted in the coding guidelines.

10. What is the correct code(s) for a patient who received an EEG for cerebral seizure with 16 channels for 48 hours?

ANSWER: 95953 × 2; The code is for each 24 hour period, so it must be multiplied times two. The EEG is specifically for 16 channels.

11. What is the correct code(s) for a patient who was given a venipuncture to test lipids?

ANSWER: 80061; The venipuncture is included in the code for the test, which is a panel test that includes several substances.

12. What is the correct code(s) for a patient who is tested for quantity of cocaine?

ANSWER: 82520; This is a quantitative test to see how much cocaine is in the body, not a qualitative test, which would be to see if cocaine was present.

13. What is the difference between hemodialysis and peritoneal dialysis?

ANSWER: With hemodialysis, waste is removed directly from the blood. In peritoneal dialysis, waste is removed indirectly through an infusion involving a fluid (dialysate) that is flushed into the peritoneum through a catheter, not through the blood. The dialysate is then removed along with the body's waste products.

14. What is the correct code(s) for a patient who received an extracapsular removal of a cataract from his left eye with capsulorrhexis and returns today for insertion of lens?

ANSWER: 66985; The insertion of the lens is done separately from the removal of the cataract, so only the insertion of lens is coded today as 66985.

15. What is the correct code(s) for a 10-year-old patient who received an IM of tetanus, diphtheria, and oral polio?

ANSWER: 90471, 90473, 90718, and 90712; Both the substance and administration are coded. The administration is coded separately as oral administration and IM.

16. What is the correct code(s) for a patient who received closure of a congenital ventricular septal

defect with a right heart catheterization with percutaneous transcatheter and implant?

ANSWER: 93580; The right heart catheterization is included in this code.

17. What is the correct code(s) for a patient who received an esophageal acid reflux test?

ANSWER: 91032; This code is selected from the Medicine section, not the testing section.

18. What is the correct code(s) for a patient who received a two-dimensional ultrasound of the chest?

ANSWER: 76604; Two dimensional is considered a B-scan.

19. What is the correct code(s) for a patient who was tested for opioid use?

ANSWER: 80101; This test is qualitative in that it is testing for presence of the opioid, but not the amount, which would be quantitative.

20. What is the correct CPT code(s) for a patient who received an intramuscular injection of penicillin?

ANSWER: 90788; This is an injection of antibiotic substance, so it is coded as an injection from the 90000 codes.

ACTIVITY 7.13

1. What is the correct code(s) for a patient who returned three days later for a lithotripsy by the same physician?

ANSWER: 50590-76; The same physician had to provide the same procedure again, so modifier 76 is attached.

2. What is the correct code(s) for a patient who had a chest tube inserted a week ago and returns today for a rotation flap on his left arm, which is 8 sq cm, by the same physician?

ANSWER: 14020-79; The patient returns for another procedure by the same physician, so modifier 79 is attached.

3. What is the correct code(s) for a patient who had a Z-plasty with fasciectomy performed on the right palm with release of the second and third digits?

ANSWER: 26123-F6 and 26125-F7; The modifiers indicate which fingers received the services, which included the Z-plasty and fasciectomy.

4. What is the correct code(s) for a patient who had two malignant lesions removed by cryosurgery in the morning from the chest, measuring 0.4 and 0.8 cm respectively, who returns later that day to have another lesion removed from the chest that also was found to be malignant, size 1.5 cm?

ANSWER: 17260, 17261, and 17262; All three lesions must be coded individually, so there is no modifier.

5. What is the correct code(s) for a patient who was seen by a different physician who provided closed treatment with manipulation of a distal radius fracture after having been treated by another physician last week for closed treatment with manipulation of an ulnar fracture?

ANSWER: 25600; This was not a return to the operating room for the previous procedure by the same physician, so it is coded separately.

6. What is the correct code(s) for a patient who was returned to the operating room for enterolysis by the same physician in the same visit?

ANSWER: 44005-76; Modifier 76 is for a repeat procedure by same physician.

7. What is the correct code(s) for a patient who received a preoperative and postoperative glucagon tolerance test?

ANSWER: 82946-91; The laboratory test was repeated, so modifier 91 is attached.

8. What is the correct code(s) for a patient who received a total abdominal hysterectomy a week after having her tubes and ovaries removed?

ANSWER: 58150-78; The total abdominal hysterectomy is a related procedure to the prior surgery, so modifier 78 is attached.

9. What is the correct code(s) for a patient who had repositioning of a permanent dual-chamber pacemaker with transvenous electrodes that was implanted one week ago?

ANSWER: No code is listed because repositioning is provided within the global period for the pacemaker.

10. What is the correct code(s) for a patient who had preoperative and postoperative chest x-rays performed with two views?

ANSWER: 71015 and 71015-76; The same procedure was repeated by the same physician, so modifier 76 is attached to the second code.

11. What is the correct code(s) for a patient who had a cerebrovascular shunt removed without replacement three days ago and returns today for a nephrectomy with partial ureterectomy by same physician?

ANSWER: 50220-79; The patient returns for a nephrectomy by the same physician, so modifier 79 must be attached.

12. What is the correct code(s) for a patient who received arthroplasties on both wrists with internal fixation?

ANSWER: 25332-50; This is a bilateral procedure, so modifier 50 is attached.

13. What is the correct code(s) for a patient who had a laparoscopic of the abdomen with collection of a specimen by brushing two weeks ago and who returns today for a direct thrombectomy of the iliac vein?

ANSWER: 34401; There is no modifier because the procedures are not related, and there is no global for the biopsy.

14. What is the correct code(s) for a patient who received an arthrotomy of both elbows?

ANSWER: 24100-50; Both elbows were operated on, so a modifier of 50 for a bilateral procedure should be attached.

15. What is the correct code(s) for a patient who had replacement of a temporary single chamber pacemaker with a dual-chamber permanent pacemaker two weeks later?

ANSWER: 33217-78; The replacement is a greater service than the implantation of the temporary pacemaker, but is related, so modifier 78 is attached.

16. What is the correct code(s) for a patient who was taken to the operating room for repair of an inguinal hernia, which was cancelled before anesthesia was administered due to patient noncompliance?

ANSWER: 49505-73; The anesthesia was not administered, so modifier 73 is attached.

17. What is the correct code(s) for a patient who had repair of the fascia and dermis for 2.1 cm and 2.2 cm lacerations on his arms?

ANSWER: 12032; No modifiers are used because with repair of lacerations, the sites are added together, so modifiers for site cannot be applied.

18. What is the correct code(s) for a patient who had an excision of a tumor from the right knee?

ANSWER: 27327-RT; The right knee is involved, so a modifier of RT is used to indicate that it is the right knee.

19. What is the correct code(s) for a patient who received a biopsy of the axillary lymph nodes, which were excised?

ANSWER: 38740; The biopsy was not an additional procedure, so no modifier is attached. Remember that biopsies with excisions are coded only as excisions.

20. What is the correct code(s) for a patient who was admitted to ASC for osteotomy of the calcaneus, which was discontinued because of the deteriorating condition of the patient once anesthesia had been administered?

ANSWER: 28300-74; Anesthesia had been administered before the procedure was discontinued, so the modifier 74 is attached.

Chapter 8: EVALUATION AND MANAGEMENT CODES (E/M)

Objectives

(1) Analyze and select proper evaluation and management (E/M) codes according to the three major components.

(2) Distinguish between ICD-9 procedural codes versus E/M codes for outpatient hospital and physician billing.

(3) Distinguish between types of E/M services and their criteria for selection.

(4) Ensure proper utilization and analysis of documentation.

(5) Apply E/M modifiers correctly.

Key Terms

Bundling—Bundling is the containment of all associated services, supplies, and materials within one main code that describes the major procedure/service provided so only one code is necessary for reimbursement purposes. It is particularly important to understand the concept of bundling with E/M codes because they are often incorporated into bundled procedures and, therefore, often cannot be billed additionally.

Documentation—Documentation is critical in the application of proper E/M codes because the process of selecting the proper code is dependent upon documentation that supports the selection of the level of service regarding the major E/M components, i.e., history, exam and medical decision making.

E/M—Evaluation and management codes are for provider services that are based on examination of patients in a variety of settings, such as office, hospital, emergency rooms, or nursing homes.

Modifiers—Modifiers are two-digit numeric codes that follow the CPT code and are separated from it with a hyphen, which can be located in Appendix A of the CPT book. Modifiers indicate when there are additional circumstances influencing the CPT code, which may alter reimbursement; therefore, they help ensure accurate payment for services.

EVALUATION AND MANAGEMENT CODES

Evaluation and management codes (E/M) are for patient care in a variety of settings, such as office, hospital, emergency rooms, or nursing homes. E/M codes reflect the time and service that facilities provide for patients' visits. With outpatient hospital coding, these codes are not used for actual physician services for which the physician would bill. It is important to remember whom you are coding for in hospital exams, you are coding for the facility, not the physician. E/Ms, like all CPTs, are never used with inpatient hospital coding

Not documented, not done is critical when using E/M codes. The reports must justify the level of service code that you select. The most distinguishing difference between coding for the physician and coding for the facility is that facilities have more flexibility when determining the level of E/M code selected. Another difference is that facilities use three levels of E/M codes—low, middle, and high. Facilities are allowed to determine their method of categorizing their E/M visits. Generally, the first two E/M levels are included in the facility low level, the middle E/M level is included in the facility middle level, and the last two levels (highest values) are included in the facility high level.

If an E/M service is provided at the same time as other services, the E/M can still be charged, if it was a significant service that was distinct from other services provided. In this case, a modifier 25 should be attached to the E/M code indicating that it is separate from other services/procedures.

Only one E/M code usually is allowed per day. Exceptions are for prolonged care and critical care codes, as well as specifying services that are provided for a different diagnosis.

MAJOR E/M CRITERIA

There are seven criteria to be considered when selecting most E/M codes. However, some E/M codes do not use these criteria, but rather use criteria such as time or age.

The seven criteria are

History

Examination

Medical Decision Making

Counseling

Coordination of Care

Nature of Presenting Problem

Time

Time

With E/M coding, facilities determine how to distinguish the various levels of E/M. Time is one important consideration with some facilities providing more time to patients than others because of the specific services provided. Facilities can use time, therefore, to determine which level of E/M code to use. Documents must support this influence. For hospital and other inpatient visits, time is defined as unit/floor, and also can include time that is not face-to-face patient care but time spent on the patient's unit or room. Unit/floor time is the time a physician spends directly with the patient or indirectly on the patient's unit caring for the patient. Unit/floor time can include observation, consultation, hospital and nursing home visits.

The Big Three

The first three criteria—history, exam and medical decision-making—are the main focus for determining E/M level. Your CPT book should list these three factors in each applicable code.

History

History is broken down into three main factors: history of present illness (HPI), review of systems (ROS), and past, family, social history (PFSH).

History of Present Illness

For the history of present illness, the first criterion is the chief complaint (CC), which always must be listed for codes that use these criteria, such as office visits.

HPI is evaluated based on elements provided in reports, such as the history and physical, which describes the patient services provided during this visit. These can include location, severity, timing, quality, duration, associated signs, context, and modifying factors. In abstracting, reports (such as history and physical reports) must be analyzed to determine how much data is contained to justify the classification being made.

HPI is divided into two levels, either brief or extended. If there are four or more elements, then the history is considered extended.

The following is an excerpt from a typical physician's report that demonstrates an extended HPI:

HISTORY OF PRESENT ILLNESS: The patient presents with complaints of stomach pain. The patient is a 57-year-old male who has been experiencing these pains for the past two hours. He says that the last time he ate was 6 hours ago, at which time he felt fine. He then watched TV until the pains began. He says that pain fluctuates but has increased in intensity and is located in the lower right quadrant. He has had no hemoptysis or diarrhea. The patient does have a fever of 100.4.

In this example there are elements for location, severity, timing, duration, associated signs, and more. Therefore, it meets the classification of extended easily.

TYPE OF HISTORY

History of Present Illness (HPI)	Review of System (ROS)	Past, Family, Social History (PSFH)	Type of History
Brief (1–3 elements)	N/A	N/A	Problem focused
Brief (1–3 elements)	Problem pertinent (1 element)	N/A	Expanded Problem focused
Extended (4+ elements)	Extended (2–9 elements)	Pertinent (1–2 elements)	Detailed
Extended (4+ elements)	Complete (10+ elements)	Complete (3 for new patient; 2 for established)	Comprehensive

Table 8-1

Review of Systems

Review of Systems (ROS) is comprised of data generally collected from the patient before the patient sees the physician which includes the patient's health history. Unfortunately, physicians often do not include enough of this information in their reports and, therefore, a higher E/M code cannot be selected. Physicians can list the various elements within ROS, which is then followed by *negative* if this is true. However, if there is positive data for an element, this can be stated as *positive,* but additional details must be provided.

ROS includes elements based on the various systems of the body such as vitals (weight, height, etc.), head, ears, eyes, nose, throat (HEENT), cardiovascular, respiratory, gastrointestinal (GI), genitourinary (GU), musculoskeletal, integumentary, neurological, psychiatric, and endocrine.

ROS is listed as either not applicable (N/A) or is divided into three categories—problem pertinent, extended, or complete. The problem pertinent usually focuses on one problem. The extended usually consists of two to nine elements and a complete usually consists of ten or more elements.

The following is an excerpt from a physician's report that demonstrates an extended ROS.

REVIEW OF SYSTEMS: HEENT: Normal. HEART: No history of angina or hypertension. RESPIRATORY: Recent bronchitis. GI: Denies abdominal pain or blood in the stool. GU: No burning or frequency. INTEGUMENTARY: Eczema in the recent past.

Past, Family, and Social History

PFSH is listed as either N/A or is divided into two categories—pertinent and complete. For new patients, all three elements must be included for a complete PSFH. With an established patient only two are needed.

The following is an excerpt from a physician's report that demonstrates a complete PFSH.

The patient does not drink or smoke. There is a history of diabetes in both the parents. The patient had an appendectomy and tonsillectomy.

Notice that all three elements are provided in the above example—the social history, family history and the patient's past history.

Selecting the Proper History Level

A ROS or PFSH from earlier reports does not need to be totally repeated but can be designated in the new report by referring to previous reports.

A physician must note in the report that he or she has acknowledged and discussed the data with the patient. This often is confirmed simply by signing his/her name to the report after confirmation with patient.

If it is not possible to obtain a history from a patient because of their condition, such as when a patient is unconscious, this can be noted and will not affect the level of the code selected.

Exam

Examination is fairly straightforward. The level of exam is based on the amount of data provided regarding the patient's physical status. This data can be categorized by either a general physical exam or by specialty areas. In a general exam, elements can include vitals (weight, height, etc.), HEENT, cardiovascular, respiratory, GI, GU, musculoskeletal, integumentary, neurological, psychiatric, and endocrine. Specialty exams do not require coverage of all previously mentioned body areas, but can focus on the specialty area that is described in more detail.

The exam is divided into four categories—problem focused, expanded problem focused, detailed, and comprehensive. Although the number of elements necessary for each category may vary, a suggested guideline is one system for a problem-focused exam, two to five for an expanded problem focused, six to eight for detailed, and nine or more for comprehensive. This depends on a number of factors, however, and it is suggested that if stricter guidelines are needed, you should research other publications and textbooks, which provide in-depth discussions on factors, such as the depth of specialty exams, that influence the extent of the exam.

TABLE OF RISK

Risk Level	Presenting Problem
Minimum	1 minor problem
Low	2 or more minor problems or 1 stable chronic or 1 uncomplicated acute
Moderate	2 or more chronic problems or new problem Acute illness or injury
High	Chronic problem with exacerbation or changes in mental status

Table 8-2

TABLE OF MEDICAL DECISION MAKING ELEMENTS

Number of Diagnoses	Morbidity and Mortality Risks	Amount and/or Complexity of Data	Type of Decision Making
Minimal	Minimal	Minimal or none	Straightforward
Limited	Low	Limited	Low complexity
Multiple	Moderate	Moderate	Moderate complexity
Extensive	High	Extensive	High complexity

Table 8-3

Stating *negative* is acceptable for elements, but when they are positive a description must be provided to explain the positive findings, for example, negative for urinary problems.

The following is an excerpt from a physician's report that demonstrates a comprehensive exam.

PHYSICAL EXAMINATION: GENERAL: The patient is a 38-year-old white female who is well-developed and well-nourished and is in no acute distress. Temperature is 98.8. Weight is 135 pounds. Blood pressure is 130/80. HEENT: PERLA. EOMs intact. Sclerae clear. NECK: Supple. No adenopathy. Thyroid not palpable. No JVD. HEART: Normal S1, S2. Normal sinus rhythm. CHEST: Lungs clear to percussion and auscultation. ABDOMEN: Soft, nontender with no organomegaly. PELVIC/RECTAL: Deferred as they have recently been done by her primary care physician. MUSCULOSKELETAL: No tenderness of the spine. INTEGUMENTARY: Some scarring on the right arm from a previous burn. NEUROLOGICAL: Nonfocal. Cranial nerves II–XII intact.

Medical Decision Making

Medical decision making (MDM) involves many factors when determining its proper level. The four levels are—straightforward, low complexity, moderate complexity, and high complexity. A Table of Risk gives a sense of the levels for MDM. For example, a minor problem, such as a cut finger that required two sutures, would have a MDM level that is straightforward. On the other hand, a patient who is experiencing chest pains and faints in the office would have a MDM level that is high complexity.

In this table, the minimum level of risk correlates with a straightforward classification, and the other classifications would follow similarly with low classification for low risk, etc.

Although this table provides some direction for choosing the correct level of MDM, there are many other elements that can be used to select the proper level, for example the number of diagnoses, amount of data, and risk of morbidity or mortality. Morbidity is disease and can be an important consideration when selecting the proper code level when that disease significantly impacts the patient's medical care.

For hospital coding, comorbidities are defined as disease conditions that increase the length of the hospital stay by at least one day for at least 75 percent of patients. These categories can contain information such as labs reviewed, data researched, consultations with other physicians, and review of old records. The following table provides some additional classifications of information and the level of decision making that would correlate with it.

The following is an excerpt form a physician's report that demonstrates a straightforward MDM.

Patient received 3 sutures for the laceration of his hand and was given a prescription for pain.

SELECTING THE PROPER E/M CODE

Once the proper level for each category is selected (history, exam, and MDM) based on the care given, they are added together to select the code. The code selected must meet either all three levels of the history, exam and medical decision making unless stated within the code's description as having to meet only two levels. For example, E/Ms for new office patients must meet all three levels, but only two levels must be met for established office visits.

TABLE OF THE BIG THREE ELEMENTS FOR DETERMINING E/M LEVEL

History	Exam	Medical Decision Making
Problem focused	Problem focused	Straight forward
Expanded problem focused	Expanded focused problem	Low complexity
Detailed	Detailed	Moderate complexity
Comprehensive	Comprehensive	High complexity

Table 8-4

Meeting the levels refers to the selection of the code in which the lowest level of the history, exam, and MDMD are all met. Remember, the higher levels naturally contain the lower levels. For example, for new patients in an office, all three levels must be met. For an established patient, only two levels must be met. This means if a new patient visit has a history that is comprehensive, an exam that is comprehensive, and a MDM that is low complexity, the level where the low complexity is located is the level that must be selected.

History, exam, and MDM are usually listed underneath the E/M code in relevant sections of the CPT book. This will help you select the proper code and is helpful particularly on the AHIMA exam, since this information is provided on the test and is your means of choosing the correct code. *Note*: On the AHIMA exam, you do not need to evaluate AHIMA's selection of levels for history, exam, and MDM. Simply abide by their choice as noted in their descriptions of these factors.

One code, 99211 for established patient office visit, does not require the immediate presence of a physician. However, the physician must be available for consultation if necessary.

HOSPITAL VISITS

Facility patients are admitted to the Emergency Room, outpatient clinic, hospital, or other facility; therefore, there is no need to distinguish between which E/M is coded for location, that is the hospital or office. Remember for inpatient hospital visits do not use CPT and E/M codes, only ICD-9 codes for diagnoses and procedures. For E/Ms, we are dealing with outpatient facilities, such as the Emergency Room and outpatient clinic visits.

EMERGENCY ROOM VISITS

Emergency Room visits are defined as occurring in a designated Emergency Room area of a hospital, which is available 24 hours per day for the purpose of providing unscheduled medical care. This definition distinguishes Emergency Room visits from urgent care, outpatient clinics, and hospital visits.

CRITICAL CARE

Critical care usually occurs in the hospital, but can be provided in an outpatient clinic. Remember, if it occurs in the hospital, the care is not coded with E/M codes. Critical care consists of unit/floor time spent with a critical care patient, which includes face-to-face time and time spent on the unit working on other issues related to the patient's care. The time does not need to be continuous but cannot be duplicated for several patients.

A physician cannot bill for the same hour for more than one patient. For example, physicians can bill only for three hours of critical care services regardless of how many patients they have cared for during that time. The time must be split up into however physicians delegate their time, for example, a half an hour with one patient, two hours with another patient and a half an hour with one more patient.

Critical care services include blood gases, gastric intubation, chest x-rays, cardiac measurements, ventilator management, and vascular access procedures and can be coded in addition to other E/M services on the same day. For critical care services that are less than 30 minutes, other appropriate E/M codes should be used. If critical care is more than an hour, but less than an additional 15 minutes, then it is coded as an hour. This is also true before an hour, i.e., if critical care is 1 hour and 43 minutes, it cannot be coded for 2 hours, but is rather 1 hour plus half an hour as the critical care codes are broken down into the first hour and then half hour segments.

Critical care for children 31 days through 24 months is coded with different codes (99293 and 99294) than regular critical care codes. Critical care for infants 30 days of age or less are coded with their own codes (99295 and 99296).

PROLONGED SERVICES

There are codes for services that extend beyond the office or hospital regarding face-to-face time with the patient (99354–99357) or indirect nonface-to-face patient care (99358–99359). These codes are based on the first hour and then subsequent half hour segments with the same rules that apply to critical care regarding time determination (less than 15 minutes does not constitute another half an hour). These codes can be attached to any level E/M code or other service.

OBSERVATION

Observations can occur in outpatient hospital facilities but not necessarily in an area designated as strictly for observation. For coding purposes, observations cannot occur in the office or for postoperative care. Currently, there are no specific time limitations for time spent in observation, but observations usually do not exceed 48 hours.

PREVENTIVE MEDICINE SERVICES

Preventive medicine services are known as well checks or ongoing medical care. They are for visits when there is no chief complaint. If, however, a problem is diagnosed and treatment initiated, an additional code for an office visit can be used to indicate the services were provided for care of a specific problem. Preventive codes are not classified by the elements of history, exam, and MDM but rather by age and new/established.

ACTIVITY 8.1

Remember, for these questions, you are coding for the facility. Code only for services provided.

1. What is the correct code for a patient admitted to an ambulatory surgical center (ASC) for an appendectomy who had received a history and physical that focused on the appendectomy? The history was detailed, the exam was detailed, and the MDM was moderate.

2. If a patient is admitted to the hospital and subsequently a temporary dual-chamber epicardial pacemaker is implanted with **documentation** demonstrating that the history was comprehensive, the exam was comprehensive, and the MDM was high, what would be the correct code?

3. What is the correct code for a 3-year-old child who is seen in the outpatient hospital clinic with documentation demonstrating that the history was expanded, the MDM was straightforward, and the exam was moderate for removal of a pencil from her left ear, which required repair of the perforated eardrum?

4. If the patient is seen in the Emergency Room for severe chest pains and the elements provided in the report include a detailed history, comprehensive exam and the MDM is moderate, would the correct code be 99204? Explain.

5. If the history in the physician's report for a new patient includes four elements for the HPI, two elements for PFSH, and ten elements for ROS, what level of history would be the proper selection?

6. What type of risk would an acute illness probably present as?

7. What is the correct E/M code for a patient who was seen in the Emergency Room for severe difficulty with breathing? The documentation demonstrates that the history and exam were expanded problem focused and the MDM was moderate. The patient developed dyspnea and was seen again by the physician for an additional 25 minutes.

8. What is the correct E/M code for a patient who was seen today in the hospital critical care unit by the physician for four and one half hours due to stroke?

9. If a new patient is seen in the outpatient hospital clinic for second-degree burns on his forearms, which were debrided and dressings applied, in addition to numerous bruises on his back, what would be the correct code? Documentation demonstrated that the history was comprehensive, the exam was comprehensive, and the MDM was moderate.

10. What are the three main factors used to select the proper level for the element of history when selecting E/M codes?

11. What is the correct code for a patient who was seen in the Emergency Room whose exam indicated the history and exam were expanded problem focused and the MDM was straightforward with focus on the fracture of the femur, which was treated with percutaneous fixation?

12. What is the correct E/M code for a patient who was seen in the Emergency Room for AIDS-related complications for which documentation demonstrates that the history was detailed, the exam was detailed, and the MDM was moderate with the physician providing an additional 50 minutes of care subsequent to the first examination?

13. If a patient was admitted to the Emergency Room for severe head injuries due to a car accident and was given critical care for three hours with documentation demonstrating that there were elements for the past history, 2 elements for the PFSH due to difficulty in obtaining information from the patient who became unconscious, 12 elements for the ROS, and 12 elements for the exam, what would be the proper code?

14. If a new patient is seen in the outpatient hospital clinic for appendicitis with documentation demonstrating an expanded problem focused history, low complexity MDM, and expanded problem focused exam, what would be the correct code for the patient's care?

15. What is the correct E/M code for a new patient who was seen in the outpatient hospital clinic for severe stomach pains with documentation demonstrating that the history was comprehensive, the exam was comprehensive, and the MDM was low?

16. What is the correct E/M code for services for a patient who is admitted to the hospital with exacerbation of her diabetes with visual disturbances and documentation demonstrates that the history was detailed, the exam was comprehensive, and the MDM was moderate?

17. What is the correct E/M code for a patient who was seen in the Emergency Room for possible pneumonia with documentation demonstrating that the history was problem focused, the exam was expanded problem focused, and the MDM was low with the patient developing severe chest pains and subsequently seen by the physician for further evaluation, which lasted 30 minutes?

18. What is the correct code for a patient who was seen in an ASC for ESWL with the history and exam described as problem focused and MDM as straightforward?

19. What is the definition of unit/floor time?

20. If a patient was admitted for critical care in the Emergency Room for 160 minutes, what would be the proper code(s)?

SUMMARY

E/M codes constitute one of the largest segments of services provided by physicians and other providers but are used for outpatient hospital services only as indications of the types of facilities and care provided to the patient. There are no physicians specified for performance of services that are billed with outpatient hospital services. Remember that inpatient facility coding does not use CPTs for billing, but instead uses Volume 3 of the ICD-9 coding book.

With outpatient hospital coding, the hospital must establish their criteria for selecting the proper level of E/M code for provided services, which contrasts with the more structured elements of physician E/M coding. E/M codes, like all codes, are determined by analyzing documentation and assessing whether the documentation supports the level of codes selected.

ACTIVITY 8.1 ANSWERS

Remember, for these questions, you are coding for the facility. Code only for services provided.

1. What is the correct code for a patient admitted to an ASC for an appendectomy who had received a history and physical that focused on the appendectomy? The history was detailed, the exam was detailed, and the MDM was moderate?

ANSWER: 44950; Only the procedure is coded.

2. If a patient is admitted to the hospital and subsequently a temporary dual-chamber epicardial pacemaker is implanted with documentation demonstrating that the history was comprehensive, the exam was comprehensive, and MDM was high, what would be the correct code?

ANSWER: 39.64; Remember that you are coding for the facility, not the physician, and this is an inpatient, so Volume 3 codes from the ICD-9s would be selected for the pacemaker.

3. What is the correct code for a 3-year-old child who is seen in the outpatient hospital clinic with documentation demonstrating that the history was expanded, the MDM was straightforward, and the exam was moderate for removal of a pencil from her left ear, which required repair of the perforated eardrum?

ANSWER: 69631-LT; The main service provided is the repair of the eardrum, so this is coded. Low levels of E/M elements indicate that the visit is included as part of the procedure.

4. If the patient is seen in the Emergency Room for severe chest pains and the elements provided in the report include a detailed history, comprehensive exam and the MDM is moderate, would the correct code be 99204? Explain.

ANSWER: No, this is an Emergency Room visit, so the code must be selected from the Emergency Department codes; therefore, the correct code would be 99284.

5. If the history in the physician's report for a new patient includes four elements for the history of present illness, two elements for PFSH, and ten elements for ROS, what level of history would be the proper selection?

ANSWER: Detailed because, the patient is new, so all three elements of PFSH must be described to have a higher level, but only two are provided, so the level is detailed.

6. What type of risk would an acute illness probably present as?

ANSWER: Moderate, as indicated in the chart provided although other factors would be considered before final determination of the level of risk.

7. What is the correct E/M code for a patient who was seen in the Emergency Room for severe difficulty with breathing? The documentation demonstrates that the history and exam were expanded problem focused and the MDM was moderate. The patient developed dyspnea and was seen again by the physician for an additional 25 minutes.

ANSWER: 99283; Although the physician did provide additional care, a code for prolonged care cannot be coded because the time was less than 30 minutes.

8. What is the correct E/M code for a patient who was seen today in the hospital critical care by the physician for four and one half hours due to stroke?

ANSWER: No E/M code is used as the patient is in the hospital. Hospitals do not use E/M codes.

9. If a new patient is seen in the outpatient hospital clinic for second-degree burns on his forearms, which were debrided and dressings applied, in addition to numerous bruises on his back, what would be the correct code? Documentation demonstrated that the

history was comprehensive, the exam was comprehensive and the MDM was moderate.

ANSWER: 99204-25 and 16015; The E/M was the primary service provided with an additional examination conducted, so both the burns and the E/M are coded with a modifier 25 attached to the E/M.

10. What are the three main factors used to select the proper level for the element of history when selecting E/M codes?

ANSWER: History of present illness, review of systems, and MDM.

11. What is the correct code for a patient who was seen in the Emergency Room whose exam indicated the history and exam were expanded problem focused and the MDM was straightforward with focus on the fracture of the femur, which was treated with percutaneous fixation?

ANSWER: 27509; The visit did not provide services other than those associated with the fracture, so no E/M code is listed.

12. What is the correct E/M code for a patient who was seen in the Emergency Room for AIDS-related complications for which documentation demonstrates that the history was detailed, the exam was detailed, and the MDM was moderate with the physician providing an additional 50 minutes of care subsequent to the first examination?

ANSWER: 99284 and 99354; Prolonged care codes must be added to the initial E/M because of the additional service provided by the physician, which consumed facility time and services.

13. If a patient was admitted to the Emergency Room for severe head injuries due to a car accident and was given critical care for three hours with documentation demonstrating that there were 4 elements for the past history, 2 elements for the PFSH due to difficulty in obtaining information from the patient who became unconscious, 12 elements for the ROS, and 12 elements for the exam, what would the proper code be?

ANSWER: 99291 and 99292 × 4; This is critical care provided in the Emergency Room, so the care is coded as critical care.

14. If a new patient is seen in the outpatient hospital clinic for appendicitis with documentation demonstrating an expanded problem focused history, low complexity MDM, and expanded problem focused exam, what would be the correct code for the patient's care?

ANSWER: 99202; Although the MDM is a higher level than the correct code, the other two categories are not high enough to pick another level for this patient who is coded as a new patient.

15. What is the correct E/M code for a new patient who was seen in the outpatient hospital clinic for severe stomach pains with documentation demonstrating that the history was comprehensive, the exam was comprehensive, and the MDM was low?

ANSWER: 99203; The MDM was not a high enough level to qualify for a higher code even though the other two elements were.

16. What is the correct E/M code for services for a patient who is admitted to the hospital with exacerbation of her diabetes with visual disturbances and documentation demonstrates that the history was detailed, the exam was comprehensive, and the MDM was moderate?

ANSWER: No code. Remember that you are coding for the facility, so no E/M is coded as this is an inpatient hospital report.

17. What is the correct E/M code for a patient who was seen in the Emergency Room for possible pneumonia with documentation demonstrating that the history was problem focused, the exam was expanded problem focused, and the MDM was low with the patient developing severe chest pains and subsequently seen by the physician for further evaluation, which lasted 30 minutes?

ANSWER: 99281 and 99354; The patient was seen for an additional 30 minutes, which does qualify for prolonged care coding.

18. What is the correct code for a patient who was seen in an ASC for ESWL with the history and exam described as problem focused and MDM as straightforward?

ANSWER: 50590; No E/M would be coded as it is not significantly distinct from the procedure, so only the procedure is listed.

19. What is the definition of unit/floor time?

ANSWER: Unit/floor time, as described in the coding notes in the CPT book, is the time the physician is present at the patient's bedside and on the floor of the unit the patient is on.

20. If a patient was admitted for critical care in the Emergency Room for 160 minutes, what would be the proper codes?

ANSWER: 99291 and 99292 × 3; The times for critical care are based on 15 minute segments, so if a patient is seen for less than 14 minutes within the next hour of care, it does not constitute a charge for another hour.

Chapter 9 — HEALTH CARE PROCEDURE CODING SYSTEM (HCPCS)

Objectives

(1) Distinguish between the application of CPT and HCPCS codes.

(2) Understand the format of the HCPCS codes and how to properly use them.

(3) Know HCPCS modifiers and their application to CPT codes.

Key Terms

DME—Durable medical equipment that is provided to a patient, such as wheelchairs.

Global—All services and materials inherent in a procedure are not coded, because their costs are assumed to be contained within the code for the procedure.

HCPCS—Health Care Procedure Coding System. HCPCS Level II is what is commonly referred to as HCPCS and consists of codes for reimbursement for services and supplies that are not represented in Level I codes.

HCPCS Level I—Current Procedure Terminology (CPT).

Modifiers—Alphanumeric digits that are attached to CPT codes to provide more information regarding services provided and that may affect reimbursement.

HEALTH CARE PROCEDURE CODING SYSTEM

CPT is known as **HCPCS Level I**. **HCPCS** Level II is commonly referred to as HCPCS and consists of reimbursement codes for supplies, and nonprovider services, including ambulance services and **durable medical equipment (DME)**, which are not represented in Level I codes. These codes are not required for the abstracting portion of the AHIMA CCS exam but can be included in the multiple-choice questions. Level III codes are local codes that are being replaced by national codes.

The use of HCPCS is mandated for Medicare claims and is preferred, not only by Medicare, but by many other payers. HCPCS is used for physician billing in outpatient settings, such as physician's offices, for services and supplies that are provided to the patient. However, only services and supplies that are not included in the **global** cost of performing other services or procedures can be charged. For example, if a physician provides services, such as removal of an ingrown toenail, to a patient in the office, a surgical tray cannot be charged because it is included in the cost of the toenail removal. This includes services performed by dentists and therapists, as well as DME. HCPCS codes are five-digit alphanumeric codes that begin with an alphabetic letter followed by four numeric digits.

Although CPT books have a code for supplies and materials (99070), the HCPCS are more specific and preferable whenever possible. Most supplies and materials are included in global procedural codes, but if they are not or if they exceed the amount normally used, they can be coded. Surgical trays are a good example of supplies and materials that are not usually coded since they are a normal part of a procedure. DME is coded using HCPCS. IVs are coded with A4305.

There are many abbreviations used in the book, which are common and accepted abbreviations in the specific fields. The meanings of the abbreviations are listed prior to the listing of the codes.

HCPCS begin with a letter—A, B, C, D, E, G, H, J, K, L, M, P, Q, R, S, or V. Letters K, G, and Q are temporarily assigned codes for services or materials while a decision is made regarding their permanent codes. There is an alphabetic index to assist in selecting the proper code.

Supplies and materials are not always listed in the alphabetic index in terms the physician may use, particularly drugs that may be known under another name. This can make it difficult to find the correct code.

When selecting codes that have measurements, you must choose a measurement that ensures the entire amount has been billed. For example, if a drug is listed as billable in 1.5 mcg, but the physician administers 2 mcg, then you must bill twice for the drug, i.e., times two.

Codes for materials associated with immunizations are listed in the CPT book. With other drugs and medicines, these are listed in the HCPCS book under the method of administration. Methods include intrathecal (IT), intravenous (IV), intramuscular (IM), subcutaneous (SC), and inhalant solution (INH), various (VAR), and other (OTH). If there is more than one method, then the most common method is listed first.

There are appendices at the back of the HCPCS book in addition to an alphabetic index. Appendices include **modifiers,** summary of changes, and table of drugs. The modifiers are composed of two digits; the first one is always a letter and the second can be either a number or a letter. However, there are one-digit modifiers composed of a letter that are used for ambulance services, which indicate the location of the service. These modifiers can be applied to CPT Level I codes, such as F5 for thumb of the right hand. These modifiers cover a wide variety of issues relating to coding and reimbursement, such as recording and storage on tape by an analog tape recorder. They are important in achieving proper reimbursement. They include modifiers such as RT for right side and LT for left side. TC also is a common HCPCS modifier used when a technical component of a service is provided, but not the professional component. The table of drugs appendix is helpful in finding other names for specified drugs, e.g., Anabolin is known as Nandrolone decanoate and can be found in the table of drugs.

It may be helpful when coding drugs from the HCPCS book to have a drug book available that can give you other names for specified drugs, routes of administration, and dosages. This information can help you select the proper code.

ACTIVITY 9.1

Provide the correct code(s)n for the following:

1. Rectal exam for screening for prostrate cancer.
2. Infusion pump for enteral feeding.
3. Sterile gauze dressing, 16 sq inches.
4. Amicar 5 g.
5. Plastic single upright KAFO with free knee.
6. Prenatal care for at-risk pregnancy.
7. Dialysis bags.
8. Kidney machine for delivery of dialysate.
9. 15 mg Cisplatin.
10. 3 needles with syringe, 3 cc.
11. What does THKAO mean?
12. What methods of administration are used for Gonadorelin HCl?
13. Venipuncture.
14. Orthosis for scoliosis that is tension based.
15. Gas permeable contact lens, bifocal, both eyes.
16. Nasal vaccine.
17. Splint for knee.
18. In what dosage(s) is Daunorubicin HCl administered?
19. Influenza vaccine administration.
20. Vaginal smear.

SUMMARY

HCPCS is composed of Level I, which are the CPT codes and Level II, which are the HCPCS codes for materials and supplies. There is a level III that is not used for the national exams. For physician billing, all materials and supplies that are an expected part of a procedure or service usually are not billable as their costs are included in the procedure or service, which is known as global. However, any other materials and supplies are billable and are, therefore, coded with HCPCS codes.

ACTIVITY 9.1 ANSWERS

Provide the correct code(s) for the following:

1. Rectal exam for screening for prostrate cancer.

ANSWER: G0102

2. Infusion pump for enteral feeding.

ANSWER: B9000

3. Sterile gauze dressing, 16 sq inches.

ANSWER: A6219

4. Amicar 5 g.

ANSWER: S0017

5. Plastic single upright KAFO with free knee.

ANSWER: L2037

6. Prenatal care for at-risk pregnancy.

ANSWER: H1000

7. Dialysis bags.

ANSWER: A4911

8. Kidney machine for delivery of dialysate.

ANSWER: E1510

9. 15 mg Cisplatin.

ANSWER: J9060 × 2; Even though the dosage is less than the 20 mg that is coded, the code would have to be multiplied by two to ensure that the correct amount was reimbursed.

10. 3 needles with syringe, 3 cc.

ANSWER: A4208 × 3

11. What does THKAO mean?

ANSWER: Thoracic-hip-knee-ankle orthosis

12. What methods of administration are used for Gonadorelin HCl?

ANSWER: Subcutaneous and intravenous.

13. Venipuncture.

ANSWER: 36415

14. Orthosis for scoliosis that is tension based.

ANSWER: L1005

15. Gas permeable contact lens, bifocal, both eyes.

ANSWER: V2512 x 2

16. Nasal vaccine.

ANSWER: J3530

17. Splint for knee.

ANSWER: L4380

18. In what dosage(s) is Daunorubicin HCl administered?

ANSWER: 10 mg

19. Influenza vaccine administration.

ANSWER: G0008

20. Vaginal smear.

ANSWER: G0141

Chapter 10 ICD-9-CM PROCEDURAL CODING

Objectives

(1) Distinguish between CPT and ICD-9-CM Volume 3 codes.

(2) Understand applications for ICD-9-CM Volume 3 codes.

(3) Use alphabetic index to locate codes.

(4) Cross-over from alphabetic to tabular list for selection of codes.

(5) Understand the structure and conventions of the ICD-9-CM tabular and alphabetic indices.

Key Terms

Alphabetical Index—There is only one Volume 3, as opposed to Volume 1 and Volume 2 of the diagnostic codes within the ICD-9-CM coding book. Both the alphabetic index and tabular index are contained within this volume. Alphabetic index provides listings of procedures and services by their name, type or anatomic site, eponym, or abbreviations in alphabetical order.

Eponym—Use of a person's name for a procedure.

Essential Modifiers—Additional information listed after a code that is required when selecting a code.

ICD-9-CM Volume 3—Volume 3 of the ICD-9-CM coding book contains the procedures for hospital-based coding.

Nonessential Modifiers—Additional information listed after a code that may or may not have an impact on the selection of the proper code but is not required.

Sequencing—The listing of codes in a selected order due to various requirements for proper reimbursement.

Tabular Index—The tabular index is contained within Volume 3 of the ICD-9-CM coding book and contains the actual numeric codes assigned to procedures and services offered within a hospital.

Volume 3 of the ICD-9-CM coding book contains the procedures for hospital-based coding. There is a tabular and an alphabetical index included in this volume. Procedures are listed in the alphabetical index (under procedure) by name, type, or anatomic site or by **eponym** (use of a person's name for a procedure) or abbreviations. The codes are three to four digits, with only two digits before the decimal and one or two after the decimal. Codes are classified according to anatomic site with specific operations listed within these anatomic sites.

Be careful of details! **Volume 3** is very specific in its notes. Check thoroughly for notes, such as excludes and see also, and follow them precisely.

ALPHABETICAL INDEX

Never code directly from the alphabetical index.

The **alphabetical index** includes names of procedures, eponyms, and conditional terms, such as manipulation, operation, removal, insertion, replacement, incision, closure, and biopsy. After eponyms, a description of the procedure will be listed. With all coding, if you cannot find the specific procedure, begin looking under related conditional terms. This makes knowledge of medical terminology and anatomy critical for proper coding.

As with other types of coding, there may be variations of a term indented underneath. When using indentations, be careful you do not end up under the wrong term.

TABULAR INDEX

The procedures are listed in the **tabular index** in ascending numerical order. The first digits of the ICD-9-CM procedure codes represent the anatomical site. For instance, codes that begin with 18, 19, and 20 represent procedures of the ear. Codes 85 and 86 represent procedures on the integumentary system.

CONVENTIONS

There are some codes in the alphabetical index that have additional codes listed after them that are contained in brackets. As with Volume 2, these two codes must be coded together.

Omit code is seen frequently in Volume 3 for procedural coding. It means that the code is not used in conjunction with the other code in which the statement omit code is listed. For example, an approach for a procedure usually is included in the code for the major procedure and is not billed separately, which may be denoted by the terms *omit code* in addition to the name of the procedure.

TERMINOLOGY

Modifying Terms

Remember, with **nonessential modifiers** (usually in parentheses), they usually are not all-inclusive, but provide additional information that may or may not have an impact on the selection of the proper code, since not all conditions related to that code will be listed within the nonessential modifiers.

Volume 3 also uses the terms *see* and *see also*. Follow these directions and check the additional information as directed by this reference. Sometimes this provides you with a choice among procedure descriptions, but sometimes it will not provide any information other than directing you to check the other term.

Watch the *details!* Read the notes. Detailed notes include *code also*, which means that you must code the additional codes that are listed when performed.

ACTIVITY 10.1

Answer the following questions.

1. What does CPAP mean?

2. What two categories can be coded for cardiac massage?

3. What term could also be used to locate administration of antitoxins?

4. On what part of the body would an Abbe operation be performed?

5. What is the code for reattachment of the retina by cryotherapy?

6. What is the code for a closure of the chest wall following open flap drainage?

7. What is the eponym for the previous operation?

8. Name another term under which you could locate TUNA in the alphabetical index?

9. Is it correct to code a radical prostatectomy as 60.96? Why or why not.

10. What else should be coded with an anoscopy?

11. What is a Sturmdorf?

12. Find another term other than Sturmdorf where this can be located within the alphabetical index.

13. How is a Mikulicz operation differentiated?

14. What is the code for a Fasanell-Servatt operation?

15. What type of procedure is described in the previous question?

16. Where else should you check for roentgenography?

17. Are stripping and incision of the bone coded the same? Why or why not.

18. Is insertion of a bone growth stimulator noninvasive? Why or why not.

19. Name three ways to perform gall bladder drainage.

20. For an anastomosis of the bladder and sigmoid colon, what else must be coded?

21. If there are multiple fusions within the spinal cord, are these each listed separately?

22. What does a CAT of the abdomen exclude?

SEQUENCING

For inpatient hospitalization, all significant procedures must be listed according to the Uniform Hospital Discharge Data Set (UHDDS), in addition to any other procedure that affects payment. A significant procedure can possess any of the following qualities: surgical in nature, has an anesthetic or procedural risk, and/or requires specialized training. The principal procedure is the procedure used for definitive treatment or to care for complications. If there is more than one principal procedure, then the procedure that is most closely linked to the primary diagnosis should be coded first. If there are several procedures closely linked to the primary diagnosis, then the procedure that is most labor intensive should be coded first.

After listing the significant procedures, therapeutic and diagnostic procedures would be coded, which is different from ICD-9-CM diagnostic codes in which therapy codes were listed first, such as V58.1 for chemotherapy. Major procedures would be considered primary over exploratory procedures. For example, if a patient is admitted for appendicitis and an appendectomy is performed in addition to amputation of the big left toe because of diabetes complications, the appendectomy would be listed first because the patient was admitted for appendicitis.

PROCEDURAL CODING

Like diagnostic codes, there are indentations with which you must be careful. Some of these indented categories will contain procedures with more information.

MULTIPLE PROCEDURES

Multiple procedures must be listed separately, unless they are an integral part of the primary procedure and, therefore, are included in the global packaging of the primary surgical code.

If the procedure does not include the term bilateral, then the code must be listed twice when coding for a bilateral procedure. If a code states that it is for either unilateral or bilateral, then the code is used only once regardless of whether it was done on both sides or not.

STAGED REPAIRS

Staged repairs are coded by the procedure performed. Some repairs that could be performed as staged are not and are coded with a code for the entire procedure, not an accumulation of several codes for the various procedures. For example, repair of congenital cardiac anomalies could be coded with 35.8, which is the repair of the genetic malformation as one total procedure, not in stages.

BIOPSIES

Biopsies are coded by method type. Biopsies can be open where the site is visualized or closed. These different biopsies may be identified by terms such as brush, endoscopic, or percutaneous. Biopsies should not be coded when the specimen is totally excised; this is coded as an excision only. However, if a procedure separate from the biopsy is performed, such as excision or approach procedure, then the biopsy can be coded also. For example, if a scope and biopsy were performed on the peritoneum, they would be coded as 54.21 and 41.32, but if the scope was part of the procedure for the biopsy, it would be coded as 44.14.

INTEGUMENTARY

Codes for integumentary systems can come from several different areas in Volume 3, including the 27s and 85–86. Grafts can be autologous (from self), allograft (from same species), or xenograft (from another species). Grafts can be free or adjacent. With adjacent grafts, a portion of the graft remains attached and the skin is stretched and shifted to cover the defected area. Split-thickness grafts include grafting the epidermis and part of the dermis. Full-thickness grafts include all of the dermis and epidermis, which complicates the ability of the donor site to be repaired as it cannot regenerate due to the complete loss of the dermis, and may require further grafting for repair.

Debridement is performed either excisionally or nonexcisionally. Excision involves tissue cutting, such as with a laser. Cutting with scissors would be considered nonexcisional.

MUSCULOSKELETAL

Musculoskeletal system procedures include fracture care, repairs, incisions, excisions, and traction. Traction is attached through skin or bone. Skin traction includes adhesives and includes Buck and Boot. Skeletal traction involves inserting pins or wires into the bone, such as the Kirschner wire. Types of skeletal traction include Russell and Dunlop. For the skull, traction devices include halos and tongs. Internal or external fixations involve pins or other devices to immobilize the bones. With external fixation, pins are attached to the bone and then to the outside frame.

There are codes for application and removal of the pins and frame. Some external devices, such as casts and splints, do not involve the application of pins into the bones. With internal fixation, pins, screws, or other devices are used, but they are implanted directly into the bone and are not attached to any external device or frame.

GASTROINTESTINAL

Calculi (stones) can be diagnosed by an intravenous pyelogram (IVP). They can be removed by lithotripsy (extracorporeal shock wave lithotripsy—ESWL), incisional (lithotomy), stents, laser, or litholapaxy (water, optic, shock, or ultrasound). When several stents are placed into the same anatomic site, only one code is assigned. Stomies are a common gastrointestinal procedure, including end-to-end and side-to-side anastomosis.

Sometimes these are coded, but sometimes not, so be sure to read the coding notes.

ENDOSCOPIC PROCEDURES

If an endoscopic procedure is converted to an open procedure, then only the open procedure is coded, and not the endoscopic procedure. Use only one code for an endoscopic procedure by selecting a code that extends as far as the procedure was completed.

CARDIOVASCULAR

For many cardiovascular procedures, cardiopulmonary bypass (extracorporeal circulation) is required and needs to be coded as indicated in the coding notes.

For cardiac bypass grafts, the number of arteries must be known for a graft from the aorta. When the graft is from a mammary artery, then it must be known if one or both mammary arteries were involved.

With cardiac catheterizations, it must be known if the catheterization was left, right, or combined. Angiographies and arteriographies are performed in conjunction with cardiac catheterizations as a means of visualizing the condition of the vessels. Angioplasties remove obstructions from occluded arteries. A catheter is inserted with a balloon on the end, which is expanded, thus clearing the obstruction.

Angioplasties can be administered percutaneously (PTCA) with thrombolytic agents infused. Vascular access devices (VADs) are catheters used to introduce substances, such as nutrition or drugs, into the body. A simple venous catheter is not considered a VAD and is coded separately as 38.93. Similar to VADs, pumps can be inserted that administer substances such as chemotherapeutic drugs.

Pacemakers can be temporary or permanent, and they are either single or dual chambered. The electrodes can be placed either on the chest (epicardial) or transvenously. With ICD-9-CM procedural codes, two codes must be used when initially inserting the pacemaker; one code for the pacemaker and one code for the leads. When revisions occur, only the service revised is coded; for example, if only the leads are revised, then only the leads are coded.

Cardioverters/defibrillators differ from pacemakers by restoring pulse when there are cardiac dysrhythmias, whereas pacemakers maintain a steady pulse. The cardioverters/defibrillators comprise both the leads and the pulse generator, and can be coded with one code.

MECHANICAL VENTILATION

Mechanical ventilation is the use of mechanical devices to maintain breathing. This is necessary when certain conditions such as pneumonia, pulmonary emboli, injuries, diseases, respiratory distress, or neurological conditions threatened a patient's ability to breathe successfully on his/her own. Mechanical ventilation is administered either continuously or intermittently. Patients can be sustained on mechanical ventilation for various lengths of time, depending on their condition.

Ventilators can be controlled with preset levels or activated by the patient's requirements. Various terminologies indicating the use of mechanical ventilation include IVVP (intermittent positive pressure breathing), PEEP (positive end-expiratory pressure), or MMV (mandatory minute ventilation). Medical complications resulting from mechanical ventilation include hypotension and decreased venous response.

Endotracheal intubation and tracheostomy are used for mechanical ventilation. Tracheostomies are more permanent and are used for long-term ventilation. A tracheostomy may be used for patients who have problems, such as severe facial trauma, burns, or tumors, using a tube. Code 96.7 is used for these categories with some excludes listed. Be sure to read these!

Mechanical ventilation codes are coded according to the number of hours the patient is maintained on the ventilation device, with the number of hours beginning when the patient receives the initial service, starting with intubation or tracheostomy. However, if the patient is admitted to the hospital and already has been intubated, then the time begins at admission. Patients initially can have an endotracheal intubation, which is followed by a tracheostomy, in which case the time begins with the intubation. No time is deducted for the changeover or for replacement time if another tube is necessary. The number of hours ends when the ventilation is stopped or when the patient dies. Weaning, which involves the use of intermittent ventilation, is included in the calculation of hours on ventilation. There are separate codes (93.90, 93.91, and 93.99) for subsequent periods of ventilation that may be required because of complications once the patient's ventilation is discontinued.

There are several important coding issues regarding the use of machines for maintenance of life. During some surgical procedures, patients are placed on cardiopulmonary bypass (extracorporeal circulation, heart-lung machine), which is coded as 39.61, to sustain them during the procedure. Extracorporeal circulation must be coded in addition to the procedure, such as replacement of heart valves (35.2).

OBSTETRICS/GYNECOLOGY

In contrast to the CPT codes, total abdominal or vaginal hysterectomies do not include removal of tubes and ovaries, which must be coded separately as detailed in the tabular list.

ICD-9-CM procedural codes for deliveries are classified according to the types of procedures used, such as forceps and episiotomies (72–75.3). Whereas abortion procedures are listed separately as abortions within the CPT codes, they are identified by the actual procedure in the ICD-9-CM procedural codes.

The ICD-9-CM procedural codes list breech extraction as characterized by the baby not being converted to a cephalic presentation first. There are codes in both the CPT and ICD-9-CM procedural codes for the external conversion of a breech to normal presentation for birth, sometimes referred to as external cephalic version.

ACTIVITY 10.2

1. A 47-year-old man was admitted to the Emergency Room after possible exposure to anthrax, so a vaccination was provided. What is the ICD-9-CM procedural code(s)?

2. Minerva jacket was applied to a 14-year-old boy, which was coded as 93.51. Is this correct? Why or why not.

3. Name two materials that can be used for insufflation of the fallopian tubes.

4. Due to the presence of a lesion, a patient received synchronous resection of the ileum with a bag attached to the outside. What is the proper ICD-9-CM procedural code(s)?

5. For this 62-year-old female who had been complaining of chest pains and difficulty breathing for the past week, a synthetic valve was used as replacement for the aortic valve. What is the correct ICD-9-CM procedural code(s)?

6. Is the following coding scenario correct for a patient receiving pinning of both ears—18.5-50? Why or why not.

7. Patient was admitted for observation for difficulty breathing. Three hours after admission, the patient experiences problems breathing and endotracheal intubation is performed. Five hours after this, a temporary tracheostomy is performed. Two hours later, the patient expires. What is the correct ICD-9-CM procedural code(s)?

8. This 23-year-old female patient had a partial-thickness removal of a chalazion from the patient's eyelid. What is the correct ICD-9-CM procedural code?

9. This 70-year-old female patient received a wedge resection of the left lung due to carcinoma of the lung. What is the correct ICD-9-CM procedural code(s)?

10. Tetralogy of Fallot and patch graft of outflow tract was repaired on this patient in one operation. The ventricular septal defect was repaired with a prosthesis. What is the ICD-9-CM procedural code?

11. Patient received a thromboendarterectomy with patch graft of the coronary artery with insertion of a stent. What is the proper ICD-9-CM procedural code?

12. The patient is a 42-year-old woman who had her ovaries, tubes, urethra, uterus, and bladder removed. What it the correct ICD-9-CM procedural code(s)?

13. If a patient received an endoscopic excision of a lesion of the pancreatic duct and it is coded as 52.21 and 51.10, would this be correct? Explain.

14. What is the correct code(s) for a patient who received cardioversion?

15. What is the correct code(s) for a patient who received a transfusion *in utero* through hysterotomy?

16. What is the correct code(s) for a patient who received an anastomosis of subclavian-pulmonary graft?

17. What is the correct code(s) for a patient who received repair of fistula of the urethra?

18. What is the correct code(s) for a patient who received an ESWL of the gallbladder?

19. What is the correct code(s) for a patient who received a treadmill stress test?

20. What is the correct code(s) for a patient who received catheterization for drainage of a cyst with adhesive substance?

ANSWERS

ACTIVITY 10.1

Answer the following questions:

1. What does CPAP mean?

ANSWER: Continuous positive airway pressure as noted in the alphabetic index.

2. What two categories can be coded for cardiac massage?

ANSWER: Closed or open. The alphabetic index provides two possible selections for cardiac massage, closed or open.

3. What term could also be used to locate administration of antitoxins?

ANSWER: Injection. Antitoxins are administered via injection, so either administration or injection can be used to locate the proper code in the alphabetical index.

4. On what part of the body would an Abbe operation be performed?

ANSWER: Either vagina or intestine. The alphabetical index notes that an Abbe operation can be either a construction of the vagina or an anastomosis of the intestine.

5. What is the code for reattachment of the retina by cryotherapy?

ANSWER: 14.52

6. What is the code for a closure of the chest wall following open flap drainage?

ANSWER: 34.72

7. What is the eponym for the previous operation?

ANSWER: Clagett

8. Name another term under which you could locate TUNA in the alphabetical index?

ANSWER: Ablation. Ablation is the procedure, so within the alphabetical index, it can be located under either TUNA or ablation. Other terms are transurethral or needle, but they are not the main terms, so ablation is more appropriate.

9. Is it correct to code a radical prostatectomy as 60.96? Why or why not.

ANSWER: No, there is an exclude note in the tabular index that says it is excluded from code 60.96.

10. What else should be coded with an anoscopy?

ANSWER: Application or administration of an adhesion barrier substance also should be coded as indicated by a note at the beginning of section 49 in the tabular list.

11. What is a Sturmdorf?

ANSWER: Conization of the cervix. This description is provided in the alphabetical index.

12. Find another term other than Sturmdorf where this can be located within the alphabetical index.

ANSWER: Operation. This is an operation, so it can be found under the term operation within the alphabetical index.

13. How is a Mikulicz operation differentiated?

ANSWER: By two stages. In the alphabetical index, a Mikulicz operation is described as either stage 1 or stage 2.

14. What is the code for a Fasanell-Servatt operation?

ANSWER: 08.35

15. What type of procedure is described in the previous question?

ANSWER: Blepharoptosis repair by tarsal technique.

16. Where else should you check for roentgenography?

ANSWER: Radiography cardiac, negative contrast.

17. Are stripping and incision of the bone coded the same? Why or why not.

ANSWER: Yes, they are coded the same as described in the alphabetical index, with stripping referring the coder to incision.

18. Is insertion of a bone growth stimulator noninvasive? Why or why not.

ANSWER: Insertion of a bone growth stimulator is described in the alphabetical index as either semi-invasive or invasive, therefore, noninvasive is not applicable to this operation.

19. Name three ways to perform gall bladder drainage.

ANSWER: Anastomosis, aspiration, or incision as noted in the alphabetical index.

20. For an anastomosis of the bladder and sigmoid colon, what else must be coded?

ANSWER: Resection of the intestine must also be coded as noted within the tabular and alphabetical indices.

21. If there are multiple fusions within the spinal cord, are these each listed separately?

ANSWER: No, they are not listed separately as there are new codes that specify and include multiple fusions.

22. What does a CAT of the abdomen exclude?

ANSWER: CAT of the kidney

ACTIVITY 10.2

1. A 47-year-old man was admitted to the Emergency Room after possible exposure to anthrax, so a vaccination was provided. What is the ICD-9-CM procedural code(s)?

ANSWER: 99.55. This code is for prophylactic administration of anthrax vaccine.

2. Minerva jacket was applied to a 14-year-old boy, which was coded as 93.51. Is this correct? Why or why not.

ANSWER: No, code 93.51 states that application of a Minerva jacket is excluded from this code, which should be coded as 93.52.

3. Name two materials that can be used for insufflation of the fallopian tubes.

ANSWER: Air, dye, gas, or saline as described in the tabular index.

4. Due to the presence of a lesion, a patient received synchronous resection of the ileum with a bag attached to the outside. What is the proper ICD-9-CM procedural code(s)?

ANSWER: 46.20 and 45.34. This is an ileostomy because a bag was placed on the outside originating from the creation of a passage from the ileum to the outside. The synchronous resection was due to the removal of the lesion, so it should be coded with 45.34.

5. For this 62-year-old female who had been complaining of chest pains and difficulty breathing for the past week, a synthetic valve was used as replacement for the aortic valve. What is the correct ICD-9-CM procedural code(s)?

ANSWER: 35.22 and 39.61. Cardiopulmonary bypass also must be coded as 39.61 in addition to the 35.22, as noted in the tabular index.

6. Is the following coding scenario correct for a patient receiving pinning of both ears—18.5-50? Why or why not.

ANSWER: Although this is bilateral and correct coding must indicate that it is bilateral, CPT modifiers cannot be applied to the ICD-9-CM procedural codes. Instead, 18.5 should be listed twice, which can be achieved by indicating 2 units on the billing form.

7. Patient was admitted for observation for difficulty breathing. Three hours after admission, the patient experiences problems breathing and endotracheal intubation is performed. Five hours after this, a temporary tracheostomy is performed. Two hours later, the patient expires. What is the correct ICD-9-CM procedural code(s)?

ANSWER: 96.71, 96.04, and 31.1. Continuous mechanical ventilation includes the time per hour for either endotracheal intubation or tracheostomy, so the hours begin when the intubation occurs, includes the tracheostomy, and ends when the patient expires. There also is a note in the tabular index that indicates that the endotracheal intubation and tracheostomy must be coded also. The tracheostomy is not permanent, so it is coded as temporary, 31.1.

8. This 23-year-old female patient had a partial-thickness removal of a chalazion from the patient's eyelid. What is the correct ICD-9-CM procedural code?

ANSWER: 08.21. Although partial-thickness is described, this is a removal of a chalazion and should not be confused with the code for partial-thickness removal of a lesion.

9. This 70-year-old female patient received a wedge resection of the left lung due to carcinoma of the lung. What is the correct ICD-9-CM procedural code(s)?

ANSWER: 32.29; Wedge resection is located within the code 32.29 for other excision or destruction of lesion of the lung.

10. Tetralogy of Fallot and patch graft of outflow tract was repaired on this patient in one operation. The ventricular septal defect was repaired with a prosthesis. What is the ICD-9-CM procedural code?

ANSWER: 35.81. Although this repair can be completed in stages, it was done in one operation, so no other codes are required as all of the procedures are included in the one operation.

11. Patient received a thromboendarterectomy with patch graft of the coronary artery with insertion of a stent. What is the proper ICD-9-CM procedural code?

ANSWER: 36.03 and 36.06. The stent also must be coded in addition to the thromboendarterectomy as noted in the tabular index.

12. The patient is a 42-year-old woman who had her ovaries, tubes, urethra, uterus, and bladder removed. What it the correct ICD-9-CM procedural code(s)?

ANSWER: 68.8; This code includes removal of the pelvic organs including the uterus and ovaries, which are not, therefore, coded as a hysterectomy.

13. If a patient received an endoscopic excision of a lesion of the pancreatic duct and it is coded as 52.21 and 51.10, would this be correct? Explain.

ANSWER: 52.21: This code includes the 51.10 for the scope.

14. What is the correct code(s) for a patient who received cardioversion?

ANSWER: 99.62; This is listed under conversion of cardiac rhythm codes.

15. What is the correct code(s) for a patient who received a transfusion *in utero* through hysterotomy?

ANSWER: 75.2 and 68.0; The hysterotomy also must be coded in addition to the transfusion.

16. What is the correct code(s) for a patient who received an anastomosis of subclavian-pulmonary graft?

ANSWER: 39.1 and 39.61; Extracorporeal circulation also must be coded as indicated by the notes in the coding book.

17. What is the correct code(s) for a patient who received repair of fistula of the urethra?

ANSWER: 58.43; The repair is coded as closure.

18. What is the correct code(s) for a patient who received an ESWL of the gallbladder?

ANSWER: 98.52; ESWL is extracorporeal shockwave lithotripsy.

19. What is the correct code(s) for a patient who received a treadmill stress test?

ANSWER: 89.41; This is a cardiac stress test.

20. What is the correct code(s) for a patient who received catheterization for drainage of a cyst with adhesive substance?

ANSWER: 52.01 and 99.77; The adhesive substance must be coded also.

Chapter 11 ABSTRACTING

Objectives

(1) Understand the concept of abstracting.

(2) Conquer your fear of abstracting.

(3) Understand the differences between inpatient and outpatient abstracting.

(4) Gain confidence in your ability to analyze and make educated guesses.

(5) To be capable of abstracting from a wide variety of documentation.

(6) Understand and recognize medical report formats.

(7) To apply your knowledge of ICD-9, CPT, and HCPCS codes when selecting proper codes.

Key Terms

Abstracting—The derivation of proper codes directly from the medical records.

Ambulatory Payment Classification (APC)—APCs are a grouping method used to determine payment for services for outpatient hospital services. APCs are based on the primary HCPCS code (which includes CPTs).

Diagnosis-Related Grouping (DRG)—DRGs are a grouping method used to determine payment for services for inpatient hospital services. DRGs are based on the principal diagnosis assigned to the patient for an inpatient visit.

Downcoding—The undesirable selection of codes that are lower in value and in criteria than services provided.

Primary Diagnosis—The diagnosis that is the main reason services are provided to patient in an outpatient setting.

Principal Diagnosis—The diagnosis that is the main reason the patient is admitted to the hospital.

Upcoding—The undesirable selection of codes that are higher in value and criteria than the documentation justifies, which may result in higher reimbursement, as well as possible fraud charges.

ABSTRACTING

Abstracting is the culmination of all of your hard work and knowledge. It brings everything together, so it can be difficult. As the term implies, it is abstract and, therefore, not memorizable. Abstracting is about using your knowledge to make educated guesses. From a common sense point of view, abstracting is about being able to prove your point, i.e., being able to explain why you did what you did. Because codes are not perfect, there are many gray areas that can be debated, and consequently changed, which is why codes undergo changes that are incorporated into the new coding books that are issued yearly. Abstracting is a valuable coding skill because it enables a coder to be their best.

Abstracting is especially important in the hospital coding exams because the principal diagnosis must be determined, so that it can be listed first. This is critical for reimbursement purposes. Inpatient hospital coding can involve a multitude of diagnoses and services since patients may remain in the hospital for several days or more. However, one code must be selected as the principal code. If more than one diagnosis is listed in the physician's report as the principal diagnosis, either one can be listed first. However, if one code is closer in relationship to the care provided, then it should be coded first.

Patients may be diagnosed with additional conditions, known as secondary conditions, while they are in the hospital or ambulatory surgical center (ASC), but secondary conditions are not the main reason for admittance so are not coded as the principal diagnosis. However, if, after inpatient (not outpatient) evaluation and testing, it is determined that there is a more definitive diagnosis for the hospital admission, this would be listed as the principal diagnosis and not as a secondary condition.

In contrast to the above scenario, if during hospitalization a more severe diagnosis is determined, but it is not related to the reason for admission, it is not listed as the principal diagnosis.

Additional diagnoses can include symptoms, histories, chronic conditions, conditions that develop during hospitalization, history of prior surgeries, and secondary conditions that may be more influential to management of the patient's care than it might typically be in a physician's office. In contrast, in a physician's office, a patient often is seen for one minor problem that is not affected by other health conditions the patient may have. In these cases one or two codes are sufficient. In a hospital setting, whether inpatient or outpatient, more time often elapses during care, and conditions may be aggravated. In addition, other diagnoses are evaluated and considered more thoroughly. For this reason, it is important to understand the disease process and what symptoms are associated with what conditions to know when a condition should be coded and when it is considered a symptom of another condition. Other codes are coded if they influence the outcome or management of a patient's care. Influence can include prolonging care or stay and extension or addition of care, including laboratory and testing services.

If there are no diagnoses or symptoms, V codes are used to indicate that there were no abnormal findings. For example, *worried well* is a V code that can be applied for visits that do not indicate any diagnosis or symptoms.

Do not code abnormal lab findings. It is the physician's job to diagnose the patient. Exceptions to this rule, where you may use labs to code, occur when a physician provides a possible diagnosis, such as strep throat, which is confirmed by laboratory results.

Report Formats

Abstracting consists of withdrawing information from the patient's file. Familiarize yourself with the format of these reports. Standard hospital reports can include operative, discharge, consultations, pathology, and history and physical reports. Other notes may include labs, physician and nursing notes, and rehabilitation care. What makes abstracting difficult is that there can be a great deal of information and no guidelines to help you: determine which information is valuable; select the proper codes; or determine their proper order. A patient may have many conditions and much information may be provided, but only information related to the current care is coded. All of this decision making is the responsibility of the coder, particularly in response to audits.

PRINCIPAL DIAGNOSIS

For inpatient hospital coding (**diagnosis-related grouping [DRG]**), the diagnosis that is the main reason the patient is admitted to the hospital is the **principal diagnosis** and should be listed first. This does not necessarily mean that this code is the patient's most serious condition as other conditions may develop during hospitalization. With outpatient hospital coding (**Ambulatory Payment Classification [APC]**), the **primary diagnosis** is the main reason services were provided.

Inpatient Abstracting

Inpatient hospital coding differs in some ways from physician coding. It involves ICD-9 diagnostic and procedural codes, not CPT codes. DRGs are used for grouping in the hospital. There are some significant differences between inpatient and outpatient/physical coding. The most noticeable difference is the amount of codes that are listed, with more codes usually coded in inpatient hospital coding. There are several reasons for this. The coding of symptoms is evaluated differently from outpatient/physician coding primarily because of extended lengths of stay and services.

In the hospital, attempts are made to determine the patient's problem and treat it while there is hospital/physician contact with the patient. Also, patients are kept in the hospital when their condition is serious, so problems often are more significant and may be exacerbated by other conditions. Conditions that influence patient care, such as increasing the length of stay and/or the treatment provided, should be coded. Conditions that require treatment in the hospital definitely should be coded. If a condition exists but is not relevant to the patient's care and is not treated, such as the presence of skin tags on the neck that are not removed, it should not be coded.

Chronic diseases may need to be coded when they affect the management of the patient's care. Chronic conditions may adversely affect services such as surgery and or conditions such as bleeding, for instance. In the hospital because of the possibility of greater stress, other conditions, such as depression, also may be important contributors to the patient's welfare. You must know when to code symptoms and when not to code symptoms, because some symptoms may be coded with inpatient coding, whereas others are not.

Usually when you apply a diagnosis code, you will not code the symptoms associated with it; however, if a diagnosis is not stated, you can code the symptoms. Also, if a symptom is not an integral component of a disease process or is influential by itself in affecting the level of care provided to the patient, then it may be coded. For example, pain is often associated with many diagnoses, but on occasion may warrant its own specific care, in which case the pain is coded. Coders must understand the disease process and what symptoms are associated with that diagnosis. It is, therefore, important with hospital coding to determine the progress of care and what conditions or diagnoses are involved. If a patient was diagnosed with chronic obstructive pulmonary disease (COPD), for example, but found to have asthmatic bronchitis after further workup, then only the asthmatic bronchitis would be coded.

Past personal, family, or surgical histories also can be important factors with inpatient hospital coding as they may indicate the type of problem or conditions exacerbating the condition of the patient and, therefore, influence patient care For example, when a patient is admitted to the hospital for severe chest pains and has had an angioplasty three months ago, this surgical history is highly relevant because it can be the source of the patient's current problem. However, not all histories are coded. If a patient is admitted to the hospital for severe chest pains and had a graft six months before to remove a lesion from his hand, this history is not relevant.

Outpatient Abstracting

Outpatient hospital coding is similar to physician billing, including the use of CPT and HCPCS codes. APCs are used for grouping and payment. Guidelines for coding symptoms are similar to physician billing, rather than inpatient hospital, since symptoms are not coded if adequate diagnoses are provided. However, if there are no appropriate diagnoses that include the symptoms, then the symptoms can be coded if they are relevant to the patient's care.

Only symptoms relevant to the patient's current care are coded if no diagnoses are provided that include the symptoms. Secondary diagnoses can be coded if they are relevant or influential to the patient's care. This would include chronic diseases and/or coexisting conditions, but these would not be coded, regardless of how serious a diagnosis unless they influence patient care. This differs from inpatient coding because outpatients rarely are seen for an extended period of time but usually are evaluated and treated within hours; therefore, certain conditions or diagnoses may not be as relevant as they are in the hospital. For example, if a patient is seen as an outpatient and receives several sutures, the fact that they have COPD probably would not influence their care and would not be coded. In this case, however, diabetes would be coded because of bleeding, which can be complicated by the diabetes.

REQUIRED CODING

Some primary and other relevant codes may require using other codes. Follow the instructions! If the codes direct you to *code also,* then code the additional required code.

RELEVANT CARE

All relevant patient care and outcomes should be coded. This is an issue particularly with hospital care because the patient generally is more at risk. Conditions such as COPD and diabetes are more relevant in the outcome of a patient's care if major surgery is involved as opposed to minor care in the office, such as removal of a wart.

SEVERITY OF INJURIES

The most severe injury is coded first, whether fractures, burns, or otherwise. Any minor integumentary injuries, such as abrasions, would not be coded additionally if there is a more severe injury at the same site. More extensive injuries, such as injuries to nerves and vessels, would be coded additionally.

SYMPTOMS

Symptoms are normally not coded if a definitive diagnosis is provided. However, there are some circumstances when symptoms are coded, for instance, when the code requires it or when the billing process requires it. For example, if a patient is admitted for vomiting and diagnosed with pneumonia and fever, the vomiting and pneumonia would be coded, but the fever would not be as it is a common symptom of pneumonia. Any symptom influencing patient care that is not associated with a coded diagnosis would be coded. Beware, however, for some conditions may be associated with each other and are both coded, since they do not meet the true definition of an inherent symptom of the diagnosis. For example, if a patient is diagnosed with septicemia with shock, the septicemia (the responsible organism) and the shock would both be coded, with septicemia listed as the principal diagnosis.

If a diagnosis is not known, then code the symptoms. Some codes will include the symptoms, which is fine.

LATE/HISTORY

Late effects and histories are coded only if they apply to the codes selected for the patient's services. Beware, not all history codes should be coded. For example, if a patient, in the office, receives two sutures in one finger and had an angioplasty three months previously, we are not concerned at this time about the angioplasty. If, however, the patient is admitted to the hospital for chest pain, the angioplasty is relevant to the care of the patient.

BEGINNINGS

First and foremost when coding abstracts, get a general overview by briefly looking at what types of reports are presented. If there are several reports, you will know the whole scenario before you begin a closer scrutiny of the reports. Reading each report separately does not provide a general overview. This creates problems if you try to code something that should not be coded and wastes time if you end up reading material you did not need to read. For example, if the first report is a well baby check, then you might begin checking for a preventive evaluation and management (E/M) code, but as you progress in the reports, you may find that the child returned and the well check is irrelevant to the new visit with the new visit being the one you need to code.

Secondly, begin with diagnostic headings. For this, a good source document to begin abstracting is the discharge summary. Headings, such as discharge diagnosis, summary of operative procedures, and admission diagnoses, are located in this type of report. Then proceed to other report headings, such as the admission diagnosis in the history and physical report or the operative procedure heading in the operative report. These can be compared to the headings in the discharge diagnosis. At this point, before continuing to other parts of the reports, it may be useful to find the diagnoses codes in the above listed headings. Write them next to their applicable description, because you cannot keep reading each report over and over, particularly during the exam. This also helps you focus, so you do not get lost in the details of complicated hospital reports, in addition to providing you with an idea of the kind of information you are seeking from the rest of the documentation. For example, if you are coding for skull base surgery, you will need to know the terms for the approach and the area of the brain where the procedure was performed. Begin with something small, like a discharge diagnosis, code it as you find it, and then proceed step by step from there, seeking the necessary documentation.

Other reports can be used to compare the diagnoses you have already coded, and discrepancies can be compared. When you first are learning how to abstract, take the time to read through the hospital reports, including irrelevant parts, such as negative histories, just to get used to report formats and documentation in the patient record. Over time, as you become more comfortable with the reports, you will be able to skim them. Beware though, as discussed earlier, skimming can be dangerous. In coding, you cannot miss vital details!

Portions of various reports are useful in finding the correct code. For CPT procedures, operative reports are critical, e.g., for catheterizations. For ICD-9 codes, pathology and laboratory reports may be crucial, indicating results from tests, e.g., infectious organisms and x-rays. However, remember, you are not the physician and you are not to diagnose the patient! Abnormal findings are not diagnoses until the physician, not you, makes the determination. Some abnormal findings usually are provided after the physician's reports are completed and considered valid, which then can be coded, such as streptococcus and other organisms.

Remember, not documented, not done. All that is lies before you. You cannot change codes based upon a physician's verbal statements. If the information is not in the file, it does not exist and cannot be added to increase costs. Physician's verbal statements can be used only to verify information that already exists in the file or as necessary clarification to select a code.

INSTRUCTIONS

Follow code instructions closely. If they instruct you to code other conditions in addition to a selected code, then you must do so. Be careful of *excludes* and *includes*. Read the details carefully!

ABSTRACTING APPLICATION RULES

The following is a summation of the abstracting process but is certainly not all-inclusive. All coding rules apply, including the many topics discussed in this book.

1. Not documented, not done.
2. Details, details, details.
3. Keep it simple.
4. Beware of skimming and assuming.
5. Code directly from tabular index, not alphabetic index.
6. Remember the order with principal diagnosis listed first.
7. Code requirements (as indicated within primary codes).

8. Code other relevant care codes (codes that affect the outcome of the care). Preexisting conditions and chronic conditions should be coded only if they affect the management of the patient's care, which may include an extended hospital stay or provision of additional services.
9. Code histories and late effects if relevant to the care of the patient.
10. Code symptoms if there is no diagnosis integrally associated with the symptoms.
11. Code acute first usually, chronic second.
12. Possible, probable, rule out, versus, either/or, etc. do not exist and are not coded for outpatient, but for inpatient, they may be coded, if relevant.
13. Code to the greatest specificity.
14. Specified better, unspecified not.
15. Remember the V codes for the unusual and unique, including histories.
16. Therapy/rehabilitation V codes listed first.
17. Do not forget late codes if needed.
18. With inpatient coding, do not forget to code for conditions that develop during the period of care and for which treatment is provided.
19. Properly linking codes justifies medical necessity where applicable.
20. Code what services the hospital provides.
21. No unbundling! Keep it within the global.
22. No upcoding, no downcoding, no overcoding, or undercoding.
23. Determine if E/M services were provided and are chargeable separate from other services provided.
24. Determine type and level of E/M if codeable.
25. Do not forget modifiers where necessary.

Code only what the physician provides. If physicians are in their offices, they can charge for all services they provide, including supplies and materials, x-rays, or medications. However, if a physician performs a surgical procedure in the office, normally expected supplies and materials cannot be billed additionally as they are included in the global. Do not unbundle them! If a procedure is listed as separate, it does not contain other service or supplies, so anything extra should be billed.

Remember, a global package for surgery contains all of the charges normally associated with the service or procedure (ancillaries). Do not unbundle any of these ancillary services by charging for them.

If a physician provides a service in a hospital, then the hospital provides all other services and supplies; therefore, the hospital is paid for the services and supplies it provides.

Upcoding is unacceptable, and can result in fraud and/or abuse charges. **Upcoding** means higher paying codes are selected than can be justified by the documentation.

Sometimes in an attempt to avoid prosecution for improper billing, as may be discovered through governmental audits, hospitals may downcode. **Downcoding** is the selection of lower paying level codes that contain less services than what was actually provided. However, auditors also consider this not acceptable.

ANSWER THE QUESTIONS FOLLOWING EACH SCENARIO.

ACTIVITY 11.1

Inpatient Hospital
HISTORY AND PHYSICAL EXAM
PATIENT NAME: Nancy Clarke
DATE: 3/2/05
DOB: 12/2/65

CHIEF COMPLAINT: Abdominal tenderness, diaphoresis, and fever.

HISTORY OF PRESENT ILLNESS: Patient states that about a week ago she began with a fever. Several days later, the abdominal tenderness began, which grew in intensity and did not subside over the next couple of days. She then began experiencing diaphoresis and subsequently sought medical care.

PAST MEDICAL HISTORY: Patient has no complaints. She is an insulin-dependent diabetic and does have hypertension, but states that she has been doing well with no problems. There is no past surgical history.

REVIEW OF SYSTEMS: Noncontributory except for diabetes and hypertension. She has no history of head, eyes, ears, nose, and throat (HEENT) problems, but does wear glasses. No history of thyroid problems. There is no history of cardiovascular problems; no murmurs, failure, or disease. There is no history of COPD or asthma. No history of abdominal masses, vomiting, diarrhea, rapid weight loss or gain, or melena. No musculoskeletal problems. Gaits are normal.

FAMILY HISTORY: The patient's mother is alive and well with diabetes. The father died in a car accident when the patient was 12 years old. He had no health problems at that time.

SOCIAL HISTORY: The patient does not smoke, but does drink on social occasions. The patient attends church regularly.

PHYSICAL EXAM: GENERAL: The patient is a well-developed, well-nourished white female who has complaints of abdominal tenderness and diaphoresis. VITAL SIGNS: Weight is 145 pounds, BP 150/90, temperature 98.6. NECK: Supple with no masses or thyromegaly. CHEST: Clear to percussion and auscultation. No difficulties breathing. HEART: Regular rate and rhythm. No gallops, murmurs, or rubs. ABDOMEN: Tenderness with palpable masses. Genitourinary (GU): Not done at this time. NEUROLOGICAL: Grossly intact. OPTHALMIC: The patient has been recently diagnosed with cataracts related to her diabetes.

IMPRESSION: Possible cholelithiasis and liver neoplasm.

CT Scan of Abdomen
PATIENT NAME: Nancy Clarke
DATE: 3/2/05
DOB: 12/2/65

TECHNIQUE: With oral contrast, non-enhanced images of the liver were obtained. With the IV injection of contrast, enhanced images of the abdomen were then obtained. Comparison is made with prior CT scans.
FINDINGS: The lung bases are clear. There is no pericardial or pleural effusion. Gallbladder enlarged. The liver does not appear cirrhotic and there are no signs of fatty infiltration of the liver.

The previously described small non-enhancing lesion present within the liver near the hepatic dome appears unchanged from previous x-rays. There are no other focal hepatic lesions.

The spleen is noted. There is no biliary dilation. The gallbladder is enlarged with apparent calculi.

The pancreas and adrenal glands appear enlarged. The kidneys are notable for parapelvic cysts.

The aorta is of normal caliber. Small retroperitoneal lymph nodes are unchanged. There is no ascites.

IMPRESSION: No change in nonspecific lesions present near the liver dome since earlier CT scans.

Operative Report
PATIENT NAME: Nancy Clarke
DATE: 3/3/05
DOB: 12/2/65

PREOPERATIVE DIAGNOSIS: Cholecystitis, cholelithiasis,

POSTOPERATIVE DIAGNOSIS: Cholecystitis, cholelithiasis, status post recent abdominal surgery.

PROCEDURE: Cholecystectomy and lysis of adhesions.

SURGEON: Fred White

ASSISTANT: Alice Marple

ANESTHESIA: General

ESTIMATED BLOOD LOSS: Minimal

COUNTS: Correct

COMPLICATIONS: None

OPERATIVE PROCEDURE: The patient was brought to the operating room where general anesthesia was achieved. The upper abdomen was prepped and draped in the usual sterile manner using Betadine solution. Used Veress needle and 3 L of CO_2 for diffuse penumoperitoneum. A trocar was inserted and numerous adhesions were observed. A second trocar was inserted down through the suprapubic area with direct visualization. Adhesions were scopically removed through a midline incision, along and between the bowel, omentum, and the anterior abdominal wall. There was no evidence of enterotomies or an injury to the bowel. The gallbladder was eventually visualized with the scope and obstruction was noted. The mass on the superior portion of the liver could not be visualized, but there was no evidence of malignancy. The gallbladder was mobilized and adhesions taken out. The cystic duct was identified, clipped and transected. The cystic artery was identified, doubly clipped, and transected in the same fashion. The gallbladder was removed with coagulation through cautery with excellent hemostasis. The wounds were copiously irrigated in all three trocar sites. The bowel was examined closely. No evidence of any antrotomy was noted. Hemostasis was excellent with no evidence of bile leak and no drains were placed. The gallbladder was removed under direct visualization, the fascia closed with 0 Vicryl and the skin with 4-0 Monocryl. Sterile dressings were applied. Sponge and needle counts were correct. The patient tolerated the procedure well without apparent complication.

Discharge Summary
PATIENT NAME: Nancy Clarke
DATE: 3/4/05
DOB: 12/2/65

ADMIT DIAGNOSES: Abdominal tenderness, diaphoresis, and fever.

DISCHARGE DIAGNOSES: Cholecystitis, cholelithiasis,

HISTORY OF PRESENT ILLNESS: Patient admitted with complains of abdominal tenderness, diaphoresis, and fever that began a week ago. PHYSICAL

EXAMINATION: GENERAL: The patient is a well-developed, well-nourished white female. VITAL SIGNS: at time of admission, weight was 145 pounds, BP 150/90, temperature 98.6. NECK: Supple with no masses or thyromegaly. CHEST: Clear to percussion and auscultation. No difficulties breathing. HEART: Regular rate and rhythm. No gallops, murmurs, or rubs. ABDOMEN: Tenderness with palpable masses.

HOSPITAL COURSE: Patient was admitted and evaluated for possible surgery. CT scan performed on patient, which revealed no changes in the liver mass from previous tests. Enlarged gallbladder was noted, so cholecystectomy was scheduled. During surgery, a trocar was inserted and numerous adhesions were observed. A second trocar was inserted down through the suprapubic area with direct visualization. Adhesions were scopically removed through a midline incision, along and between the bowel, omentum, and the anterior abdominal wall. There was no evidence of enterotomies or an injury to the bowel. The gallbladder was eventually visualized with the scope. The mass on the superior portion of the liver could not be visualized, but there was no evidence of malignancy. The gallbladder was mobilized and adhesions taken out. The cystic duct was identified, clipped, and transected. The cystic artery was identified, doubly clipped, and transected in the same fashion. The gallbladder was removed with coagulation through cautery with excellent hemostasis. No evidence of any antrotomy was noted. Hemostasis was excellent with no evidence of bile leak and no drains were placed. The gallbladder was removed under direct visualization, the fascia closed with 0 Vicryl and the skin with 4-0 Monocryl. Patient tolerated the procedure well and was dismissed with discharge medications.

1. What is the principal diagnosis?

2. Why did you select the previous code as the principal diagnosis?

3. Were there two existing concurrent conditions, either of which could have been the principal diagnosis?

4. Should the liver mass be coded at this time? Why or why not?

5. Why was the liver mass not the principal diagnosis?

6. Should the liver mass be coded as malignant or benign?

7. What is the proper code for the liver mass?

8. What would you code for abdominal tenderness, diaphoresis, and fever?

9. Should any history code(s) be selected? If so, what is it?

10. Should the diabetes be coded? Why or why not?

11. If coding for diabetes, what is the proper code? Why did you select this code?

12. Should any other codes be selected? Why or why not? If there are any other codes, what are they?

13. Is code 51.22 the correct code for the procedure? Why or why not?

14. What is the correct code for lysis of adhesions for this scenario?

15. Are any other services coded? Why or why not? If so, what are their codes?

16. Should any other procedural codes be coded? Why or why not?

ACTIVITY 11.2

Emergency Room
PATIENT NAME: Terese Rawlings
DATE: 9/16/04
DOB: 10/14/77

ADMISSION DIAGNOSIS: Femoral hernia.

DISCHARGE DIAGNOSIS: Femoral hernia.

HISTORY OF PRESENT ILLNESS: Patient states that she had been feeling fine until today when she was bowling, when she suddenly experienced acute abdominal pain. She presented to the Emergency Room, where she was diagnosed with non-reducible hernia.

PAST MEDICAL HISTORY: Remarkable for cardiopathy, COPD, and essential hypertension.

FINDINGS: This femoral hernia was incarcerated. On opening the sac, the incarcerans was no longer present. The fact that it reduced indicates it probably was not infarcted bowel. The Cooper's ligament repair was performed.

DESCRIPTION OF PROCEDURE: The patient was brought to the operating room, placed in the supine position, administered a general anesthetic. The right groin was prepped and draped in the usual sterile fashion.

A curvilinear incision was made, beginning just superior to the pubic tubercle, through the skin and subcutaneous tissue, Scarpa's to external oblique. The external oblique was additionally infiltrated with 0.5% Marcaine and then opened in the direction of its fibers through the external ring. The cord was small and atrophic and it was removed.

Dissection was carried down to the posterior wall of the inguinal canal, along the Cooper's ligament. The mass was also isolated, dissected outside of the external oblique. It was opened and then transected near the fascial defect. The remainder of the sac could then be brought through the defect and was either excised or placed back toward the peritoneal cavity. The remaining fibers passing through the defect were removed. A Cooper's ligament repair was performed, coapting the conjoined tendon, first to the pubic tubercle, then with interrupted sutures of 0 Prolene to Cooper's ligament. There was a transition suture up to the per-venous adventitial tissue, back up to the inguinal ligament and two sutures placed lateral to this, between conjoined tendon and the inguinal ligament with insertion of mesh. A relaxing incision was made along the linea alba. The external oblique was then closed with a running 3-0 Vicryl. The subcutaneous closure was with 3-0 Vicryl, and the skin closure with subcuticular 4-0 Monocryl suture. Sterile dressing was applied.

Patient was returned to postanesthetic recovery (PAR) where she was doing fine and she had no problems with her asthmatic bronchitis. Sponge, needle, and instrument counts were reported correct.

1. What is the correct code(s) for the hernia?

2. Why did you select the specific code that you did in the previous question?

3. What does *non-reducible* mean?

4. What does incarcerated mean?

5. How would you code for cardiopathy and hypertension?

6. What is the correct code for COPD?

7. Which diagnosis is the principal diagnosis?

8. In the previous question, why did you select the diagnosis that you did as the principal diagnosis?

9. Is code 53.29 the correct code for the procedure?

10. What other codes should be used for the services?

11. What code should be used to code for the mesh?

12. What code did you use for the ligament repair?

ACTIVITY 11.3

Inpatient Hospital
PATIENT NAME: George Karan
DATE: 7/23/05
DOB: 1/5/59

CHIEF COMPLAINT: Chest pain, jaundice, edema, and nausea.

HISTORY OF PRESENT ILLNESS: This gentleman was admitted to the Emergency Room with recurrent episodes of severe retrosternal chest pain radiating to the back and to the right arm. This started when he was sitting in a chair watching TV. He became sick to his stomach and felt bloated. He took a nitroglycerin and the symptoms subsided. Then, about 10 minutes later, the symptoms came back as more severe, associated with some shortness of breath. He called 911 and was found to be in mild congestive heart failure. He was admitted with a combination of heart failure and angina pectoris. He was given IV Lasix and more nitroglycerin and morphine, and his symptoms decreased.

The patient has a past history of arteriosclerotic heart disease but never had myocardial infarction. He has a past history of congestive heart failure. One year ago he had an angioplasty. He has had chronic atrial fibrillation.

PHYSICAL EXAMINATION: The patient is a pleasant gentleman who looks younger than his stated age. Vital signs demonstrated blood pressure of 110/70, pulse 68, and respirations 22. HEENT showed head normocephalic and PERLA. Neck was supple with good range of motion. Fair carotid upstrokes with no bruits. Chest revealed few crackling rales bilaterally. Heart with PMI not felt. There was no heave, lift, or thrill. There is a II/VI systolic murmur along the left sternal border. Abdomen revealed no organomegaly and was soft with no masses. Extremities had good range of motion. Peripheral pulses were present. There was no peripheral edema.

LABORATORY FINDINGS: 12-lead EKG revealed atrial fibrillation, rate controlled, with nonspecific diffuse ST segment changes. Negative for cirrhosis.

The patient was diuresed. He was found to be in atrial fibrillation. Rates went from 40 to 100. He was on Cardizem, which was decreased and his Capoten was increased. On that program, heart rate stabilized. He felt better and was discharged on the same medications as admission. The Cardizem was decreased to CD 120 mg daily, Capoten was increased to 12.5 mg b.i.d. He was continued on Premarin .3 mg daily, Lasix 40 mg daily, Digoxin .125 mg daily, Slow-K one b.i.d.

and Coumadin 1 mg daily. He will be followed in the office in one week.

FINAL DIAGNOSES:
1. Congestive heart failure.
2. Arteriosclerotic heart disease.
3. Angina pectoris.
4. Atrial fibrillation
5. History of osteoporosis with compression fractures of the thoracic spine.

1. What is the principal code?

2. Should any other cardiac codes be listed? If so, what are they?

3. If you coded more cardiac codes in the previous question, why?

4. What is the correct code for the edema?

5. Are there any other conditions that should be coded?

6. Should ICD-9 procedural codes or CPT procedural codes be used to code for procedures and services provided? Why?

7. Are there any service or procedural codes? If so, what are they?

8. If there was a service listed in the previous questions, was the professional component provided? Explain.

ACTIVITY 11.4

Emergency Room Visit

PATIENT NAME: Becky Jones
DATE: 9/18/04
DOB: 9/23/95

CHIEF COMPLAINT: Left forearm and wrist pain, possible fracture.

HISTORY OF PRESENT ILLNESS: This 10-year-old child slammed a school door on her arm and has some pain over her wrist and her left hand. She has no numbness or tingling. She denies any other injury.

PAST MEDICAL HISTORY: Not significant. No past surgeries or illnesses.

MEDICATIONS: None

ALLERGIES: None

REVIEW OF SYSTEMS: Negative

PHYSICAL EXAM: Vital signs show pulse 67, respiratory rate 18, pulse oximetry 96% on room air, which is considered normal for this patient, temperature 97.2. Left wrist and forearm showed some pain down into the wrist area and the dorsum of the hand with swelling. Range of motion, vascular, sensation are intact.

EMERGENCY ROOM COURSE: She was sent for an x-ray of the left wrist, forearm, and hand with the finding of a green stick fracture of the ulnar styloid. Manipulation was used and the arm placed in a sling. The patient is to schedule a follow-up in one week.

1. What is the principal code?

2. What is the code for the services provided to the patient?

3. Is the procedure coded as open or closed? Why or why not?

4. Is the fracture coded as open or closed?

5. What is the code for application of the cast in this scenario?

ACTIVITY 11.5

Emergency Room Visit

PATIENT NAME: Ann Bannick
DATE: 1/7/05
DOB: 3/3/73

CHIEF COMPLAINT: Vomiting.

HISTORY OF PRESENT ILLNESS: This white female has a chief complaint of vomiting for three days, which revolves around post-traumatic stress syndrome. She states that she has also been feeling shaky. She has been queasy with the vomiting episodes. She has not taken her usual medications, which include Clonazepam and basically is in some distress. She cannot relate it to any dietary change, indiscretion, etc.

PAST MEDICAL HISTORY: Post-traumatic stress disorder, hysterectomy, and coophorectomy in June of last year due to trauma.

SOCIAL HISTORY: The last time I saw this patient she was here with her son, who has attention deficit/hyperactivity disorder (ADHD), and she states that he has since been taken out of her care. Patient does not smoke, but does continue to drink heavily on occasion.

ALLERGIES: Penicillin.

PHYSICAL EXAMINATION: The patient is an alert and oriented female who does not appear to be in distress. Vital signs showed O2 to be 98%. She does seem a bit anxious and is a bit shaky. HEENT is benign with normal fundi. Neck is supple without jugular venous distention. Chest shows lungs that are clear in all fields. Heart sounds are normal with no murmurs. Abdomen is soft and nontender. There is no organomegaly. She has a scar secondary to her aforementioned surgery. Extremities show no clubbing, cyanosis, or edema. Neurological shows cranial nerves II through XII intact with motor and sensory function in all four extremities.

DIAGNOSTIC DATA: CBC is benign. Liver function tests and amylase are benign as well. Basic metabolic panel shows a potassium of 3.1. Otherwise, data is within normal limits.

EMERGENCY ROOM COURSE: An IV was placed. She received 25 mg of Benadryl, which helped her nausea but she still was not feeling well. We gave her one mg of Lorazepam and the vast majority of her symptoms almost immediately settled down.

ASSESSMENT: She may well be having a bit of withdrawal from Benzodiazepine. This was discussed with the patient and she is probably in agreement with that. She was offered something for nausea so she would be able to keep her medications down and states when she gets home she will begin her Clonazepam and other medications again.

DISCHARGE DIAGNOSIS: Vomiting, possible Benzodiazepine withdrawal syndrome.

1. What is the principal code?

2. Why did you select the previous code as the principal diagnosis?

3. What E code should be used to code for the drug effect?

4. What importance does the drug scenario have regarding the coding?

ACTIVITY 11.6

Inpatient Hospital

HISTORY AND PHYSICAL EXAM
PATIENT NAME: Joanie Hunter
DATE: 5/18/04
DOB: 3/7/42

CHIEF COMPLAINT: Menorrhagia with severe cramping with pain radiating into the leg, possible dysmenorrhea.

HISTORY OF PRESENT ILLNESS: This female, G6, P7, has had a very long history of having severe and worsening dysmenorrheal, which has been completely refractory to all conservative medical care. With the patient's persistent worsening problem that is severely affecting her ability to function, with continued and worsening pain over the past two years, again completely refractory to all conservative measures, the patient has elected for a definitive procedure, i.e., a transvaginal hysterectomy.

PAST MEDICAL HISTORY: Significant for anemia, arthritis of the back, migraines, and hypertension.

PAST SURGICAL HISTORY: Significant for tubal ligation.

SOCIAL HISTORY: Negative for alcohol and tobacco.

MEDICATIONS: Iron replacement therapy, Verapamil, Clonidine, hydrochlorothiazide.

ALLERGIES: No known drug allergies.

FAMILY HISTORY: Noncontributory

REVIEW OF SYSTEMS: Reviewed and found to be noncontributory.

PHYSICAL EXAMINATION: Vital signs are stable and patient is afebrile. HEENT and neck were completely within normal limits. No thyroid is palpable. No palpable thyroid nodules. Lungs are clear to percussion and auscultation in all quadrants. Breasts are nontender without masses. Heart shows regular rate and rhythm without murmur. Back is normal. Abdomen is nontender. GU shows vault and vulva normal with a first-degree cystourethrocele, first-degree rectocele, a boggy, and approximately an 8-week size uterus. No adnexal masses felt. Rectal was normal without palpable masses or tenderness. Confirms pelvic exam. Extremities are normal without deformity. Cranial nerves II through XII grossly intact. Deep tendon reflexes are brisk and symmetrical. Motor and sensory intact and symmetrical. Skin without lesions. Lymph showed no palpable adenopathy.

IMPRESSION: Severe dysmenorrhea and some menorrhagia, completely refractory to all conservative medical care. The patient is strongly desirous of a definitive procedure.

PLAN: Transvaginal hysterectomy.

COUNSELING NOTE: The patient has been fully informed of the risks of surgery, to include, but not exclusive of, infection, bowel, bladder, ureteric or other great vessel injury that could require more surgery. She has been informed of the risk of bleeding to the point of requiring blood transfusion, with approximately 1 in 400,000 risk of AIDS, and also approximately 1 in 50 chance of hepatitis. The patient has also been informed of the risk of anesthesia. The patient appears to understand these risks and agrees to the proceed with the proposed procedure.

Operative Report

PATIENT NAME: Joanie Hunter
DATE: 5/18/04
DOB: 3/7/42

Surgeon: Greg Jones

DISCHARGE INSTRUCTIONS: 1. Regular diet. 2. Pelvic rest and no physical exertion. 3. Follow up in two weeks. 4. Vicodin as needed for pain.

DISCHARGE DIAGNOSIS: Severe dysmenorrheal refractory to conservative medical therapy.

1. What is the principal code?

2. Did the pathological results affect the coding of the principal code? Why or why not?

3. What directions does the alphabetical index provide regarding the coding of the pathological results?

4. Should any other diagnostic codes be coded? Why or why not? If yes, what are the codes?

5. Are ICD-9 or CPT codes used to code for the procedure? Why?

6. What is the correct code(s) for the procedure?

7. If you coded more than one code, why?

8. What additional operative circumstances influenced the choice of codes?

ACTIVITY 11.7

Emergency Room Visit

PATIENT NAME: Donna Vanderhoff
DATE: 1/29/04
DOB: 7/31/42

CHIEF COMPLAINT: Abdominal distention and hypoactive bowel sounds, possible peritonitis.

HISTORY OF PRESENT ILLNESS: The patient states that within the last several days she has experienced significant bloating and distention in the abdominal area, in addition to dull pain. She says that it has been continuous and does not seem to be aggravated by anything nor alleviated by any medications.

PAST MEDICAL HISTORY: Patient states that she has occasional abdominal complaints, including reflux. Patient had a PEG tube inserted two weeks ago. She called Dr. Miller's office and they told her that she needed an x-ray, that something was wrong with the stomach and not with the tube.

PHYSICAL EXAMINATION: Vital signs show hypotensive blood pressure at 98/56. Pulse is 102 and respirations are 18. Temperature is 96.6.

EMERGENCY ROOM COURSE: On arrival to the Emergency Room, I deflated the balloon of the patient's PEG tube and removed the tube, placed a Foley in place of the tube temporarily. I found the tube to be plugged. The tube was irrigated. The tube was then replaced and functioned perfectly. The family was given an opportunity to demonstrate that the tube functioned perfectly. We did hydrate the patient with 16 ounces of fluids through the PEG tube. Lidocaine jelly was used to facilitate the procedure. The patient was discharged to home.

1. What is the principal code?

2. In the previous question, how did you find the code?

3. What types of services are included in the code you selected in the previous questions?

4. What does the abbreviation PEG mean?

5. What is a PEG commonly known as?

6. Are there other diagnostic codes that should be coded? If so, what are they and why are they coded?

7. The codes for these procedures are selected from what code book?

8. What is the primary code for this visit?

9. How did you find this code in the alphabetical index?

10. Should any other procedural codes be listed? Why or why not?

11. Does the code for the procedure include a scope? Why or why not?

ANSWERS

ACTIVITY 11.1

1. What is the principal diagnosis?

ANSWER: 574.10; Note that there is a fifth digit of 0 because obstruction is mentioned.

2. Why did you select the previous code as the principal diagnosis?

ANSWER: It was the main reason for admission after evaluation and testing were completed.

3. Were there two existing concurrent conditions, either of which could have been the principal diagnosis?

ANSWER: No, code 574.10 included both the cholelithiasis and cholecystitis.

4. Should the liver mass be coded at this time? Why or why not?

ANSWER: Yes, it should be coded at this time because services were provided for it.

5. Why was the liver mass not the principal diagnosis?

ANSWER: The liver mass was not based on the reason for admission, which was the tenderness, diaphoresis, and fever, which later, after evaluation and testing, were found to be cholelithiasis and cholecystitis.

6. Should the liver mass be coded as malignant or benign?

ANSWER: Neither, the behavior was not specified within the CT scan report, so uncertain must be selected.

7. What is the proper code for the liver mass?

ANSWER: 235.3

8. What would you code for abdominal tenderness, diaphoresis, and fever?

ANSWER: These would not be coded as they are symptoms of the cholecystitis and cholelithiasis.

9. Should any history code(s) be selected? If so, what is it?

ANSWER: Yes, there is a recent history of abdominal surgery, which may be pertinent to the patient's diagnosis and care. The code is V15.2.

10. Should the diabetes be coded? Why or why not?

ANSWER: There is surgery involved in the hospital, so the diabetes should certainly be coded as it is relevant to the outcome of the patient's care and prognosis.

11. If coding for diabetes, what is the proper code? Why did you select this code?

ANSWER: 250.51 and 366.41. A fourth digit of 5 must be used, in addition to two codes because the patient has ophthalmic manifestations from the diabetes. The diabetes is insulin-dependent, so a fifth digit of 1 must be selected.

12. Should any other codes be selected? Why or why not? If there are any other codes, what are they?

ANSWER: Yes, the hypertension should be coded as 401.9. It is not specified as benign or malignant, so a fourth digit of 9 should be selected for unspecified.

13. Is code 51.22 the correct code for the procedure? Why or why not?

ANSWER: No, 51.22 is for a cholecystectomy without a scope, but a scope was used. Code 51.23 would be correct as it includes the scope.

14. What is the correct code for lysis of adhesions for this scenario?

ANSWER: Lysis of adhesions for the bladder are not coded as noted within the exclusions of code 54.5 because it is part of the global procedure for the cholecystectomy.

15. Are any other services coded? Why or why not? If so, what are their codes?

ANSWER: No, all other services are included in the procedural code.

16. Should any other procedural codes be coded? Why or why not?

ANSWER: No, all other services are included in the global procedural code.

ACTIVITY 11.2

1. What is the correct code(s) for the hernia?

ANSWER: 552.00

2. Why did you select the specific code that you did in the previous question?

ANSWER: It is incarcerated which means that it is obstructed and it should be coded as 552.00, and not 553.00.

3. What does *non-reducible* mean?

ANSWER: Non-reducible means that it cannot be placed back into the normal position.

4. What does incarcerated mean?

ANSWER: Incarcerated means strangled or constricted, such as with obstruction.

5. How would you code for the cardiopathy and hypertension?

ANSWER: 429.9 and 401.9. These would not be coded as hypertensive heart disease (402), as the physician has not stated that they are linked.

6. What is the correct code for the COPD?

ANSWER: 493.90. The COPD is not coded as 496, because it is stated in the report that the patient has asthmatic bronchitis.

7. Which diagnosis is the principal diagnosis?

ANSWER: Incarcerated femoral hernia.

8. In the previous question, why did you select the diagnosis that you did as the principal diagnosis?

ANSWER: The hernia is the main reason for the services provided.

9. Is code 53.29 the correct code for the procedure?

ANSWER: No, this is an Emergency Room visit, so CPT codes should be used.

10. What other codes should be used for the services?

ANSWER: 49553 is the correct code, the only one, not the ICD-9 procedural code.

11. What code should be used to code for the mesh?

ANSWER: Coding of mesh is only admissible for incisional or ventral, so the mesh would not be coded.

12. What code did you use for the ligament repair?

ANSWER: None, the repair is part of the hernia repair.

ACTIVITY 11.3

1. What code would be listed first?

ANSWER: 428.0

2. Should any other cardiac codes be listed? If so, what are they?

ANSWER: 414.00, 413.9, and 427.31

3. If you coded other cardiac codes in the previous question, why?

ANSWER: Although they are all cardiac-related, the codes are not commonly associated with other cardiac codes but can occur on their own; they can occur together and may affect each other, but they are not integral symptoms of each other.

4. What is the correct code for the edema?

ANSWER: The edema should not be coded, as it is part of the heart problems.

5. Are there any other conditions that should be coded?

ANSWER: Yes, the jaundice should be coded, as it is not related to the heart problem but is a consideration when treating the patient. The services provided at this visit to treat the jaundice also should be coded.

6. Should ICD-9 procedural codes or CPT procedural codes be used to code for procedures and services provided? Why?

ANSWER: ICD-9 procedural codes should be used, because services were provided in the hospital.

7. Are there any service or procedural codes? If so, what are they?

ANSWER: 93000

8. If there was a service listed in the previous questions, was the professional component provided? Explain.

ANSWER: Yes, the professional services were provided, which is indicated by *supervision and interpretation*. In the report with the lab results, results are described verifying the professional component.

ACTIVITY 11.4

1. What is the principal code?

ANSWER: 813.43

2. What is the code for the services provided to the patient?

ANSWER: 25650

3. Is the procedure coded as open or closed? Why or why not?

ANSWER: The procedure is coded as closed, because there is no mention of it being either closed or open, so it must be coded as closed.

4. Is the fracture coded as open or closed?

ANSWER: Greenstick indicates that it is a closed diagnosis.

5. What is the code for the sling in this scenario?

ANSWER: There is no code for the sling, because this is included in the global code for the fracture care.

ACTIVITY 11.5

1. What is the principal code?

ANSWER: 787.03

2. Why did you select the previous code as the principal diagnosis?

ANSWER: The only code that is valid is the symptom of vomiting since the drug effect was described as possible.

3. What E code should be used to code for the drug effect?

ANSWER: None, the drug is not listed.

4. What importance does the drug scenario have regarding the coding?

ANSWER: The benzodiazepine is described as possible and may be, so it is not coded since this is an Emergency Room and not the hospital. Remember that conditional statements, such as possible, are not coded with physician or outpatient coding, but they are with hospital coding.

ACTIVITY 11.6

1. What is the principal code?

ANSWER: 625.3

2. Did the pathological results affect the coding of the principal code? Why or why not?

ANSWER: No; The metaplasia of the cervix should not be coded.

3. What directions does the alphabetical index provide regarding the coding of the pathological results?

ANSWER: Under metaplasia, the alphabetical index instructs to either omit the code for the cervix or to see condition for the squamous metaplasia, which are derived from the pathological report.

4. Should any other diagnostic codes be coded? Why or why not? If yes, what are the codes?

ANSWER: 401.9; Hypertension should be coded because it is stated as a condition and it can certainly affect the management of the patient's care.

5. Are ICD-9 or CPT codes used to code for the procedure? Why?

ANSWER: ICD-9 codes are used because the services were provided as inpatient.

6. What is the correct code(s) for the procedure?

ANSWER: 68.59 and 65.63

7. If you coded more than one code, why?

ANSWER: Yes, code 68.59 instructed the coder to code the removal of the ovaries and tubes.

8. What additional operative circumstances influenced the choice of codes?

ANSWER: Both ovaries and tubes were removed and a scope was used as indicated within the operative report.

ACTIVITY 11.7

1. What is the principal code?

ANSWER: V55.4

2. In the previous question, how did you find the code?

ANSWER: V55.4 is for *attention to artificial openings* and has a category for *other artificial openings of digestive tract*, which lists a PEG.

3. What types of services are included in the code you selected in the previous questions?

ANSWER: Adjustment, closure, passage of sounds, reforming, removal, replacement, or cleansing.

4. What does the abbreviation PEG mean?

ANSWER: Percutaneous endoscopic gastrostomy.

5. What is a PEG commonly known as?

ANSWER: Feeding tube

6. Are there other diagnostic codes that should be coded? If so, what are they and why are they coded?

ANSWER: The hypotension should be coded, as indicated in the report, with code 458.9.

7. The codes for these procedures are selected from what code book?

ANSWER: From the CPT book, as services were provided in the Emergency Room.

8. What is the primary code for this visit?

ANSWER: 43760

9. How did you find this code in the alphabetical index?

ANSWER: Under change of gastrotomy tube.

10. Should any other procedural codes be listed? Why or why not?

ANSWER: No, there are no directions or statements to code anything else.

11. Does the code for the procedure include a scope? Why or why not?

ANSWER: No, it does not include a scope, because it is not described in the report.

Chapter 12 REIMBURSEMENT

Objectives

(1) Determine proper rules and regulations for reimbursement.

(2) Understand insurance billing terminology and applications, particularly in reference to Medicare and Medicaid.

(3) Be able to complete and interpret CMS-1500 and CMS-1450 billing forms to produce a clean claim.

(4) Understand how reimbursement and coding intertwine.

(5) Be able to audit coding and billing scenarios and make appropriate corrections.

Key Terms

CMS-1450 form—Generic billing form created by the Centers for Medicare and Medicaid Services that is accepted for reimbursement for facilities.

CMS-1500 form—Generic billing form created by the Centers for Medicare and Medicaid Services that is accepted by most payers of medical services, including the government.

Clean claim—A billing form that is acceptable for payment which is completed properly and codes are used correctly.

Downcoding—The undesirable selection of codes that are lower in value and in criteria than services provided.

Linkage—Joining ICD-9-CM codes with CPT/HCPCS codes through referencing to demonstrate medical necessity for services.

Superbill—An office form on which ICD-9-CM, CPT, and HCPCS codes and other information are listed for selection by provider to report the types of service provided. These forms are then entered into an office database and processed for reimbursement.

Upcoding—The undesirable selection of codes that are higher in value and criteria than the documentation justifies, which may result in higher reimbursement, as well as possible fraud charges.

REIMBURSEMENT WITHIN CODING

Coding culminates in the reimbursement billing process. Because of this link, coders play an important role in ensuring that facilities receive fair and just reimbursement for services, which ensures that patients receive proper care. Reimbursement enables coders to demonstrate their importance to facilities by communicating with facilities about the steps that must be taken for proper reimbursement. Proper report formatting and demonstrating medical necessity are two areas that coders oversee and audit to ensure proper reimbursement.

The national coding exams focus on basic coding principles and do not concentrate on the idiosyncrasies of any one insurance company or program.

Not documented, not done. Not signed, not done. That is absolute, whether on reports or on the billing form, it must have been signed and signed legibly!

Health Care Financing Administration (HCFA) is now known as Centers for Medicare and Medicaid Services (CMS). Their forms, the **CMS-1500** and the **CMS-1450** (UB-92), are used primarily for billing for physician services and hospital services, respectively. These forms are included on the national exams. Therefore, it is important that you understand the basic billing practices on both forms and know when each form is used for hospital billing.

INSURANCES

For the hospital coding exams, it is important to know extensive billing information, such as types of insurances and groupers as well as other medical information.

Know your insurance terminology! A *beneficiary* is the person receiving the medical care. If a child has health insurance coverage by both parents' policies, then the primary insurance is considered the policy belonging to the parent who has the earliest birthday within the year (not by birth year). This is known as the Birthday Rule. A *third-party payer* is the insurance or government agency responsible for payment. Medicare is the largest third-party payer. *Guarantor* is the person responsible for payment of health care services, other than the third-party payer. The guarantor is usually the person who is the insurance policy holder, such as a parent. *Coinsurance* is the dollar amount for medical services not covered by insurance. Coinsurance, usually 20 percent with insurance often covering the 80 percent, must be paid by the patient. A *deductible* is the dollar amount that a patient pays for health care, usually on an annual basis. Patients must pay their insurance deductible before the insurance policy will pay for any medical costs. After the deductible is reached, then insurance coverage would apply. For example, a patient may have a $500 deductible, which the patient must pay first. After the deductible is paid, medical charges that are covered by the insurance company will be determined and paid, usually by the 80/20 split between the insurance company and the patient. Deductibles are tracked by the insurance company, since individual health care providers do not know what charges have been incurred by other facilities or physicians, and therefore, would not have access to a patient's deductible information.

Copayments are standard charges that patients must pay for each visit, e.g., $25. *Premiums* are the dollar amount paid on a regular basis to obtain and maintain insurance coverage. *Providers* are the persons or agencies providing medical care, such as physicians. Providers sign a participating provider agreement (PAR) with an insurance company or government program that states the conditions for their providing medical services to the insureds. *Accepting assignment* is a contracted agreement in which the provider agrees to accept payment in full from the insurance company or government agency and will not charge the patient additional costs.

CORRECT CODING INITIATIVE (CCI)

In 1996, CMS (formerly known as HCFA) instituted the National Correct Coding Initiative to assist in proper coding and billing of health care services, including issues regarding bundling. The CCI establishes rules for the correct methods to combine codes. CCI edits inform the coder when combinations are not acceptable, such as the coding for an episiotomy when charging for vaginal delivery, which would be included in the cost of the vaginal delivery (global). The CCI means that you must be current and know the field of coding and its applications through study programs, seminars, research materials, publications, etc. CCIs are issued on a quarterly basis (January 1–March 31; April 1–June 30; July 1–September 30; and October 1–December 31).

UNIFORM HOSPITAL DISCHARGE DATA SET (UHDDS)

Uniform hospital discharge data set (UHDDS) was originally developed by several national organizations, but later adopted by the federal government as the standard documentation requirements. UHDDS requires that certain information be maintained on each patient. This includes ethnicity, sex, payer, identification number, admission and discharge dates, physician identification numbers, discharge disposition of patient, diagnoses, and procedures.

TYPES OF INSURANCE GROUPS

GENERAL INSURANCE

Preferred Provider Organization (PPO) is a group of physicians and/or other medical care providers who agree to provide services for a group of patients at discounted rates. Health Maintenance Organizations (HMOs) are organizations in which a majority of services are provided by the same medical group of physicians and health care providers. A primary care doctor is the gatekeeper who directs patients to other health care providers as necessary. HMOs can be either a Staff Model or an Individual Practice association (IPA) model. In the Staff Model, the HMO employs the physician or health care provider. In the IPA, the physician provides services for a contracted fee from the organization. In HMOs, fees usually are paid on a regular basis, such as monthly, with the patient being able to access the physician or provider when needed. If proper medical care cannot be provided by the contracted physicians or providers, the patient can be referred outside of the organization, but usually the cost to the patient will be greater and requires prior authorization from the primary care physician and the organization's approval board.

With these types of insurance groups, coverage usually is provided when the insured pays a set dollar amount each month. Some groups may pay physicians a fixed fee for service whereas others groups may have contracts with the physicians who receive a set amount of money each month from the group for providing health care as needed. This is known as *capitation*. With many types of insurance coverage, the health care provider (e.g., physician) must obtain preauthorization or approved referral before providing services to ensure that services are covered. Without this preauthorization or referral, the health care provider risks failure of reimbursement.

TRICARE is the health insurance plan for military personnel and their families, which was formerly known as CHAMPUS (Civilian Health and Medical Program of the Uniformed Services). Bills must be submitted to the contractor in the area where services are provided, not the military base from which the insured originally enlisted. Services must be obtained first from a military treatment facility, if available. CHAMPVA (Civilian Health and Medical Program of the Veterans Administration) provides health insurance coverage to veterans, including disabled veterans, and their families (including families of soldiers killed in action). CHAMPVA usually is the secondary payer, with other health insurance policies paying first, except for Medicaid.

MEDICARE

Know Medicare issues because many other payers adopt similar policies and procedures.

The Department of Health and Human Resources (DHHS) is responsible for the administration of the Medicare program. They contract with fiscal intermediaries, such as insurance companies, for billing services. These insurance companies interact with patients and physicians. Medicare provides services for people 65-years and older and to people with disabilities (including the blind), and those with kidney failure.

Part A of Medicare coverage is for hospital services (which would be coded using ICD-9-CM and Diagnosis-Related Groups [DRGs]), and Part B is for physician services and durable medical equipment (DME), which are not covered under Part A. While coverage for Part A is automatic for eligible persons, coverage for Part B is subscription-based and requires a monthly premium. Part B Medicare covers 80 percent of covered costs. Patients must pay the other 20 percent. Remember—there are deductibles for Part A and B services.

Covered services are determined primarily by medical necessity and include all related services, such as hospital room and board, supplies, and labs. Medically necessary is defined as services that are appropriate to the patient's diagnoses and are not elective or experimental. Part A includes medically necessary services, including semiprivate hospital rooms, care in a Medicare-certified skilled nursing facility (SNF), hospice, and eligible home health care dependent upon physician approval as medically necessary (which would be coded using ICD-9-CM, CPT, and HCPCS codes). Coverage for Part A hospitalization is provided for the first 60 days, which includes a deductible. There is an additional 30 days, known as coinsurance days, in which the patient is responsible for paying the coinsurance amount. Then there are an additional 60 days available, known as lifetime reserve, which are usable after a patient has been in the hospital for more than the 90 days of the combined regular inpatient care and coinsurance days.

For SNF visits, there is a coinsurance payment for days 21 through 100. Medicare considers the spell of illness to consist of 60 consecutive days after the patient is dismissed from the facility. Any services incurred during this 60-day period are considered to be part of the previous visit and additional deductibles are not collected, even if the services are provided for another reason.

Part B includes physician services, laboratory, therapies, outpatient mental health care, second opinions, home health care, outpatient hospital services (ambulatory surgery centers [ASCs]), DME, ambulance services, Pap smears, mammograms, prostate screenings, and vaccines for pneumonia, hepatitis B and influenza. Blood components are allowable under Part A and B, however, the patient is responsible for the first three pints. Patients can use replacements for the first three pints, provided either by themselves or by someone else on their behalf. If they provide their own replacement, they will not be charged for the first three pints.

Medicare does not pay for medically unnecessary services or routine preventive health services (except mammograms, Pap smears, screenings for prostate cancer, and vaccines for pneumococcal pneumonia, hepatitis B, and influenza), routine dental, eye or ear exams, most foot care, and cosmetic surgery. Advance Beneficiary Notices (ABNs) should be signed by patients before treatment is provided that may not be covered by Medicare. If the proper signatures on the proper forms are not obtained before services are provided, the care provider may not be reimbursed. For this reason, preauthorization is important and required by most payers in certain circumstances.

Providers can participate in the Medicare program or not participate in the program. There are distinct differences between participating and nonparticipating providers. For PAR, payment is sent to the provider directly, but not to nonparticipating providers. There is an extra 5 percent reimbursement to participating providers for services. In addition, claims are processed faster and patients are advised of the benefits associated with using participating providers. For nonparticipating physicians (non-PAR), there is a limiting charge that sets a limitation on the fees that PARs can charge. Non-PARs are allowed to charge above a certain percentage (i.e., 15 percent in 2004) of the Medicare allowable charge.

Medigap insurance is insurance provided by private carriers for a fee that provides reimbursement for costs not paid by Medicare, such as deductibles, uncovered services, and coinsurance. Medigap insurance is available only to patients who have the original Medicare coverage and is not necessary for some policies, such as managed care plans.

Sometimes Medicare is the secondary payer (MSP), such as when an eligible person is employed and has health insurance through their current employer or their spouse's employer (who has more than 20 employees), has disability insurance through an employer (who has more than 100 employees), or has coverage through another insurance, such as worker's compensation or automobile. In addition, Medicare is a secondary payer for veterans, patients with black lung disease, and patients with end-stage renal disease (ESRD). Medicare also may pay as the secondary payer if services are denied or partially paid by the primary payer. In the case where there are multiple payers with Medicare as a secondary payer, the primary payer will forward the bill to the secondary payer after the primary payer has paid or denied its portion of the bill. Proper filing of insurance claims to multiple payers is known as *coordination of benefits* (COB).

The *Federal Register* is the publication of federal rules and regulations and is a good source for information on what is happening in the medical field concerning Medicare issues.

MEDICAID

Medicaid is a government program established under the Social Security Act that requires state governments to administer health insurance programs to needy individuals. Eligibility requirements are determined by federal and state regulations. The program is funded by both state and federal governments but is administered by the state, therefore, there are some differences in how programs are administered differently according to individual state regulations, and the coverage provided is dependent on each state. Federal funding agencies require a wide variety of services to be provided, including inpatient and outpatient hospital services, family planning services, Emergency Room services, medically necessary physician services, prenatal care, home health and SNF care. Under Medicaid programs, children are eligible for preventive health care, known as *Early and Periodic Screening, Diagnosis, and Treatment* (EPSDT), which includes regular checkups and immunizations. Medicaid is known as the payer of last resort. If there is any other health insurance coverage, it will be charged first, and Medicaid charged last.

Some patients may be eligible for both Medicare and Medicaid benefits, which is known as *Medi-Medi*. Claims for these patients are sent first to Medicare, and then to Medicaid. Some Medicaid programs pay for Medicare deductibles for Part A services and premiums for Part B. Some Medicaid coverage also may pay for services that are not payable under Medicare.

ACTIVITY 12.1

MATCHING

A. UHDDS

B. Insured

C. MSP

D. Coinsurance

E. Beneficiary

F. PAR

G. Staff Model

H. Third-party payer

I. Capitation

J. HMO

K. Assignment

L. TRICARE

M. Guarantor

N. Copayment

O. Advanced Beneficiary Notice

P. Medicare Part B

Q. Deductible

R. Coordination of Benefits

S. PPO

T. IPA

1. Select organization of medical care providers who provide services based on standardized and regularly scheduled payments

2. Person who is eligible for benefits

3. Standard fee, for which insured is responsible before services are rendered

4. Medical care provider who has chosen to participate with Medicare in providing services

5. Fixed monetary amount paid by insurer on a regular basis for each patient

6. Insurer

7. Formerly known as CHAMPUS

8. When a payer other than Medicare is the primary payer

9. Medicare payments for physician services and DMEs

10. Individual medical care providers

11. Responsible party for the policy

12. Process and order by which claims are reimbursed by multiple payers

13. Person responsible for payment

14. Signed statement verifying patient's acknowledgement that claim may not be paid by third-party payer

15. Select organization of medical providers providing medical services based initially on standard prepayments and in which the providers are employees

16. Specified amount for which insured is responsible before coverage begins

17. Defined set of patient information for hospitals

18. Specified amount to be paid by insured for services

19. Group of affiliated medical care providers who provide services to members

20. Agreement to accept payment in full for medical services that are billed according to specific allowed charges

PROSPECTIVE PAYMENT SYSTEMS

The government created Prospective Payment Systems (PPS) to help establish regulated payment systems for health care. DRGs are for inpatient hospital payments and Ambulatory Payment Classifications (APCs) are outpatient hospital payments. Health insurance prospective payment system (HIPPS) is for services provided at SNFs. APCs took root in 1986 when Congress mandated the OPPS (Outpatient Prospective Payment System). OPPS involves predetermined amounts that are paid by the selected APC or DRG code. DRGs are based on the principal diagnosis assigned to the patient for an inpatient visit. APCs are based on the primary HCPCS code (which includes CPTs). Basic unit of payment for a DRG is admission/discharge; for APC it is the visit for each day. These methods are meant to simplify the billing process, limit fraud and abuse, and ensure proper medical care for patients.

DIAGNOSIS-RELATED GROUPS

DRGs are the methods used to code for inpatient hospital services. DRGs are based on the facility's case mix and correlate to the hospital resources consumed. As stated earlier, DRGs are based on a single primary ICD-9-CM diagnosis for admission; the basic unit of payment is admission/discharge. It must be reiterated that the DRG selected is based on the principal diagnosis, not other conditions that may develop and be treated while a patient is in the hospital (DRGs 468, 476, and 477). Contained within each DRG code may be numerous ICD-9-CM codes. However, some ICD-9-CM codes can be coded additionally, as they may contribute to the need for increased medical care (e.g., diabetes, chronic obstructive pulmonary disease [COPD], cellulitis, cardiomyopathy, renal failure, and urinary tract infection). Other factors also may affect the DRG assignment, such as age, sex, birth weight, discharge status or complications/comorbidities.

Selecting appropriate DRGs based on selected ICD-9-CM codes generally is performed by computerized coding systems, known as *groupers*. It is crucial that the correct ICD-9-CM code is selected as the primary code for admission, as it can make a significant difference in reimbursement amounts.

There are almost 500 DRGs, which are divided into categories called *major diagnostic categories* (MDCs). There were 25 MDCs in 2003. These categories are based on body systems, in addition to some general categories for conditions that may affect numerous systems, such as infections. The DRG is based on the principal diagnosis and comorbidities and complications. There are also DRGs for diagnoses that are not well specified (469) or inaccurate (470) and, therefore, cannot be placed into the proper DRG category.

An important concept to understand with DRGs is the hospital's case mix upon which reimbursement is based. Case mix is based on a variety of factors that influence the types of patients (cases) that the hospital is serving, which ultimately influences reimbursement. Reimbursement for DRG is based on the DRG weight and facility rate. These factors include the chosen DRG, patient's age, and exacerbating conditions (such as burns, myocardial infarction, newborn), volume of same types of cases, hospital costs, patient types, types of cases, discharge status, and geographic area. It should be noted that the case mix is not a measure of the difficulty of patients' medical conditions or cases that a hospital has, but is rather a reimbursement issue, namely what types of costs are associated with particular services.

It must be remembered that services may require costs or time that are greater than the typical costs or time for a specific procedure. These services are known as *outliers* and additional charges can be made for them with the proper documentation that documents and demonstrates the increase in services.

AMBULATORY PAYMENT CLASSIFICATIONS

APCs are similar to DRGs in that they both use the concept of groupers, but APCs are used to bill for outpatient hospital services and visits, such as ASCs, (but not physician services) and Emergency Room visits, whereas DRGs are used to code for inpatient hospital services. There are about 450 APC groups, which include clinic and Emergency Room visits, surgical procedures, diagnostic procedures, supplies, and preventative services. ICD-9-CM, CPT, and HCPCS are used to code for APCs, in addition to patient's age, sex, and other factors. APCs are based on groupings of similar HCPCS (including CPT codes). For example, ophthalmological services are contained in one APC group.

Multiple groups may be selected with APCs as payment is made by each line item on the claim, which is in contrast to DRGs, which are based on a single primary code. APCs are based on packages, including nursing services, anesthesia, certain drugs, operating room, and associated supplies. Services excluded from APC reimbursement include inpatient, prosthetics, ESRD dialysis services, clinical laboratory, screening mammography, DME, ambulance services, therapies, and outpatient services at SNFs, all of which would have to be billed additionally. Institutions that use APCs include hospitals, SNFs, home health agencies, community mental health centers, hospice programs, and comprehensive outpatient rehab facilities.

APCs are a fixed payment for services provided in hospital outpatient settings, as previously described. The fixed amount for APCs are based on facility rate, APC weight, number of units, and discount. Remember, CPTs are based on Resource Based Relative Value Scale (RBRVS) as determined by the relative value unit of the service, geographical location, and a conversion factor. This provides the hospital with the ability to, for instance, focus on improving management practices, thereby reducing costs and waste. By doing this, the hospital can begin to make a profit or to increase profits. Whereas in the past, hospitals often were not reimbursed fully for their costs, with APCs they can now better manage their medical services.

Policies similar to both APCs and physician practices include the concept of global packaging and reimbursement discounting for multiple procedures or discontinuation of a service. With APC discounting, the highest APC service is paid at the normal rate with each additional service paid at 50 percent, if a payment status indicator *T* is applied, indicating multiple procedures (discussed later in this book). Global packaging, in which one procedure is coded that encompasses many different services that are not billed separately, is a component of APCs. APC global services include surgical trays, operating room, related supplies, IVs, preparation for surgery, casts/splints, venipuncture, and drugs, as well as the visit on the same day. APCs do not include preoperative and postoperative services, e.g., evaluation and management (E/M) visits, because patients are seen for immediate services and treatment since this is a hospital-based facility and there are no long-term patients.

There are a number of payment status indicators associated with each HCPCS code within APCs, which indicates how the code is paid. It is important to understand the payment status indicators for they are similar to CPT modifiers. The following is a list of the indicators and their meanings:

- (A) indicates that the services are paid under another method of payment.
- (C) indicates that Medicare will not pay for the service as outpatient but only as an inpatient for safety reasons based on the invasiveness of the service, physical condition of the patient, and the need for postoperative care, which must be at least 24 hours in length.
- (D) indicates the code has been deleted.
- (E) indicates that payment is not allowed under the hospital outpatient payment system.

(F) indicates separate payment for corneal tissue acquisition.
(G) indicates payment is transitional pass-through for a drug.
(H) is used for transitional pass-through for devices.
(K) is not for transitional pass-through, but does indicate the drug will be paid separately.
(N) indicates services that are incidental, so the cost is included in the global package.
(P) indicates that these codes will be only paid if incurred during partial hospitalization.
(S) demonstrates that the code is for a significant procedure, which is paid at full rate of reimbursement and not reduced as a multiple procedure.
(T) indicates that the service is reimbursed only if incurred as hospital outpatient, but the reduction in reimbursement for multiple procedures does apply.
(V) indicates that reimbursement is for services, which are incurred in a clinic or emergency departments.
(X) indicates ancillary services, for which payment is allowed under the payment system.

Because of new technology, many of the procedures that were once considered inpatient only have been changed to outpatient with these types of changeovers continuing as technology grows.

OUTPATIENT CODE EDITOR

Outpatient code editors (OCEs) are computerized systems that process code assignment for groupers, i.e., APC and DRG. They provide information on how clean the claim is regarding data and they assist with proper grouping and assignment of groupers. OCEs provide information on denial or rejection of claims. Errors can include conflicts with codes and age or sex or invalid codes.

AMBULATORY SURGERY CENTERS

With advancements in the medical field, hospitals now offer many services in an outpatient setting, so patients no longer have extended stays in the hospital. These centers can be freestanding or part of the hospital. Freestanding indicates that the center is independent of the hospital with separate tax identification numbers. However, like a hospital, ASCs bill for their services and not the medical care provider (physicians bill for their own services), unless the medical care provider is an employee of the center. ASCs bill in the same manner as outpatient hospital departments, using revenue codes and HCPCS.

For ASC billing purposes, there are two lists—one list consists of surgeries that can be performed at an ASC, and the other list is for surgeries that cannot be performed at ASCs. Each list contains nine groups, with each group receiving the same rate of payment with a geographic adjustment. (Geographic adjustment means that more or less will be charged depending on where the services are provided within the country. So, for example, more might be charged for services in New York than in Omaha because of the differences in cost-of-living expenses). However, if the ASC is attached to the hospital, then hospital costs also determine the payment. For surgeries not approved for performance at ASCs, only hospital costs are considered for determining payment.

If a procedure is terminated for medical reasons after it has begun, but before anesthesia, then ASC services will be paid at 50 percent. If the procedure is terminated for nonmedical reasons after it has begun, but before anesthesia, then there is no payment. If the procedure is terminated after anesthesia, then ASC charges are paid at full rate.

The surgeries on the ASC approved lists are global, meaning that charges that are considered to be a normal part of a service are included in the primary code and would not be billed separately, similar to physician coding. This is not, however, applicable to hospital coding in which services are charged unless the coding books state otherwise. Services normally included in a global payment at a freestanding ASC include general supplies, anesthesia, recovery room, diagnostic services, and simple laboratory tests. Services that are not considered part of the global service provided by the hospital and which can, therefore, be billed separately include professional fees, ambulance services, prosthetic devices, and DME.

There are temporary APC codes for transitional pass-through items/devices and new technology, which can be used for no more than three years. Within those three years, new APCs must be created and calculated for these services. The purpose of having temporary codes is to allow for billing of items that are not listed appropriately in the current APC list. For transitional pass-throughs, drug and device reimbursement as provided for within the APC system is inadequate. The pass-through payment allows additional billing charges for a calculated difference between the amount that would be paid under the APC system and the remaining additional cost. Remember, as with HCPCS coding, if the units administered are greater than the amount listed in the HCPCS book, then additional units must be charged to ensure coding of the entire amount of drugs or devices administered to the patient. Materials and supplies typically associated with a procedure, which are not coded additionally, cannot be considered transitional pass-throughs. A device that is considered a transitional pass-through must be approved by the Food and Drug Administration and considered medically necessary.

New technology differs from transitional pass-throughs in that there may not be a coding group for them. In these cases, new APC special coding groups are created, which are based on clinical documentation and monetary midrange value. For example, if a new technology costs $400 to provide, it is placed in a new APC category, which may have values ranging from $200 to $500. Similar to transitional pass-throughs, additional reimbursement can be paid for outliers, which are services whose costs exceed by more than 2.5 the amount typically reimbursed.

Partial hospitalization services are covered under APCs. Partial hospitalizations are an alternative to inpatient care and are charged per day. The APC charges are for the facility's charges and not the professional charges, such as charges for psychologists or physician assistants, who would bill for their own services.

For Medicare, observation services can now be charged with APCs, if they meet certain criteria for patients with chest pain, asthma, or congestive heart failure. Criteria include peak expiratory flow rate or pulse oximetry for asthma. For chest pain, at least two sets of CPK, troponin, or cardiac enzymes in addition to two sequential electrocardiograms are required criteria for chest pain. A chest x-ray and electrocardiogram with pulse oximetry are performed each time for an observation visit when diagnosing congestive heart failure and should be charged accordingly. Other visits, including Emergency Room or critical care, can be charged on the same day or the day before the observation visits. There is a minimum of 8 hours and a maximum of 48 hours allowed for observation services, which are billed per hour. There are three levels of E/M codes in APCs—low (level 1 and 2 of the CPT codes), mid (level 3), and high (level 4 and 5). APC facilities can use their own method for determining what level of care was provided for these services. This includes classifying current E/M codes into the three levels of low, middle, and high, instead of distinct separate codes.

ACTVITIY 12.2

1. Which of the following is not a PPS?
   A. DRG
   B. MDS
   C. APC
   D. HIPPS

2. Which of the following is true about APCs?
   A. Basic unit of payment is admission/discharge.
   B. Based on a single primary code.
   C. Fixed amount is based on facility rate, APC weight, number of units, and discount.
   D. Fixed amount is based on relative value units, geographic area, facility rate.

3. A zero claim means:
   A. Discount provided.
   B. Claim is dirty.
   C. No claim is filed.
   D. No payment is expected.

4. In APCs, what is an outlier?
   A. Additional services that are not part of the global package.
   B. Services whose costs exceed by more than 2.5 the amount typically reimbursed.
   C. Facility costs that are too high or too low.
   D. Costs associated with overcoding.

5. Which of the following are not true about Medicare payment for transitional pass-throughs?
   A. Monetary adjustment to ensure proper payment.
   B. Temporary codes.
   C. Can be applied to all materials and supplies.
   D. Can only be applied to medically necessary procedures.

6. Which of the following is a description for APC payment status indicator C?
   A. Payment is transitional pass-through for a drug.
   B. Services are incidental so the cost is included in the global package.
   C. Payment is not allowed under the hospital outpatient payment system.
   D. Medicare will not pay for the service as outpatient, only as an inpatient.

7. Which of the following are not considered for determining reimbursement for DRGs?
   A. Relative value factor.
   B. Patient's age.
   C. Exacerbating conditions.
   D. Types of cases.

8. What factor would not be considered in the determination of DRGs for a facility?

   A. Patients.

   B. Case mix.

   C. Facility resources.

   D. Number of physicians in the facility.

Use this information to answer the following questions:

MDC 1: Nervous System
MDC 4: Respiratory System
MDC 5: Circulatory System
MDC 6: Digestive System
MDC 8: Musculoskeletal System
DRG 107: Coronary bypass with cardiac catheterization
DRG 127: Heart failure and shock
DRG 139: Cardiac arrhythmia and conduction disorder
DRG 517: Percutaneous cardiovascular procedure with coronary artery stent
DRG 154: Stomach esophageal and duodenal procedures
DRG 161: Inguinal and femoral hernia procedures

9. Under what MDC would you locate the DRG code for right heart catheterization?

10. What is the correct DRG for inguinal hernia repair with graft?

11. What is the correct DRG for vagotomy?

12. What is the correct MDC for an esophagogastroplasty?

13. What is the correct DRG for a coronary arteriogram?

14. What is the correct DRG for an ablation of a lesion of the heart?

Use this information to answer the following questions:

CPT code	11200	11600	13121	16010	25530
APC code	0013	0019	0025	0016	0044
Weight	1.51	4.56	3.71	3.31	2.73

15. Using descriptive terms (not codes) to identity the procedure, what procedure would cost the most? Explain.

16. What is the APC code for treatment of burns?

17. Using descriptive terms (not codes) to identity the procedure, what procedure would cost the least? Explain.

Use this information to answer the following questions:

CPT code	32400	35473	774210	76604	93510
APC code	0005	0081	0276	0266	0080
Weight	6.71	22.04	1.63	1.67	32.2

18. If the reimbursement rate is 50.30 per unit, what is the cost for the procedure for repair of the arterial blockage?

19. Which APC code classification pays the highest? Explain.

20. Under which APC classification would a needle biopsy of the chest be listed?

BILLING FORMS

Bad habits are born of superbills. **Superbills** are the forms used in offices to simplify the coding and billing process by allowing a list of codes to be selected. Superbills make lazy habits. They do not ensure that proper justification is included in the files nor do they account for services that are not listed on the superbill. Superbills also may not be regularly audited and changed as services change and codes change. They also do not guarantee that follow up is provided, so that staff and physicians are ensured of staying knowledgeable and educated about proper code selection. The problem is not always the superbill, but the failure of staff or physicians to provide proper documentation. Superbills and documentation must be audited and analyzed regularly by a qualified coder or manager. Superbills are only as good as the coder or coding manager who audits them.

CLEAN CLAIM

It is critical that coding and billing are properly executed to ensure prompt reimbursement. A **clean claim** means there are no errors on the bill and it can be processed immediately. The following discussions will examine the types of errors that may occur on the 1500 billing form and how they can be corrected. Computer software can be used to check claims before submitting them to ensure that they are properly completed.

CODING WITH THE CMS-1500 FORM

The CMS-1500 form sometimes is used in conjunction with the CMS-1450 for billing professional and facility services. However, for most outpatient services, the CMS-1450 (UB-92) form is used. For auditing and coding purposes, we will examine only the bottom half of the CMS-1500 form.

Blocks 17 and 17A must be filled out if another physician referred a patient. This part of the form is used when a patient is referred by a physician for services offered by another physician. However, when a patient is referred to another physician for a consultation, blocks 17 and 17A must be completed. Be sure to include the referring physician's tax identification number (UPIN).

Block 18 must be filled out if the patient is in the hospital for the services charged. This box must link to block 24B, which indicates the place of service. If a patient is in the hospital, 21 must be entered in block 24b to indicate that services were provided in the hospital.

Block 21 is used to list ICD-9-CM codes with up to four codes allowed. If there are more diagnoses listed than four, you must select the four most important and relevant codes. Remember, some codes will indicate that other codes must be coded additionally; this means if you list the one code, you must list the other required code. This takes up two lines, which can be limiting if you have many diagnoses. But a required code must be listed in addition to the code that requires it.

ICD-9-CM codes then must be linked properly to CPT/HCPCS codes as listed in block 24D. This **linkage** is crucial in proving medical necessity and ensuring prompt reimbursement. The linkage of ICD-9-CMs and CPTs in blocks 21 through 24 justifies the services charged. Block 24E is used to denote this linkage. Line numbers from block 21 are used in block 24E to link the ICD-9-CMs and CPTs. The line number of the relevant ICD-9-CM code is selected and then applied in block 24E, which corresponds with the service provided. Do not list the codes themselves in block 24E, only line numbers as indicated in block 21. Several line numbers may be used to demonstrate multiple links, such as a patient who has diabetes and a laceration, both of which are relative to the care of the wound; therefore, the line numbers for both codes can be listed in block 24E.

Block 24 includes several different blocks, i.e., date(s) of services, place of service, type of service, CPT/HCPCS code, diagnosis code, charges and units within block 24. Place of service uses specifically designated numbers: 11 is used to indicate an office visit; 21 for inpatient hospital; 12 is the patient's home; 22 an outpatient hospital; 23 the Emergency Room; 31 a SNF; 34 is hospice. Not all carriers use type of service, which can include: 1 for medical care, 2 for surgical care, 3 for consultation, 4 for x-rays, 5 for laboratory, 6 for radiation therapy, and 7 for anesthesia.

The CPT/HCPCS codes listed in block 24D are required. Up to six CPT/HCPCS codes may be included on this form. Remember, both ICD-9-CMs and CPT/HCPCS must be listed in order to be reimbursed. You cannot charge for services without a code, even when results are normal and there is no chief complaint. Modifiers must be included after the code, as indicated on the form. Multiple modifiers may be applied. Do not unbundle CPT/HCPCS codes by listing codes that are included in the services of another already listed code. Check CCI edits.

Do not forget to include charges and units in blocks 24F and G. Charges are charged by the unit and then totaled in block 28. Do not charge for additional units if a code does not permit coding for each unit separately and additionally. Some codes will include charges for a multitude of units, such as skin tag code 11200, which includes up to 15 tags.

The tax identification number, whether a social security number (SSN) or federal employer identification number (EIN), must be included to receive reimbursement. The SSN or EIN box must be checked.

In block 27, the Accept Assignment box should be checked.

Total charges, amount paid, and balance due should be completed as necessary.

Signature and date must be included in block 31 for reimbursement, as well as address and phone number in block 33.

Billing errors, whether purposeful or not, may be interpreted as fraud and/or abuse with accompanying fines and possible imprisonment. Coders are a good defense for medical practices against such punishments if coders apply their expertise and knowledge. A coder is essential in ensuring prompt reimbursement. Coders also can provide follow up for unpaid claims by investigating why claims were rejected and not paid.

CODING WITH THE CMS-1450 (UB-92) FORM

Hospitals charge for their inpatient and outpatient services using the CMS-1450 (UB-92) form. Other facilities that use the CMS-1450 form include rehabilitation centers and laboratory services. As with the 1500 form, the 1450 is often completed electronically. The electronic completion of CMS-1450 forms usually is completed through a technological tool known as the Chargemaster. The Chargemaster is a computerized program that contains coding and billing information, such as codes and their associated costs. An additional technological tool that also may be used is a scrubber that evaluates claims and determines if they are properly completed. A properly complete claim is known as a *clean claim*. An example of this type of technology is the OCE, which Medicare uses to review claims. These types of technology search for errors such as unbundling global codes and inadvertent use of mutually exclusive codes, which cannot be used together as described in the coding manuals.

Because of the size of hospitals, coders usually perform only coding and do not interact with the billing system, as coders in a physician's office might. However, it is still important, similar to conditions with the CMS-1500 form for physician billing, that coders understand the form and the billing process to ensure proper reimbursement.

CMS-1450 (UB-92) Indicators

Indicators on the CMS-1450 form, formerly known as UB-92, can be numeric, alphanumeric, or alphabetic. There are 86 blocks, or form locators (FLs) on the CMS-1450 form. Some are general information and some indicate specific information about the claim. There are ten of them that are not assigned a specific task, which are for either state (FL 2, 11, 56, and 78) or national use (FL 31, 49, and 57). Three are used for specific purposes, which are the provider name, address, and phone number for FL 1, internal control number (FL37), and responsible party name and address (FL38).

The following sections provide information about the FLs; however, not all possibilities are listed. For complete listings of FL possibilities, textbooks about hospital billing with the CMS 1450 (UB-92) form should be consulted.

Dates/Times

Be careful of dates. Some FLs require single dates, whereas others require from-through dates. Some formats also may differ as to days, months, and years. The proper dates are important to ensure that entries in other FLs are correct. If dates do not coincide with each other or with services, payment may be denied. There are some FLs in which the dates may be reflected as MM/DD/YY, such as for discharge (FL 6); however, remember that in these instances, the actual count of days will not include the discharge date, such as in FL 7, which is for days in the hospital that are covered by insurance. Remember, too, that for Medicare hospital care cannot exceed 150 days, which could be composed of the first 60 days for hospitalization, 30 days for coinsurance days, and 60 days for lifetime reserves. For SNFs, coinsurance days consist of days 21 through 100. FL 8 would be used to list days in the hospital that are not covered. FL7 and FL 8 should reflect the from-through time period listed in FL 6. FL 9 is for the number of Medicare coinsurance days.

For a patient's birth date, however, the format is different from the previously mentioned fields. In FL 14, the patient's birth date would be listed as MM/DD/CC/YY, in which CC stands for century, so the year of birth is a four-digit number, not a two-digit number as previously seen in FL 6, 7 and 8. The birth date should not be left vacant, even if the birth date is unknown, but should be filled in with zeros to indicate that an error has not been made.

It should be noted that the date format might change depending on whether the claim is filed electronically or mailed. If it is filed electronically, some of the date may use the format that includes the century, whereas the paper claim would not.

Times should be listed in military time and without minutes, but instead with zeros following the hour. For example, 1:45 a.m. would be listed as 01:00 and 1:45 p.m. would be listed as 13:00. Note that there also are two digits for the hour, with a zero preceding a single number. These time formats area applied in FLs, like admission or discharge hour.

Names

Names are reported by last name, first, and middle. Suffixes, such as *Jr.*, are displayed as part of the last name without a comma.

Bill Type

FL 4 is for bill type and uses a three-digit numeric indicator. The first digit is for the type of facility, e.g., hospital or SNF. The second digit is for type of care for billing purposes, e.g., inpatient or clinics. The third digit is for the bill frequency, e.g., admit, nonpayment, continuing, adjustment, replacement, or late charges.

Admission/Discharge

FL 19 is for type of admission, which is indicated by single numbers. One (1) is for admission to the Emergency Room, 2 is for urgent, 3 is for elective, 4 is for newborn, and 5 is for a trauma center.

FL 20 is for type source of admission or how the patient is referred to the hospital. This FL includes single numeric or alphabetic numbers, such as 1 for physician referral, 2 for clinical referral, 3 is for HMO referral, 4 is for transfer from a hospital, 5 is for transfer from a SNF, 7 is a transfer from an Emergency Room, and A is for transfer from a critical access hospital. If FL 19 is for a newborn, then FL 20 is different since the newborn was born at the hospital and not transferred from somewhere else. Possible selections for FL 20 include 1 for normal delivery, 2 for premature delivery, 3 for sick baby, and 4 for extramural birth (birth outside of nonsterile area). Discharge status is indicated in FL 22 and includes 01 for discharge to home, 02 for discharge to another inpatient hospital, 03 for discharge to SNF, etc.

Condition Codes

Condition codes are in FLs 24–31. They can be a two-digit numeric or alphanumeric code, which explains special circumstances about services provided. This includes insurance codes (01 through 16), such as 01 for military service-related, 06 for patient with ESRD in the first 18 months in employer's health insurance plan, or 10 for a patient employed with no employer health insurance coverage. Codes 17 through 35 are for situations that are particular to the patient, such as 17 for homelessness, 18 for retention of maiden name, 25 for non-U.S. resident, 28 for employer health insurance coverage which is secondary to Medicare, and 31 through 34 are for student status. Codes 36 through 45 indicate room services, such as 39 for a private room as medically necessary and 40 for same-day transfer. Codes 46 through 65 are used for TRICARE or SNF visits. These include 55 for SNF bed not available at the time of discharge, which was subsequently delayed. Codes 66 through 69 indicate other services, such as 68 in which the patient wishes to use lifetime reserve days with Medicare (LTR). Codes 70 through 76 are for dialysis services, including 71 for full care in the hospital, 72 for self-care in the hospital, and 74 for home care. Codes 77 through 99 are for other situations, such as 78 for nonimplementation of new coverage by an HMO. Codes A0 through B2 are for special programs, such as A1 for EPSDT, A2 for physically handicapped children, A4 for family planning, A6 for pneumoccal pneumonia vaccine, and A9 for second opinion regarding surgery. Codes C0 through C6 indicate approval status, such as C1 for approved as billed, C3 for partial approval, and C6 as admission preauthorization. Codes D0 through G0 are for changes in claims, such as D0 for changes in service dates, D1 for changes in charges, D2 for changes in revenue codes, and E0 for changes in the status of the patient.

Occurrence Codes

Occurrence codes indicate a significant event that is related to the health care services being provided. The codes are listed in FL 32–25 and must be accompanied by an occurrence date. Each code can only be listed once. Codes 01 through 08 are for accidents, for example, 01 is for an auto accident, 04 is for employment-related accidents, and 06 is for a crime victim. Codes 09 through 15 are for conditions related to an illness or condition, for example, 10 indicates when the last menstrual period was and 11 indicates when an illness or condition began. Codes 16 through 39 provide additional insurance information, for example 16 is the date of last therapy, 24 is the date insurance was denied, 29 is the date outpatient physical therapy was established or reviewed, 31 is the date when a patient was informed that future services would not be covered under their insurance plan (which must be at least three days before provision of services), and 38 is for the date when home IV therapy began. Codes 40 through 99, and including A1 through Z9, are for situations related to a patient's visit, such as 40 is the date of admission, 42 for date of discharge, 43 is the date when surgery was cancelled, A1 for birth date, and A3 for date of exhaustion of benefits.

It should be noted that FL 36 is for conditions that span a space of time (from-through), whereas FL 32 through 35 are specific dates. If more space is necessary for additional conditions that cover a span of time, FL 32 through 35 can be used. Codes for 36 range from 70 through 99 and M0 through Z9, and include span of times for noncovered services (code 74), SNF care (75), and for when the patient assumes liability for payment of charges (76).

Value Codes

Value codes justify specific charges and are located in FL 39 through 41. Value codes include a code but also include numeric amounts. They are used for charges, and number of visits/units, and must have two digits to the right of the decimal/delimiter. For example, if there were 5 visits, it would be recorded as 5.00. The codes can include codes and charges for semiprivate rooms, combination billing, blood deductibles (including furnished and replaced), professional charges billed within hospital services, Medicare lifetime reserve and coinsurance amounts for first and second calendar years, MSP, veteran, disabled, patient liability, hour of accident, hemoglobin or hematocrit readings, and rehabilitation visits (e.g., physical, occupational, cardiac, speech), weight of newborn, home health codes, and deductibles. A listing of six zeros is an indication to Medicare that condition payment is requested, which may occur when other insurance plans are involved, such as employer's, automobile, liability, or worker's compensation.

Revenue Codes

A major element of the CMS-1450 is the revenue codes, which are listed in FL 42 and are three digit codes. They indicate the service or materials provided to the patient. They must be listed numerically, although 001 is listed last because it is the total of all of the charges. Descriptions can be provided in FL 43 but are not required by all payers, including Medicare except for several exceptions including certain cancer drugs.

All codes must have a date completed in FL 45. Number of units, FL 46, also must be completed. Units can be based on pints, days, visits, miles, etc. There are 23 lines available, but additional pages can be used— up to 450 lines for one bill. Revenue codes indicate services provided by the hospital, both inpatient and outpatient. Some revenue codes must be used in conjunction with the appropriate level of HCPCS codes (remember these include CPT codes). The HCPCS codes are listed in FL44, if required. Both HCPCS and revenue codes should be used by the hospital when billing outpatient hospital Medicare claims. Often, hospital computerized software programs provide the code when the description is entered.

Codes 10X through 20X apply to institutional accommodations. 12X is for room and board in a semi-private room with two beds. 14X is for a deluxe private room. 17X is for the nursery, 20X is for intensive care. The X is replaced by a properly selected code indicating the specific department where the patient was located—1 is for surgical, 2 is for obstetrics, 3 is for pediatric, 7 is for oncology, and 9 is for other. Third digits differ in some codes, such as newborns, which are classified as levels, ranging from routine care to constant care.

Other services charged by the hospital that use revenue codes include pharmacy (25X), IV therapy (26X), DME (29X), laboratory (30X), radiology (32X and 33X), nuclear medicine (34X), anesthesia (37X), respiratory services (41X), physical therapy (42X), occupational therapy (43X), Emergency Room (45X), cardiology (48X), and ambulance (54X). The third digit varies between categories depending on applicable specifics.

Payer Information

The payer is the third-party agency responsible for payment. With all of the payer FLs, such as 50 through 66, there are three lines, so information is provided for each payer listed. FL50 is for payer. Several lines are provided so that several payers can be listed, such as secondary payers or gap providers. The payers should be listed in order according to order of payment, with the primary payer listed first. The remainder of the FLs, 52 through 66, also provide three lines so that requested information can be provided for each payer.

FL51 is for the provider number that refers to the provider of health services for which payment is requested, such as the hospital or nursing home.

FL52 indicates a signed statement from the patient consenting to release of medical information to appropriate others. A "Y" means "yes," "N" means "no," and "R" means "restricted" meaning that the release is limited as to who can have access.

FL53 indicates if an assignment of benefits form has been signed with "Y" indicating "yes" and "N" indicating "no." Assignment of benefits means that payment is made directly to the payer and not the patient.

In FL54, the amount already paid by either the patient or other payers is listed according to their order in FL50. The final line is described as "Due from patient" and is used to list the amount the patient has paid. This can include amounts for deductibles and coinsurance.

Estimated amount due is listed in FL55. Up to ten digits can be provided, which includes two ending digits for cents and a final digit to indicate if it is credit. Note that there is a space for "Due from patient" also.

FL58 is used to list the insured's name, who is the person who has the insurance. The name is listed with the last name first. Remember that if the name is not spelled exactly the way it is on the insurance card, the claim will be denied, since no coverage will be indicated under the misspelled name. FL59 corresponds with FL58 as it indicates the relationship of the actual patient to the insured who is listed in FL58. FL64 indicates the insured's (FL58) employment status and FL65 indicates the employer's name. FL66 lists the employer's address. This is particularly important when the insured has both Medicare and insurance through an employer, making Medicare the secondary payer. Numbers indicating employment status include 1 for employed full time, 2 for employed part time, 3 for not employed, and 5 for retired. This information can include the patient, a spouse, child, etc. FL60 requires the insured's identification number as assigned by the payer agency, which may be the insured's social security number or health insurance claim number (HICN) from Medicare. The name of the group or plan is listed in FL61 with the identification number for this group listed in FL62. Medicare payers do not have identification group numbers. For military claims, "ACT" for active or "RET" for retired is listed in this space. An alphanumeric indicator is listed in FL63 to denote that services have been authorized by the payer.

Diagnosis And Procedure Codes

FL67 through 81 are important locators as the diagnostic and procedural codes are listed here. Decimals should not be used when listing the diagnostic codes, so 250.00 would be listed as 25000. Remember, for FL67, which is for the principal diagnosis, the primary reason the patient is admitted to the hospital is listed after evaluation, not more severe diagnoses that may develop during hospitalization. For outpatient services, the primary reason services are provided is selected as the principal diagnostic code. Up to eight additional ICD codes can be listed, in addition to the principal code. FL76 is for the diagnosis that was the initial reason the patient was admitted. FL76 may differ from the principal diagnosis in FL67, because a patient may be admitted to the hospital with one diagnosis and later be diagnosed with another condition after evaluation. For example, a patient may be admitted for chest pain, which would be listed in FL76, but be diagnosed with myocardial infarction (MI) after tests are completed. The MI would be listed in FL67. FL77 is for an E code.

FL80 and FL81 list procedural codes. Up to five additional procedural codes can be listed in addition to the principal procedural code. Dates are included with the codes. FL79 lists which coding system was used for the services provided, with 4 indicating CPT-4 codes were used, 5 indicating HCPCS were used, and 9 indicating ICD-9-CM procedural codes were used. FL80 is for the principal procedure, which may be the most closely linked service to the principal diagnosis.

Physician Information

FL82 and 83 are for the provider's name and/or number. FL82 is for the attending physician. Identification number is the UPIN, which is assigned by CMS. A new identification number, known as the national provider identifier (NPI), may replace the UPIN in the future. FL83 indicates the physician responsible for providing care for the principal procedure if that physician differs from the attending physician listed in FL82. For both FL82 and 83, the name and identification number should be listed in the lower line, which is below the box where the FL number and description is listed. The final FLs are for signature, date, and remarks.

HCPCS/Rates

HCPCS, including CPT codes, are placed in FL 44 for outpatient services. HIPPS codes for SNFs also are placed there. Rates for inpatient services could be listed here. These codes are listed in addition to the revenue codes in FL 42. Supplies, drugs (not chemotherapeutic), and ESRD services do not have to be recorded here for Medicare. The description of the services and/or supplies may be listed in FL 43, although this FL is not required by all payers. All codes must have a date completed in FL 45. Units can be based on pints, days, visits, or miles, etc.

When using the CMS-1450 form, it is not necessary to use the modifier 26 for the professional and technical component, because the revenue code indicates this information. If, however, the code itself denotes a professional component, use the code; it is only the modifier that does not need to be used with revenue codes.

Diagnostic Codes

Diagnostic codes from Volume 1 of the ICD-9-CM book are listed in FL67 for the principal diagnosis, and FL68–75 for other pertinent diagnostic codes. FL76 is for the diagnostic code, which is the reason for the visit.

REIMBURSEMENT ERRORS

CODES

(1) Undercoding
(2) Overcoding
(3) Wrong or improperly applied code, particularly regarding the number of required digits
(4) Improperly applied modifier
(5) Improper linkage of ICD-9-CM and CPT/HCPCS codes
(6) Unbundling
(7) Failure to demonstrate medical necessity
(8) Lack of necessary information to validate codes
(9) Transposed or wrong codes

BILLING

(10) Billing for services not rendered
(11) Billing for services provided to another patient
(12) Billing for services provided by another provider
(13) Billing for services not covered or for patients who are not covered
(14) Billing for services by unauthorized personnel
(15) Duplication of charges
(16) Duplication of services
(17) Billing for more units of services than allowed or in improper increments
(18) Overcharging, particularly Medicare patients
(19) Failure to bill payers in proper order, i.e., primary and secondary
(20) Improper mathematical determination of units and accompanying cost
(21) Wrong physician name, address, or ID number

DOCUMENTATION

(22) Altering report information to receive higher reimbursement, either before or after reports have been completed
(23) Missing or inadequate documentation
(24) Lack of proper signatures
(25) No referring physician with ID, particularly for consultations
(26) Wrong dates or locations
(27) Improper information, such as place of service

ACTIVITY 12.3

1. What does JCAHO mean?

2. In what FL on the CMS-1450 form is the principal diagnosis listed?

3. What is the definition of a principal diagnosis for inpatient visits?

4. What does NPI mean and what is its purpose?

5. On the CMS-1450 form, how many procedural codes can be listed and in what FLs?

6. On the CMS-1500 form, how many diagnostic codes can be listed and where on the form?

7. Why is modifier 26 not listed on the CMS-1450 form?

8. Which of the following value codes is not correct?

   A. 10.50 gm
   B. 3.00 days
   C. 2.50 ounces
   D. 3 visits

For questions 9 and 10, answer the following question: If a CMS-1500 billing form had code 95813 listed in block 24D and code 427.1 listed in block 21 for services that were provided for 45 minutes, why would this bill be rejected?

9.

10.

For questions 11 and 12, answer the following questions: On a CMS-1450 form, what would the diagnostic and procedural code be for an inpatient visit for removal of a toy, without an incision, from the pharynx of a 6-year-old, and where would they be located?

11.

12.

13. If a patient is seen in the Emergency Room after referral from a physician, what blocks on form CMS-1500 should indicate this?

14. If on CMS-1450, FL 67 was listed as 614.6 and FL 80 is coded as 65.81 and 65.89, why would this be rejected?

15. Must all revenue codes be listed numerically on the CMS-1450 form?

16. Which of the following would not be an acceptable measure of units for coding?

   A. Pints
   B. Days
   C. Degrees
   D. Visits

17. Is FL 58 used to list the name of the patient? Explain.

18. What would be wrong if on the CMS-1500 form in block 21, code 807.0 was listed, and in block 24d, code 21805 was listed times three? Explain.

19. What would be wrong if on the CMS-1500 form in block 21 code 871.5 was listed, and in block 24d, code 68801-50 was listed?

20. What days are coded from the 61st day through the 90th for inpatient visits?

SUMMARY

Reimbursement lies at the heart of the medical practice, for without money medical services could no longer be realistically provided and continued advances in medicine would be curtailed. There are many reasons incorrect reimbursement and coding play a critical role in the reimbursement process. It is, therefore, critical that coders, as a part of the process, understand their role in ensuring proper reimbursement. A coder's role can include auditing bills and follow up, in addition to continually educating physicians and staff regarding errors or limitations. Because of their knowledge and expertise, coders are a critical part of proper and full reimbursement for a medical practice.

ANSWERS

ACTIVITY 12.1

A—17
B—11
C—8
D—18
E—2
F—4
G—15
H—6
I—5
J—1
K—20
L—7
M—13
N—3
O—14
P—9
Q—16
R—12
S—19
T—10

ACTIVITY 12.2

1. Which of the following is not a PPS?
   A. DRG
   B. MDS
   C. APC
   D. HIPPS

Answer B: MDS is minimum data set and not a PPS (Prospective Payment System). DRGs are Diagnosis-Related Groups, APCs are Ambulatory Payment Classifications, and HIPPS are Health Insurance Prospective Payment Systems.

2. Which of the following is true about APCs?
   A. Basic unit of payment is admission/discharge.
   B. Based on a single primary code.
   C. Fixed amount is based on facility rate, APC weight, number of units, and discount.
   D. Fixed amount is based on relative value units, geographic area, facility rate.

Answer C: Basic unit of payment for APC is the visit for each day. APCs may assign multiple groups as it is paid by each line item on the claim, which is in contrast to DRGs, which are based on a single primary code. Fixed amount is based on facility rate, APC weight, number of units, and discount is correct.

3. A zero claim means
   A. Discount provided.
   B. Claim is dirty.
   C. No claim is filed.
   D. No payment is expected.

ANSWER D: Zero claims are claims for which no payment is expected, but are filed to provide the basis for submission to another payer.

4. In APCs, what is an outlier?
   A. Additional services that are not part of the global package.
   B. Services whose costs exceed by more than 2.5 the amount typically reimbursed.
   C. Facility costs that are too high or too low.
   D. Costs associated with overcoding.

Answer B: Outliers are services that cost more than the usual and customary amount that is usually reimbursed.

5. Which of the following are not true about Medicare payment for transitional pass-throughs?
   A. Monetary adjustment to ensure proper payment.
   B. Temporary codes.
   C. Can be applied to all materials and supplies.
   D. Can only be applied to medically necessary procedures.

Answer C: Transitional pass-throughs do not apply to materials and supplies associated with procedures. Remember that services that are not medically necessary are not paid for.

6. Which of the following is a description for APC payment status indicator C?
   A. Payment is transitional pass-through for a drug.
   B. Services are incidental so the cost is included in the global package.
   C. Payment is not allowed under the hospital outpatient payment system.
   D. Medicare will not pay for the service as outpatient, but only as an inpatient.

Answer D: Payment status indicator C indicates that Medicare will only pay for the service as inpatient due to health concerns.

7. Which of the following are not considered when determining reimbursement for DRGs?
   A. Relative value factor.
   B. Patient's age.
   C. Exacerbating conditions.
   D. Types of cases.

Answer A: Reimbursement for DRGs is based on the DRG weight and facility rate. These factors include the selected DRG, patient's age and exacerbating conditions (e.g., burns, MI, newborn), volume of same types of cases, hospital costs, types of patients, types of cases, discharge status, and geographic area.

8. What factor would not be considered when determining DRGs for a facility?
   A. Patients.
   B. Case mix.
   C. Facility resources.
   D. Number of physicians in the facility.

ANSWER D: DRGs are based on the correlation between case mix and facility resources. Case mix is the type of patient the facility treats, so the more complex the patients' care is, the more costly. Facility resources include human, spatial, and financial resources.

Use this information to answer the following questions:
MDC 1: Nervous System
MDC 4: Respiratory System
MDC 5: Circulatory System
MDC 6: Digestive System
MDC 8: Musculoskeletal System
DRG 107: Coronary bypass with cardiac catheterization
DRG 127: Heart failure and shock
DRG 139: Cardiac arrhythmia and conduction disorder
DRG 517: Percutaneous cardiovascular procedure with coronary artery stent
DRG 154: Stomach esophageal and duodenal procedures
DRG 161: Inguinal and femoral hernia procedures

9. Under what MDC would you locate the DRG code for right heart catheterization?

ANSWER: 5; MDC 5 is for circulatory system.

10. What is the correct DRG for inguinal hernia repair with graft?

ANSWER: 161; DRG 161 is for inguinal and femoral hernia procedures.

11. What is the correct DRG for vagotomy?

ANSWER: 154; DRG 154 is for stomach esophageal and duodenal procedures, which includes vagotomy.

12. What is the correct MDC for an esophagogastroplasty?

ANSWER: 6; MDC 6 is for diseases and disorders of the digestive system.

13. What is the correct DRG for a coronary arteriogram?

ANSWER: 107; DRG 107 is for coronary bypass with cardiac catheterization.

14. What is the correct DRG for an ablation of a lesion of the heart?

ANSWER: 517; DRG 517 is for percutaneous cardiovascular procedures with coronary artery stent, which includes the ablation.

Use this information to answer the following questions:

CPT code	11200	11600	13121	16010	25530
APC code	0013	0019	0025	0016	0044
Weight	1.51	4.56	3.71	3.31	2.73

15. Using descriptive terms (not codes) to identity the procedure, what procedure would cost the most? Explain.

ANSWER: Removal of skin lesions. This procedure (code 11600) would cost the most, because it has the highest weight by which the cost of reimbursement is determined.

16. What is the APC code for treatment of burns?

ANSWER: 0016; Code 16010 is the procedural code for burns.

17. Using descriptive terms (not codes) to identity the procedure, what procedure would cost the least? Explain.

ANSWER: Removal of skin tags. Code 11200 has the smallest weight of 1.51, which is what the cost of reimbursement is based on.

Use this information to answer the following questions:

CPT code	32400	35473	774210	76604	93510
APC code	0005	0081	0276	0266	0080
Weight	6.71	22.04	1.63	1.67	32.2

18. If the reimbursement rate is 50.30 per unit, what is the cost for repair of the arterial blockage?

ANSWER: $11086.12; Multiply 50.30 times 22.04.

19. Which APC code classification pays the highest? Explain.

ANSWER: 0080; This code classification is for diagnostic cardiac catheterization and has the highest weight.

20. Under which APC classification would a needle biopsy of the chest be listed?

ANSWER: 0005; This is the classification for needle biopsy/aspiration except bone marrow.

ACTIVITY 12.3

1. What does JCAHO mean?

ANSWER: Joint Commission on Accreditation of Healthcare Organizations.

2. In what FL on the CMS-1450 form is the principal diagnosis listed?

ANSWER: FL67

3. What is the definition of a principal diagnosis for inpatient visits?

ANSWER: With inpatient visits, the principal diagnosis is the main reason for admission, which is determined after evaluation. For outpatient visits, the principal diagnosis is the main reason services were provided.

4. What does NPI mean and what is its purpose?

ANSWER: National provider identifier (NPI), which is used to identify the medical care provider.

5. On the CMS-1450 form, how many procedural codes can be listed and in what FLs?

ANSWER: Six, including the principal procedure, which are located in FLs 80 and 81.

6. On the CMS-1500 form, how many diagnostic codes can be listed and where on the form?

ANSWER: Four, which are located in block 21.

7. Why is modifier 26 not listed on the CMS-1450 form?

ANSWER: The revenue codes provide the additional information needed for coding and billing, so this modifier for professional services is not necessary.

8. Which of the following value codes is not correct?

A. 10.50 gm
B. 3.00 days
C. 2.50 ounces
D. 3 visits

ANSWER: D. Value codes must have a decimal and two values to the right of the decimal for correct format.

For questions 9 and 10, answer the following question: If a CMS-1500 billing form had code 95813 listed in block 24D and code 427.1 listed in block 21 for services that were provide for 45 minutes, why would this bill be rejected?

9.

ANSWER: An EEG does not link properly to cardiac dysrhythmias.

10.

ANSWER: Code 95813 is for more than one hour of service.

For questions 11 and 12, answer the following questions: On a CMS-1450 form, what would the diagnostic and procedural code be for an inpatient visit for removal of a toy, without an incision, from the pharynx of a 6-year-old, and where would they be located?

11.

ANSWER: The code in FL 67 for the diagnosis would be listed as 933.0.

12.

ANSWER: The code in FL 80 would be listed as 09.13.

13. If a patient is seen in the Emergency Room after referral from a physician, what blocks on form CMS-1500 should indicate this?

ANSWER: Blocks 17 and 17a.

14. If on CMS-1450, FL67 was listed as 614.6 and the FL80 is coded as 65.81 and 65.89, why would this be rejected?

ANSWER: 65.81 is for removal of adhesions from the ovaries and the tubes, and the diagnosis is also for adhesions in the ovary and tubes, so only one procedural code is necessary.

15. Must all revenue codes be listed numerically on the CMS-1450 form?

ANSWER: No, they must be listed numerically except for 001, which is listed last, because it is the total of all of the charges.

16. Which of the following would not be an acceptable measure of units for coding?

A. Pints
B. Days
C. Degrees
D. Visits

ANSWER: Degrees may be used for burns, such as first and second, but it is not used as a measure of units for coding.

17. Is FL58 used to list the name of the patient? Explain.

ANSWER: No, FL58 is for the name of the insured, which is the person under whose name the insurance policy is listed.

18. What would be wrong if on the CMS-1500 form, in block 21, code 807.0 was listed, and in block 24d, code 21805 was listed times three? Explain.

ANSWER: There is no error. The code can be times three, because it states that it is for each rib. The diagnosis of rib fracture, which is closed, does not affect the coding of the procedure.

19. What would be wrong if on the CMS-1500 form, in block 21, code 871.5 was listed, and in block 24d, code 68801-50 was listed?

ANSWER: Code 871.5 is for foreign object in the eye, not the lacrimal punctum that is listed in the procedural code.

20. What days are coded from the 61st day through the 90th for inpatient visits?

ANSWER: Coinsurance days. The first 60 days are for hospitalization, the next 30 days for coinsurance days, and next 60 days for lifetime reserves.

Appendix A   CPC-H MOCK EXAM 1

The following appendix provides a mock example of the Certified Professional Coder-Hospital (CPC-H) exam offered by the American Academy of Professional Coders (AAPC).

ICD-9 (including Volume 3), CPT, and HCPCS coding books are allowed for this exam.

The CPC-H exam is divided into two sections. The first section contains 20 questions on medical terminology, 20 on ICD-9, 20 on billing, 20 on outpatient coding and 20 on surgery coding. Section 2 contains 50 case studies. This includes information related to APCs, and the UB-92 billing form. Sequencing and linkage of codes are important issues for this coding exam.

For the following scenarios, select the best answer. (Some scenarios may provide only ICD-9s or only CPTs, and not necessarily both. Select the correct one.)

1. Streptococcus is a(n)
   A. Rod-shaped bacterium
   B. Organism that grows in clusters
   C. Berry-shaped bacterium
   D. Chain-like virus

2. Which of the following is not part of the integumentary system?
   A. Lipocytes
   B. Sweat glands
   C. Paronychium
   D. Dendritic cell

3. Analyte is
   A. Substance measured
   B. Tissue that is submitted for pathological examination
   C. Withdrawal of fluid
   D. An antibody

4. Protrusion of the urinary bladder through the vaginal wall is
   A. Ileostomy
   B. Perineorrhaphy
   C. Cystocele
   D. Meningioma

5. Ganglion is
   A. Membrane lining the alimentary canal
   B. Cluster of nerve cell bodies
   C. Inflammation of the genital tract
   D. Varices

6. TNM is not
   A. A type of delivery system
   B. A method of classifying malignancy
   C. Inclusive of lymph nodes
   D. Concerned with metastasis

7. A LEEP is
   A. Loupe electrocautery excision procedure
   B. Lumbar extracorporeal excision paresis
   C. Lumbar extracorporeal excision procedure
   D. Loop electrocautery excision procedure

8. Eversion means
   A. Turned inward
   B. Lying down
   C. Turned outward
   D. Lying on the back

9. Hypofunctioning of the adrenal cortex is known as
   A. Cushing disease
   B. Hamman disease
   C. Addison disease
   D. Tay-Sachs disease

10. Receptor cells of the eyes are known as
    A. Macula
    B. Cones
    C. Cornea
    D. Retina

11. Jaundice is not caused by
    A. Hemolysis
    B. Pneumoconiosis
    C. Choledocholithiasis
    D. Liver disease

12. Tetralogy of Fallot does not include
    A. Cirrhosis
    B. Stenosis
    C. Septal defect
    D. Hypertrophy

13. Achondroplasia is also known as
    A. Hirsutism
    B. Epstein-Barr
    C. Cretinism
    D. Dwarfism

14. Formation of connection between two parts of the colon is known as
    A. Colostomy
    B. Colocolostomy
    C. Cholecystectomy
    D. Cholostomy

15. Which of the following is not affected by the condition of pancytopenia?
    A. Eosinophils
    B. Antibodies
    C. Erythrocytes
    D. Platelets

16. Ventral hernia is
    A. Hernia in the area of the abdominal wall
    B. Hernia in the area of the umbilicus
    C. Hernia in the area of the groin
    D. Hernia of the upper abdomen

17. Which of the following would not be a tumor?
    A. Polyp
    B. Neoplasm
    C. Mass
    D. Pustule

18. Uveitis is
    A. Inflammation of the small tissue located at the back of the throat
    B. Inflammation of the iris
    C. Resistance to UV lights
    D. Abnormal condition of the chorioid

19. To fill a body part with gas is
    A. Thoracentesis
    B. Insufflated
    C. Venipuncture
    D. Bronchiectasis

20. Which of the following bones is not included in the face?
    A. Parietal
    B. Vomer
    C. Zygomatic
    D. Lacrimal

21. What is the correct code(s) for a patient who was treated for burns on his right leg and chest area, which covered 33% of the body with 10% second-degree and the rest third-degree with infection?
    A. 948.32, 958.3, 93.57
    B. 942.20, 945.20, 93.56
    C. 948.33, 93.57
    D. 948.33, 958.3, 86.59

22. What is the correct code(s) for a patient who was given CAPD for chronic kidney disease?
    A. 403.91, 39.95
    B. 593.9, V56.8, 54.98
    C. 585 54.98
    D. 585, 39.95, V56.8

23. What is the correct code(s) for a patient who was treated for pneumonia due to Pneumoniae pneumonia, streptococcal?
    A. 481
    B. 482.30
    C. 482.9
    D. 480.9

24. What is the correct code(s) for a patient who was treated for an acute bucket handle tear of the meniscus with an arthroscopy performed?
    A. 717.0, 80.36
    B. 836.0, 80.36
    C. 836.0, 80.26
    D. 717.0, 80.20

25. What is the correct code(s) for an unborn child of a mother who has idiopathic thrombocytopenia, which is affecting the child?
    A. 776.1
    B. 666.3, 287.5
    C. 666.3
    D. 287.5

26. What is the correct code(s) for a patient who had severe coronary atheroma and received a double mammary-coronary bypass graft with extracorporeal circulation?

   A. 414.01, 36.10
   B. 414.00, 36.16
   C. 414.00, 36.16, 39.61
   D. 414.01, 39.61, 36.10

27. What is the correct code(s) for a patient with Hodgkin's granuloma of the spleen and axillary lymph nodes who had biopsies?

   A. 201.18, 40.11, 41.32
   B. 201.90, 41.32
   C. 201.17, 201.14, 41.32, 40.11
   D. 201.17, 41.32

28. What is the correct code(s) for a patient who was treated for arthritis due to severe cuts on the wrists six months ago, which involved tendons?

   A. 716.13, 905.8
   B. 713, 905.2
   C. 716.13, 906.1
   D. 713

29. What is the correct code(s) for a patient who was treated for pneumonitis due to varicella?

   A. 480.8
   B. 480.8, 052.8
   C. 052.1
   D. 484.8, 052.1

30. What is the correct code(s) for a patient who was treated for excision of sequestra from the patella as a result of a staph infection?

   A. 730.16, 041.1, 77.06
   B. 730.17, 77.16
   C. 730.16, 041.1, 77.66
   D. 730.06, 77.06

31. What is the correct code(s) for a patient who was treated for carcinoma of the bronchus, which was removed with bronchoplasty?

   A. 197.0, 32.01
   B. 162.9, 32.09, 33.48
   C. 162.9, 32.09
   D. 197.0, 33.0, 33.48

32. What is the correct code(s) for a patient who was treated for acute and chronic obstructive bronchitis?

   A. 493.9, 491.20
   B. 496, 491.21
   C. 496
   D. 491.21

33. What is the correct code(s) for an insulin-dependent diabetic patient who was treated for hypoglycemic shock?

   A. 251.1
   B. 250.81
   C. 250.31
   D. 250.30

34. What is the correct code(s) for a patient who was treated for CVA with hemiparesis?

   A. 436, 434.91, 342.9
   B. 438.20, 436, 907.0
   C. 436, 342.9
   D. 434.91, 342.9

35. What is the correct code(s) for a patient who was treated for hemorrhage after CABG with PTCA and insertion of a stent?

   A. 410.9, 36.05
   B. 996.72, 36.05, 36.06
   C. 410.9, 36.05, 36.06
   D. 996.72, 36.05

36. What is the correct code(s) for a pregnant woman who was injured in a car accident and received a severe laceration of the scalp requiring 8 stitches?

   A. 873.1, V22.2, 86.59
   B. 851.9, 86.51
   C. 851.17, V22.2, 02.92
   D. 873.1, V22.2, 02.92

37. What is the correct code(s) for a patient who delivered a single live birth at 42 weeks with a third-degree perineal laceration with surgical rupturing of membrane after delivery had begun?

   A. 645.21, 664.21, 73.09
   B. 664.21, V27.0, 73.6
   C. 645.11, V27.0, 73.01
   D. 645.11, 664.21, V27.0, 73.09

38. What is the correct code(s) for a patient who was treated with chemotherapy for carcinoma of the muscle of the right leg?
    A. V58.1, 171.3
    B. 195.5, V58.1
    C. 239.2
    D. V58.1, 195.5

39. What is the correct code(s) for a patient who was suffering from seizures after having ingested her husband's prescription for Benzedrine in addition to a large quantity of alcohol?
    A. 969.7, E939.7
    B. 780.39, 980.9, E939.7, E947.8
    C. 969.7, 980.9, 780.39
    D. 436, E939.7

40. What is the correct code(s) for a patient who received a herniorrhaphy for a direct inguinal hernia with graft?
    A. 550.90, 53.01
    B. 550.00, 53.02
    C. 550.90, 53.03
    D. 551, 53.11

41. Third-party payer is
    A. The medical office
    B. The patient
    C. The guarantor
    D. The organization responsible for payment

42. Which is not true about a PAR?
    A. Assignment of benefits is not necessary
    B. Payment is sent directly to PAR
    C. Additional percentage of reimbursement
    D. Patients are advised who PARs are

43. With CHAMPUS, the following is true
    A. Now known as TRICARE
    B. Bills are submitted based on insured's home base
    C. CHAMPUS is always the secondary payer
    D. None of the above

44. Which of the following is not a correct billing procedure?
    A. Writing clearly and legibly
    B. Obtaining proper signatures
    C. Whiting out numbers
    D. Providing patient name and address

45. Fiscal intermediary is
    A. Guarantor
    B. Company responsible for payment of Medicare bills
    C. Person responsible for payment for medical services
    D. Collection agency

46. Lifetime reserve is
    A. Coinsurance
    B. An additional 30 days of hospital coverage
    C. An additional 60 days of hospital coverage
    D. A monetary value that can be applied to medical services

47. Which of the following is information that does not need to be collected during registration?
    A. Deductible
    B. Insurance company
    C. Copayment
    D. Coinsurance

48. If a Medicare patient is injured while at work at Microsoft, what would be the proper procedure for reimbursement?
    A. Send bill to employer's insurance company for their employees first
    B. Send bill to Worker's Compensation insurance carrier first
    C. Send bill to Worker's Compensation second
    D. Send bill to Medicare first

49. UHDDS is not
    A. Used by long-term care facilities
    B. Inclusive of patient diagnoses
    C. Inclusive of E codes
    D. The standard for hospital patient information

50. Which of the following would not be considered an occurrence code on the CMS-1450 form?
    A. Comorbidities
    B. Illness
    C. Auto accident
    D. Last menstrual period

51. Guarantor is
    A. The insurance company responsible for payment
    B. The individual responsible for payment
    C. The insured
    D. The physician providing services

52. CCI is not
    A. Incorporated within OCEs
    B. Applicable to correct coding combinations
    C. Composed of various edits
    D. A determining factor in the RBRVS

53. What is the proper APC billing method for outpatient observation services?
    A. Never bill for outpatient observation services
    B. Bill for no more than 24 hours
    C. Bill in increments of one hour
    D. Do not bill with other services that have a status indicator of T

54. Allowable charge is based on charges that are
    A. Usual
    B. Medically necessary
    C. Customary
    D. Reasonable

55. Medicare does not provide coverage for
    A. Senior citizens
    B. Indigent
    C. Patients with kidney failure
    D. Blind

56. With APCs, a status indicator of C indicates that
    A. This service is a transitional pass-through
    B. Code has been deleted
    C. This is a multiple procedure
    D. These charges will only be paid as inpatient by Medicare, not outpatient

57. If a patient was seen in the Emergency Room in Los Angeles, California, for appendicitis, where would the bill be sent for reimbursement if the patient enrolled in the U. S. Air Force in Tucson, Arizona, and were presently living in Butte, Montana?
    A. Tucson
    B. Butte
    C. Los Angeles
    D. None of the above

58. What practices can a medical facility perform in order to collect payment for services rendered?
    A. Call during the day and early evening
    B. Contact employer about patient's delinquency in payment
    C. Threaten the patient with a lawsuit
    D. Call only during the day

59. Form locators are
    A. Blocks to be completed on the CMS-1500 form
    B. Blocks to be completed on the CMS-1450 form
    C. Blocks on billing forms for diagnostic codes
    D. Blocks on billing forms indicating where services were provided

60. The CMS-1450 form may be referred to by some people as:
    A. CMS-1500 form
    B. HCFA-1500
    C. UB-92
    D. Physician billing form

61. What is the correct code(s) for a patient who was seen again in the Emergency Room for another rigid esophagoscopy for removal of a foreign object, which was performed by the same physician?
    A. 43200
    B. 43215-76
    C. 45332
    D. 43215

62. Radioimmunoassay tests are found in what section?
    A. Chemistry
    B. Diagnostic Ultrasound
    C. Nuclear Medicine
    D. Clinical Pathology

63. What is the correct code(s) for a patient who is admitted to the Emergency Room for fracture and dislocation of the right bimalleolar ankle with manipulation without anesthesia?
    A. 27810
    B. 27808
    C. 27810, 27830
    D. 27808, 27830

64. Which one of the following coding scenarios for services provided within an ASC is valid?
    A. 99203, 99202-24
    B. 99281, 99202
    C. 99281-27, 99283-27
    D. 99203, 99211

65. What is the correct code(s) for a patient who received photodynamic therapy during bronchoscopy for 1 hour for removal of abnormal tissue?
    A. 31641
    B. 31641, 96570, 96571 × 2

C. 96570, 96571 × 2

D. 31640

66. Which of the following is not a proper way to code analytes?

A. Report separately for multiple specimens from different sources

B. Report separately for specimens obtained at different times

C. Must be combined into one comprehensive code that includes all analytes

D. Analytes can be collected from any source

67. For E/M codes, the levels of history for Emergency Room visits do not include

A. Problem focused

B. High complexity

C. Comprehensive

D. Detailed

68. What is the correct code(s) for a patient who was treated with hot packs and diathermy for her injured knee?

A. 97010, 97024

B. 97024

C. 97010

D. 97039

69. What is the correct code(s) for a patient who was tested for presence of cocaine?

A. 82520

B. 80101

C. 80100

D. 86160

70. Administration of contrast material would be coded "with contrast" when

A. Oral

B. Injected

C. Rectal

D. Infused

71. What would not be proper when coding for mapping for intracardiac electrophysiological procedures?

A. Used to code for tachycardia origination site

B. Use of multiple catheters

C. Coded in addition to the procedure

D. Standard mapping should be reported in addition to 3-D mapping

72. If a patient is seen in the Emergency Room for complaints of severe chest pain with expanded problem focused history and exam and moderate medical decision making complexity, which level of code would be used for billing with APCs?

A. 0600

B. 99283

C. 99203

D. 99204

73. Which of the following conditions would be true if code 90919 is used?

A. Patient is age 14

B. Billed for a full month of services

C. Patient is in the hospital

D. Non-dialysis services were provided

74. What is the correct code(s) for a patient who received a B-scan, which included the mediastinum?

A. 76506

B. 71275

C. 76604

D. 76645

75. What are the differences between coding for critical care in the hospital for an adult and for a child 2 years old?

A. Pediatric critical care is coded by days, not hours

B. Pediatric critical care is coded by age

C. Pediatric critical care is coded by initial and subsequent

D. Both A and C

76. If surgery for a patient is cancelled prior to administration of anesthesia and surgical preparation, how should this be coded?

A. With modifier 74

B. With modifier 73

C. No code

D. With modifier 52

77. What is the correct code(s) for a patient who received a unilateral duplex scan of the aorta?

A. 93979

B. 93880-52

C. 93978-52

D. 93882

78. The unit of service for surgical pathology is based on
    A. Size
    B. Specimen
    C. Anatomic site
    D. Both B and C

79. What is the correct code(s) for a patient who is tested for *H. pylori* by enzyme immunoassay?
    A. 87338
    B. 78267
    C. 83014
    D. 87339

80. Which modifier cannot be applied to both physician and ASC procedures and services?
    A. 52
    B. 57
    C. 58
    D. 78

81. When coding for cardiac catheterizations, the code should be coded to
    A. Include an additional code for introduction
    B. Include catheters
    C. The highest order
    D. Type of catheter

82. What is the correct code(s) for a patient who received a CABG with single venous and single arterial grafts including the saphenous with endarectomy?
    A. 33517, 33533, 33572
    B. 33533
    C. 33517, 33572
    D. 33510, 33533, 33572

83. What is the correct code(s) for a patient who received wound repair for a 6.2-cm laceration of the abdomen, which required considerable time and enlargement of the wound for adequate treatment including coagulation of various vessels extending into the muscle and fascia and repair of nerve?
    A. 13101, 35221
    B. 20102, 64872
    C. 20102, 35221, 64872
    D. 13101, 35206, 64872

84. What is the correct code(s) for a patient who received a thoracotomy for insertion of permanent epicardial pacemaker?
    A. 33206
    B. 33200
    C. 33210
    D. 33201

85. Which of the following statements is not true about percutaneous fixation?
    A. X-rays can be used
    B. Implants are included
    C. Is always coded as open
    D. Can be inserted into the bone

86. What is the correct code(s) for a patient who received a mastoidectomy and resection with decompression and an infratemporal post-auricular approach for an intradural resection of the neoplasm of the cavernous sinus with graft with transection of the carotid artery?
    A. 61591, 61608
    B. 61591, 61607
    C. 61608
    D. 61591, 61608, 61609

87. What is the correct code(s) for a patient who received care for treatment of hallux valgus that included fusion and transplantation of the metatarsal bone?
    A. 28290
    B. 28294
    C. 28296
    D. 28290, 28296

88. What is the correct code(s) for a patient who received a nephrectomy and total ureterectomy with the same incision?
    A. 50543
    B. 50548
    C. 50234
    D. 50548, 50548

89. What is the correct code(s) for a patient who had a biopsy of the transverse colon with partial excision with anastomosis through abdominal approach?
    A. 44100, 45114
    B. 44140
    C. 44025, 45114
    D. 44140, 45100, 45114

90. What is not true about coding for wound repair?
    A. Debridement is not coded
    B. Complex repair includes closure

C. Simple repair includes suturing

D. The lengths of the wounds are added together for each anatomic group listed

91. What is the correct code(s) for a patient who received I&D with incisional removal of glass of three fingers at the metacarpophalangeal joint?

    A. 26075, 25028

    B. 25248, 10180

    C. 26075

    D. 26070 × 3

92. What is the correct code(s) for a patient who had repair of a closed unicondylar fracture of the proximal tibia with internal fixation?

    A. 27759, 20690

    B. 27756

    C. 27535

    D. 20690

93. What is the correct code(s) for a patient who received a flexible colonoscopy with removal of a tumor by snare including insertion of a stent?

    A. 45384

    B. 50605

    C. 45385, 45387

    D. 43256

94. What is the correct code(s) for a patient who received curettement of 4 benign lesions?

    A. 17000, 17003 × 3

    B. 11056

    C. 17000, 17003

    D. 11056 × 4

95. Which of the following is not a HCPCS code?

    A. I1020

    B. 67316

    C. B9000

    D. None of the above

96. What is the correct code(s) for a patient for whom bone marrow was obtained through fine needle aspiration?

    A. 10021

    B. 38220

    C. 10022

    D. 10021, 38220

97. What is the correct code(s) for a patient who received selective catheterization of the right common artery and right and left middle cerebral artery and introduction?

    A. 36217, 36218

    B. 36215

    C. 36215, 36217, 36218

    D. 36217, 36100

98. What is the correct code(s) for a patient who received repair of a strangulated ventral hernia with mesh with repair of the intestinal obstruction?

    A. 49507

    B. 44615

    C. 44615, 49561, 49568

    D. 44615, 49568

99. What is the correct code(s) for a patient who received a 2-cm Z-plasty with destruction of a malignant lesion of the scalp?

    A. 15120

    B. 14020, 17272

    C. 14020

    D. 15120, 17262

100. Which of the following codes do not include the use of a microscope?

    A. 34001

    B. 20955

    C. 17304

    D. 44970

101. What are the correct code(s) for a patient who has occlusive disease of the coronary arteries with a possible MI? The patient has a significant history of ongoing care for COPD and malignant hypertension as well as asthmatic bronchitis. Patient received a retrograde catheterization through the femoral vein of the left main coronary artery with cineangiography of the left coronary arterial system, which was then followed by insertion of a catheter into the right coronary artery with monitoring of arterial pressure, which included a bypass graft.

    A. 411.81, 493.00, 496, 93501, 93545, 93556

    B. 411.81, 401.0, 493.90, 93526, 93540, 93545, 93556

    C. 410.90, 401.9, 493.90, 93510, 93526

    D. 410.90, 401.9, 496.00, 493.00, 93510, 93501, 93540, 93545

102. What are the correct code(s) for a patient who presents with complaints of severe chest pains and was found to have severe coronary artery disease with left ventricular dysfunction and ejection fraction of 30%? CABG was performed times three which included the left internal mammary artery to distal left anterior descending artery and sequential WYE saphenous vein graft with the main branch from the aorta to the posterolateral artery and the WYE branch from the mid portion of this graft to the posterior descending artery.

    A. 414.01, 33518 × 3

    B. 429.2, 33533 × 3

    C. 414.01, 33533, 33518

    D. 429.2, 33533, 33511

103. What are the correct code(s) for a pregnant woman with aseptic meningitis due to adenovirus?

    A. 647.83, 049.1, 322.9

    B. 647.83

    C. 647.83, 049.1

    D. 049.1, V22.2

104. What are the correct code(s) for a patient who was diagnosed with cellulitis of the left foot with gangrene?

    A. 682.7

    B. 785.4

    C. 682.7, 785.4

    D. 785.4, 682.7

105. What are the correct code(s) for a patient who received trephination for drainage of brain abscess due to TB?

    A. 324.0, 013.30, 61150

    B. 324.0, 013.30, 61105

    C. 324.0, 61105

    D. 013.30, 61150

106. What are the correct code(s) for a patient who presented with symptoms of glomerulonephritis with renal necrosis?

    A. 593.9, 580.4, 584.9

    B. 583.9

    C. 580.9, 584.9

    D. 580.4, 593.9

107. What are the correct code(s) for the newborn with a mother with incompetent cervix?

    A. 654.4

    B. 622.5

    C. 622.5, V22.2

    D. 761.0

108. What are the correct code(s) for a patient who received photocoagulation with an additional cryotherapy for repair of retinal detachment of the left eye with defects?

    A. 361.00, 67101

    B. 361.2, 67105

    C. 361.2, 67105, 67101

    D. 361.02, 67105

109. What are the correct code(s) for a patient who was seen in urgent care for pneumonia as a result of measles? Patient is currently being treated for arthritis. Frontal and lateral chest x-rays were taken and patient provided prescription for antibiotics. History and exam were problem focused and medical decision making was low.

    A. 055.1, 99201, 71020

    B. 055.9, 99202, 71020

    C. 055.9, 486, 99201, 71023

    D. 055.1, 486, 99212, 71023

110. What are the correct code(s) for a patient who received thoracentesis to remove fluid in the right lung with ultrasound in association with small cell carcinoma of the upper lobe?

    A. 191.1, 511.9, 32002

    B. 162.3, 511.8, 32000

    C. 162.3, 32000

    D. 239.1, 511.9, 32002

111. What are the correct code(s) for a patient who had 8 sections of hypertrophied skin removed on both sides of the body?

    A. 701.9, 11200

    B. 733.9, 11200 × 8

    C. 701.9, 11200-51

    D. 733.9, 11200-50

112. What are the correct code(s) for a patient who received a 50 sq cm porcine graft for scarring on both arms due to third-degree burns sustained two months ago?

    A. 709.2, 15000-50, 15100-50

    B. 934.30, 15400-50

    C. 709.2, 906.7, 15000, 15400

    D. 943.30, 15400

113. What are the correct code(s) for a patient who was seen in the Emergency Room for blackouts and vomiting due to intentional overdose with benzodiazepines? History and exam were problem focused, although the patient was not able to provide much information and the medical decision making was low.

   A. 780.2, E950.3, 99282
   B. 780.02, 969.4, E950.3, 99281
   C. 780.02, 99282
   D. 969.4, 780.2, E950.3, 99281

114. What are the correct code(s) for a patient who received a T11–T12 laminectomy with decompression and T11–T13 posterior spinal fusion with posterior instrumentation using screws for lumbar instability, lumbar degenerative disc disease, and chronic low back pain? This included structural femoral head allografting, and intrathecal Duramorph injection. The patient was found to have left sided hemiparesis from a past CVA.

   A. 722.72, 436, 342.90, 63170, 22840, 20930
   B. 722.51, 438.20, 63003, 22610, 22614, 22840, 20931
   C. 722.72, 63003, 22630, 20931, 22610
   D. 722.51, 438.20, 436, 22630, 22840, 29030

115. What are the correct code(s) for a patient who was diagnosed with osteomyelitis due to infection by *Salmonella* of the first metacarpal with amputation and a flap?

   A. 730.2, 003.24, 26910
   B. 730.2, 26952
   C. 003.24, 26910
   D. 003.24, 26952

116. What are the correct code(s) for a patient who received an arthroplasty with allograft for traumatic left intertrochanteric fracture due to a fall at home? The patient has also been diagnosed with adult onset diabetes mellitus with neuropathy, hypertension, hypercholesterolemia, hyperglyceridemia, and intestinal diverticulosis.

   A. 820.21, 250.40, 583.81, 401.9, 272.2, 562.00, 27130
   B. 820.8, 250.00, 401.9, 272.0, 272.1, 562.00, 27125
   C. 820.21, 250.40, 272.2, 562.01, 27130, 27130, 29031
   D. 820.8, 250.40, 583.81, 401.9, 272.0, 272.1, 27130

117. What are the correct code(s) for a patient who had a 0.5 cm lesion removed from the upper frenulum of her lip, which was found to be malignant?

   A. 140.0, 11600
   B. 140.3, 40819
   C. 140.3, 11600
   D. 140.0, 40819

118. What are the correct code(s) for a patient who had complicated repair of a 3.5-cm laceration of the right cheek and 4.6-cm complete laceration of the nose, which all required sutures?

   A. 873.51, 873.30, 13132, 13152
   B. 879.9, 13132, 13133
   C. 873.30, 873.51, 12053
   D. 873.51, 873.30, 13132, 13133

119. What are the correct code(s) for a patient who had a tube inserted into the lungs for drainage of fluid? Patient has been diagnosed with metastatic carcinoma of the right lung from the esophagus and is currently receiving chemotherapy.

   A. V28.1, 162.9, 43267
   B. 150.9, 197.0, 43267
   C. V28.1, 511.9, 162.9, 32020
   D. 150.9, 511.9, 197.0, 32020

120. What are the correct code(s) for a patient who was diagnosed with severe external hemorrhoids, for which a rigid protoscopy was performed for visualization in addition to open complete hemorrhoidectomy with a Harmonic scalpel, which were found to be irreducible?

   A. 455.6, 46230, 46600
   B. 455.3, 46230
   C. 455.5, 46250
   D. 455.3, 46230, 46600

121. What are the correct code(s) for a patient who had subtemporal craniectomy of ganglion cysts?

   A. 727.43, 61450
   B. 727.41, 61440
   C. 727.43, 20612
   D. 727.41, 61450

122. What are the correct code(s) for a patient who developed ulcerative proctitis due to radiation for endometrial carcinoma for which she had a vaginal hysterectomy? Today, patient received rigid proctoscopy with topical treatment.

   A. 601.0, 46600

B. 990, 556.2, 46600
C. 990, 46604
D. 556.2, 990, 46600

123. What are the correct code(s) for a patient who had a first dorsal compartment decompression fasciotomy, carpal tunnel release and flexor tenosynovectomy for tendonitis of the left wrist of the first dorsal compartment, carpal tunnel syndrome of the left wrist, and left wrist chronic flexor tenosynovitis?

A. 354.0, 727.05, 726.90, 25115, 25118, 64721
B. 354.0, 727.00, 726.00, 64721, 25118
C. 354.0, 727.05, 25020
D. 354.0, 727.00, 25115, 25020

124. What are the correct code(s) for a patient who presents with hyperparathyroidism due to renal disease as associated with diabetes? Patient underwent a subtotal parathyroid resection.

A. 588.81, 250.40, 60240
B. 252.0, 250.40, 60500
C. 250.40, 583.81, 588.81, 60500
D. 250.40, 588.81, 60240

125. What are the correct code(s) for a patient who had intradural resection of neoplasm of the parasellar area with infratemporal approach with a cranioplasty with the use of a synthetic graft?

A. 61607, 61591
B. 61607, 61590, 61618
C. 61607, 61591, 61618
D. 61590, 61607

126. What are the correct code(s) for a patient who presents today with complaints of dyspnea, chest pain, and nausea? He received an ECG that demonstrated an old MI and enlarged left atrium due to hypertension.

A. 412, 402.90, 89.52
B. 410.90, 429.3, 89.51
C. 410.90, 429.3, 89.52, 401.9
D. 412, 89.51, 402.90, 401.9

127. What are the correct code(s) for a patient who has a history of gallstones and presents today again with pancreatitis with cholelithiasis? A cholecystectomy was performed for removal with an intraoperative cholangiogram.

A. 577.1, 574.20, 47605
B. 577.0, 575.0, 47563
C. 577.1, 575.0, 47563
D. 574.20, 47605

128. What are the correct code(s) for a patient who had abnormal findings of an AIDS test, but the patient is not exhibiting any symptoms at this time?

A. V01.7
B. V08
C. 042
D. 795.71

129. What are the correct code(s) for a patient who had a graft implanted for repair of an abdominal aneurysm and aorta with endarterectomy?

A. 442.84, 34800, 33572
B. 441.2, 35091
C. 440.4, 34800
D. 442.84, 35092

130. What are the correct code(s) for a patient who was seen in the urgent care for shortness of breath? She has a 40-year history of smoking two packs a day but quit two months ago. She was diagnosed as having an acute exacerbation of her COPD as complicated by her emphysema with bronchitis. She was given oxygen to continue at home in addition to prescriptions for Flovent and prednisone. The history was expanded problem focused, the exam was detailed and the decision making was low.

A. 491.21, V15.82, 99202
B. 496, 492.8, 305.1, 99282
C. 496, V15.82, 99282
D. 496, 491.21, 305.1, 99203

131. What are the correct code(s) for a diabetic patient who was seen with an infected tib-fib fracture, which was fixated with screws with casting of the right lower extremity in addition to transient hypertension?

A. 823.80, 823.81, 401.9, 27752, 29405
B. 823.92, 958.3, 796.2, 27758
C. 823.82, 958.3, 796.2, 27532, 27759
D. 823.92, 401.9, 27532, 27780, 29405

132. What are the correct code(s) for a patient who had ECE of the right eye due to a cataract associated with glaucoma, which included lens implant?

A. 366.31, 66830
B. 366.9, 365.9, 66882-RT
C. 365.9, 366.31, 66984-RT
D. 36631, 365.9, 66982-RT

133. What are the correct code(s) for a patient who was seen for complaints of distention and progressively severe abdominal pain, which were diagnosed to be associated with carcinoma of the descending colon, which had invaded through the bowel wall into the subserosa into the lymphatics in that area for which she received a sigmoid colectomy with anastomosis?

   A. 154.0, 196.2, 44140
   B. 198.9, 44140
   C. 154.0, 198.9, 44150
   D. 154.0, 55140

134. What are the correct code(s) for a patient who delivered a single liveborn at 36 weeks who developed severe hypertension and albuminuria after delivery and who had been under physician care throughout her entire pregnancy?

   A. 650, V27.0, 50400
   B. 642.21, 642.52, V27.0, 59409
   C. 650, 644.21, 401.9, 646.2, 59400
   D. 642.51, 644.21, V27.0, 59400

135. What are the correct code(s) for a patient who received a right shoulder hemiarthroplasty due to complaints of right shoulder pain within the glenohumeral area, which is complicated with inflammation? She had a cortisone injection 3 weeks ago for the pain but it did not relieve the pain. She is also ESRD with peripheral vascular disease with the shoulder pain attributable to fragmentation with sequestrum of the humeral head due to osteonecrosis with osteomyelitis.

   A. 730.11, 585, 443.9, 23470
   B. 733.40, 443.9, 585, 24400
   C. 730.11, 733.40, 23470
   D. 730.11, 733.40, 585, 23472

136. What are the correct code(s) for a patient who was diagnosed with ovarian cysts including pelvic adhesions and endometriosis of the tubes, for which her tubes were removed and she was given a TAH in addition to a Burch bladder suspension?

   A. 620.2, 614.6, 615.9, 58210
   B. 620.2, 58150, 617.2, 58140
   C. 620.2, 614.6, 617.2, 58152
   D. 620.2, 568.0, 58150, 58140

137. What are the correct code(s) for a patient who was experiencing severe hypoxemia with oxygen saturations of 82% with decreased level of consciousness when seen in the Emergency Room with possible pulmonary embolism due to COPD? The history and exam were detailed and the medical decision making was high complexity. The patient has a history of long time tobacco abuse with COPD, which is contributing to the hypoxia and panic attacks. She is currently being treated for chronic urinary incontinence. Her fingernails are cyanotic.

   A. 496, 788.30, 300.01, V15.82, 99284
   B. 411.89, 496, V15.82, 300.01, 625.6, 99284
   C. 411.89, 625.6, 300.01, 99285
   D. 799.0, 496, 788.30, 99203

138. What are the correct code(s) for a patient who was trimming a callus from his left foot, which has ulcerated as complicated by his diabetes and extends into the muscle fascia, which is excised today in the outpatient clinic at the hospital?

   A. 707.15, 11011
   B. 250.80, 707.15, 11043
   C. 707.15, 250.00, 11011
   D. 250.70, 707.15, 11043

139. What are the correct code(s) for a patient who was seen in an outpatient care clinic for an EKG with three leads and two-view chest x-ray due to complaints of rapid pulse and shortness of breath? Patient had angioplasty six months ago and is currently under treatment for hypertension. X-ray demonstrated cardiomegaly. Patient admitted to hospital.

   A. 429.3, 401.9, 71020, 93040
   B. 402.9, 429.3, 93041, 71030
   C. 402.9, 71020, 93040
   D. 429.3, 71020, 93041

140. What are the correct code(s) for a patient who received an ankle arthroplasty with implant, syndesmosis arthrodesis, harvesting of distal tibial bone graft, removal of external fixator under anesthesia for previous fracture with which the patient has continued to have problems due to nonunion?

   A. 824.8, 27700, 27870, 20670
   B. 824.8, 27700, 20900, 20670
   C. 733.82, 905.4, 27702, 20900, 27870, 20694
   D. 733.82, 905.4, 27702, 20900

141. What are the correct code(s) for a 4-year-old patient who received repair of an umbilical hernia, which was incarcerated with mesh, and release of the omentum?

   A. 552.1, 49582
   B. 553.1, 49580

C. 552.1, 49582, 49568

D. 553.1, 49587, 49568

142. What are the correct code(s) for a patient who was admitted to the hospital for diaphoresis and purpura with possible thrombocytopenia due to use of prescription drugs? The patient has been treated for chronic hemolytic anemia in addition.

A. 287.4, 287.2, 283.0

B. 287.3, 283.9

C. 287.4, 283.9

D. 287.2, 287.3

143. What are the correct code(s) for a 14-year-old patient who was diagnosed with hypertrophy of the adenoids, which were excised a week ago, but returns today for excision of his tonsils, which were found to be inflamed and enlarged at the patient's follow-up visit?

A. 474.11, V45.89, 42821

B. 474.00, V45.89, 42826-78

C. 47400, 42821-78

D. 474.11, 42826

144. What are the correct code(s) for a patient who was seen in the Emergency Room for lacerations sustained when he fell through a glass window? The lacerations consisted of a 2.8 cm heavily contaminated wound of the right cheek that required single layer closure, layered closure of the superficial fascia of the chin that is 3 cm, and a layered closure of the superficial fascia of a neck wound that was 3.3 cm and required ligation of muscular blood vessels and removal of glass.

A. 873.8, 12052, 12013, 13132

B. 874.8, 12054

C. 874.9, 873.44, 873.41, 12052, 12013

D. 874.9, 873.44, 873.41, 12053, 12042

145. What are the correct code(s) for a patient who had a loose screw removed from her leg for fixation of a fracture?

A. 996.4, 27500

B. 996.4, 20680

C. V52.8, 20680

D. V52.8, 20650

146. What are the correct code(s) for a patient who received a left heart catheterization due to persistent chest pains, although the stress test was negative with possible MI? The pain is left-sided and not related to exertion or associated with shortness of breath. The angiocath was placed in the right radial artery and into the left main, LAD, and circumflex, with angiographies of the right coronary artery and left ventricle. Findings were normal coronary arteries and ventricular function.

A. 786.50, 93510, 93545, 93543, 93555, 93556

B. 414.00, 93510, 93543

C. 410.9, 93510, 93556

D. 786.50, 93510, 93555, 93556

147. What are the correct code(s) for a patient who received debridement with anesthesia for 7 days for burns sustained over 32% of his body with 15% as third-degree burns with infection and the remainder as second-degree?

A. 948.31, 958.3, 16015 × 7

B. 949.3, 949.2, 16025

C. 949.2, 949.3, 16015

D. 948.31, 16015 × 7

148. What are the correct code(s) for a patient who received a bunionectomy with proximal metatarsal osteotomy and internal fixation with wire for metatarsus primus varus, hallux valgus, and bunion?

A. 28296, 28740

B. 735.0, 735.1, 28292

C. 727.1, 735.0, 735.1, 28296

D. 735.0, 28292

149. What are the correct code(s) for a patient who had two adjacent malignant skin lesions removed from the chest that were 1.2 cm and 1.4 cm with one excision, which was then covered with a Z-plasty?

A. 195.111603, 14000

B. 174.9, 11603

C. 195.1, 11602 × 2, 14000

D. 195.1, 14000

150. What are the correct code(s) for a patient who was seen in the Emergency Room for urinary tract infection with sepsis due to *Pseudomonas*, for which the patient was provided antibiotics? History and exam for this visit were expanded problem focused and the medical decision making was straightforward.

A. 995.91, 99282

B. 599.0, 041.7, 99281

C. 038.9, 041.7, 99201

D. 995.91, 99281

ANSWERS

CPC-H MOCK EXAM 1

ICD-9 (including Volume 3), CPT, and HCPCS coding books are allowed for this exam.

The CPC-H exam is divided into two sections. The first section contains 20 questions on medical terminology, 20 on ICD-9, 20 on billing, 20 on outpatient coding, and 20 on surgery coding. Section 2 contains 50 case studies. This includes information related to APCs, DRGs, and the UB92 billing form. Sequencing and linkage of codes are important issues for this coding exam.

For the following scenarios, select the best answer. (Some scenarios may provide only ICD-9s or only CPTs, and not necessarily both. Select the correct one.)

1. Streptococcus is a(n)
   A. Rod-shaped bacterium
   B. Organism that grows in clusters
   C. Berry-shaped bacterium
   D. Chain-like virus

ANSWER C: Streptococcus is a berry-shaped bacterium that grows in twisted chains. Staphylococci is also berry-shaped bacterium but grows in clusters.

2. Which of the following is not part of the integumentary system?
   A. Lipocytes
   B. Sweat glands
   C. Paronychium
   D. Dendritic cell

ANSWER A: Dendritic cells are found in neurons of the nervous system and are not part of the integumentary system.

3. Analyte is
   A. Substance measured
   B. Tissue that is submitted for pathological examination
   C. Withdrawal of fluid
   D. An antibody

ANSWER A: Analyte is the substance measured for laboratory tests. Tissue that is submitted for pathological examination is a specimen.

4. Protrusion of the urinary bladder through the vaginal wall is
   A. Ileostomy
   B. Perineorrhaphy
   C. Cystocele
   D. Meningioma

ANSWER C: Cyst means bladder and cele means hernia. A protrusion is a hernia.

5. Ganglion is
   A. Membrane lining the alimentary canal
   B. Cluster of nerve cell bodies
   C. Inflammation of the genital tract
   D. Varices

ANSWER B: Ganglions are groups of nerve cell bodies, which are located outside of the brain and spinal cord in various areas of the body.

6. TNM is not
   A. A type of delivery system
   B. A method of classifying malignancy
   C. Inclusive of lymph nodes
   D. Concerned with metastasis

ANSWER A: TNM means tumor, node, and metastasis and is a method for classifying the malignancy of a tumor

7. A LEEP is
   A. Loupe electrocautery excision procedure
   B. Lumbar extracorporeal excision paresis
   C. Lumbar extracorporeal excision procedure
   D. Loop electrocautery excision procedure

ANSWER D: Loop is the correct spelling. Loupe is a convex magnifying lens. LEEP is a procedure that uses a loop that provides electrical stimulation to remove tissue.

8. Eversion means
   A. Turned inward
   B. Lying down
   C. Turned outward
   D. Lying on the back

ANSWER C: The "e" at the beginning of the word indicates that this is a turning outward.

9. Hypofunctioning of the adrenal cortex is known as
   A. Cushing disease
   B. Hamman disease
   C. Addison disease
   D. Tay-Sachs disease

ANSWER C: Addison disease is a deficiency in the secretion of cortical steroids as produced by the adrenal glands.

10. Receptor cells of the eyes are known as
    A. Macula
    B. Cones
    C. Cornea
    D. Retina

ANSWER B: Cones and rods are the receptor cells of the eyes and are located on the retina.

11. Jaundice is not caused by
    A. Hemolysis
    B. Pneumoconiosis
    C. Choledocholithiasis
    D. Liver disease

ANSWER B: Pneumoconiosis is an abnormal condition of the lungs caused by dust in the lungs and not related to intestinal conditions as the other possible answers are.

12. Tetralogy of Fallot does not include
    A. Cirrhosis
    B. Stenosis
    C. Septal defect
    D. Hypertrophy

ANSWER A: Tetralogy of Fallot is a congenital malformation of the heart, of which cirrhosis certainly would not be included. It includes hypertrophy of the right ventricle, shift of the aorta to the right, ventricular septal defect, and pulmonary artery stenosis.

13. Achondroplasia is also known as
    A. Hirsutism
    B. Epstein-Barr
    C. Cretinism
    D. Dwarfism

ANSWER D: Achondroplasia is a congenital condition in which the bones of the arms and legs do not grow properly.

14. Formation of connection between two parts of the colon is known as
    A. Colostomy
    B. Colocolostomy
    C. Cholecystectomy
    D. Cholostomy

ANSWER B: The connection is between two parts of the colon so it would be termed colo and colo to indicate that the two parts are from the colon and not somewhere else. Colostomy is from the colon to the outside. Remember that for the colon, colo is used, not cholo.

15. Which of the following is not affected by the condition of pancytopenia?
    A. Eosinophils
    B. Antibodies
    C. Erythrocytes
    D. Platelets

ANSWER B: Antibodies are protein substances that are produced within the body to fight infections and is not one of the main blood cell types. Eosinophils are a granular leukocyte, so they are affected by the condition of pancytopenia.

16. Ventral hernia is
    A. Hernia in the area of the abdominal wall
    B. Hernia in the area of the umbilicus
    C. Hernia in the area of the groin
    D. Hernia of the upper abdomen

ANSWER A: Hernia of the groin is known as inguinal. Hernia of the upper abdomen is known as hiatal hernia. Hernia in the area of the umbilicus is known as epigastric hernia.

17. Which of the following would not be a tumor?
    A. Polyp
    B. Neoplasm
    C. Mass
    D. Pustule

ANSWER D: Pustule is an elevation of the skin with pus and is not a tissue growth.

18. Uveitis is
    A. Inflammation of the small tissue located at the back of the throat
    B. Inflammation of the iris
    C. Resistance to UV lights
    D. Abnormal condition of the chorioid

ANSWER B: Itis refers to inflammation. The tissue situated at the back of the throat is known as the uvula, which is not the combing form used in the term uveitis. Uveitis refers to the uvea, which includes the iris, ciliary body, and choroids.

19. To fill a body part with gas is
    A. Thoracentesis
    B. Insuffulated
    C. Venipuncture
    D. Bronchiectasis

ANSWER B: Insufflated means to fill a body part with vapor or gas.

20. Which of the following bones is not included in the face?
    A. Parietal
    B. Vomer
    C. Zygomatic
    D. Lacrimal

ANSWER A: The parietal bone is located in the skull, not the face.

21. What is the correct code(s) for a patient who was treated for burns on his right leg and chest area, which covered 33% of the body with 10% second-degree and the rest third-degree with infection?
    A. 948.32, 958.3, 93.57
    B. 942.20, 945.20, 93.56
    C. 948.33, 93.57
    D. 948.33, 958.3, 86.59

ANSWER A: The infection must be coded in addition to the burn. The fifth digit of the burn code is for the amount of third-degree burn, while the fourth digit is for total amount burned with all degrees. The application of dressing for the burn should also be coded.

22. What is the correct code(s) for a patient who was given CAPD for chronic kidney disease?
    A. 403.91, 39.95
    B. V56.8, 593.9, 54.98
    C. 585 54.98
    D. 585, 39.95, V56.8

ANSWER B: The patient has chronic renal disease, not failure, which is coded as 593.9. There is no mention of hypertension, so this is not coded. Continuous ambulatory peritoneal dialysis is provided, so this is coded with 54.98. The visit for the dialysis should be coded with a V code.

23. What is the correct code(s) for a patient who was treated for pneumonia due to Pneumoniae pneumonia, streptococcal?
    A. 481
    B. 482.30
    C. 482.9
    D. 480.9

ANSWER A: The code 481 is specifically for the pneumonia caused by the organism Pneumoniae pneumonia.

24. What is the correct code(s) for a patient who was treated for an acute bucket handle tear of the meniscus with an arthroscopy performed?
    A. 717.0, 80.36
    B. 836.0, 80.36
    C. 836.0, 80.26
    D. 717.0, 80.20

ANSWER C: This is an acute derangement so it is coded from the 836 codes. The arthroscopy must be coded.

25. What is the correct code(s) for newborn of a mother who has idiopathic thrombocytopenia, which is affecting the child?
    A. 776.1
    B. 666.3, 287.5
    C. 666.3
    D. 287.5

ANSWER A: Remember to code for the child, not the mother.

26. What is the correct code(s) for a patient who had severe coronary atheroma and received a double mammary-coronary bypass graft with extracorporeal circulation?
    A. 414.01, 36.10
    B. 414.00, 36.16
    C. 414.00, 36.16, 39.61
    D. 414.01, 39.61, 36.10

ANSWER C: The atheroma is coded from the artherosclerosis codes as noted in the coding book. The procedure code is for a double mammary-coronary graft. In addition the heart-lung machine must also be coded.

27. What is the correct code(s) for a patient with Hodgkin's granuloma of the spleen and axillary lymph nodes who had biopsies?
    A. 201.18, 40.11, 41.32
    B. 201.90, 41.32
    C. 201.17, 201.14, 41.32, 40.11
    D. 201.17, 41.32

ANSWER C: The granuloma must be coded twice, as it is located within the spleen and the axillary lymph nodes. Although there is a fifth digit for multiple sites, it is only for lymph nodes, so the spleen must also have a code. Biopsies indicate that they were taken from the lymph nodes and the spleen.

28. What is the correct code(s) for a patient who was treated for arthritis due to severe cuts on the wrists six months ago, which involved tendons?

    A. 716.13, 905.8
    B. 713, 905.2
    C. 716.13, 906.1
    D. 713

ANSWER A: The arthritis is coded as traumatic, since it occurred as a result of trauma. The late effect must also be coded.

29. What is the correct code(s) for a patient who was treated for pneumonitis due to varicella?

    A. 480.8
    B. 480.8, 052.8
    C. 052.1
    D. 484.8, 052.1

ANSWER C: The code 052.1 includes both the varicella and the pneumonitis, so no other codes should be listed.

30. What is the correct code(s) for a patient who was treated for excision of sequestra from the patella as a result of a staph infection?

    A. 730.16, 041.10, 77.06
    B. 730.17, 77.16
    C. 730.16, 041.1, 77.66
    D. 730.06, 77.06

ANSWER A: The sequestra are coded under osteomyelitis. The staph infection should also be coded. The sequestra are excised from the knee area, so code 77.06 should be used.

31. What is the correct code(s) for a patient who was treated for carcinoma of the bronchus, which was removed with bronchoplasty?

    A. 197.0, 32.01
    B. 162.9, 32.09, 33.48
    C. 162.9, 32.09
    D. 197.0, 33.0, 33.48

ANSWER B: The carcinoma is primary, so it must be coded as 162.9. The excision was performed with a scope, so this also must be coded. The bronchoplasty must be coded in addition to the excision of the lesion.

32. What is the correct code(s) for a patient who was treated for acute and chronic obstructive bronchitis?

    A. 493.9, 491.20
    B. 496, 491.21
    C. 496
    D. 491.21

ANSWER D: The acute and chronic are coded together with the code 491.21, which is for the obstructive bronchitis, not asthma and not COPD. Acute means acute exacerbation so it should be included in the code.

33. What is the correct code(s) for an insulin-dependent diabetic patient who was treated for hypoglycemic shock?

    A. 251.1
    B. 250.81
    C. 250.31
    D. 250.30

ANSWER B: The patient is type I diabetic, who is not reported as uncontrolled, so a fifth digit of 1 is used. The patient is diabetic and has hypoglycemic shock, so the code 250.8 must be used, which includes the coma.

34. What is the correct code(s) for a patient who was treated for CVA with hemiparesis?

    A. 436, 434.91, 342.9
    B. 438.20, 436, 907.0
    C. 436, 342.9
    D. 434.91, 342.9

ANSWER D: The CVA is described as an acute condition that is currently being treated, so it is not coded as a late condition.

35. What is the correct code(s) for a patient who was treated for hemorrhage after CABG with PTCA and insertion of a stent?

    A. 410.9, 36.05
    B. 996.72, 36.05, 36.06
    C. 410.9, 36.05, 36.06
    D. 996.72, 36.05

ANSWER B: This is a surgical complication, so it is coded from the 996 codes. PTCA is percutaneous transluminal coronary angioplasty, which must be coded in addition to the insertion of the stent.

36. What is the correct code(s) for a pregnant woman who was injured in a car accident and received a severe laceration of the scalp requiring 8 stitches?

   A. 873.1, V22.2, 86.59
   B. 851.9, 86.51
   C. 851.17, V22.2, 02.92
   D. 873.1, V22.2, 02.92

ANSWER A: The laceration is of the skin and does not extend into the brain or skull, so it is coded as a wound. The repair is suturing of the skin, so it is coded as closure of skin. The fact that the woman is pregnant should be coded with an incidental pregnancy code, V22.2.

37. What is the correct code(s) for a patient who delivered a single live birth at 42 weeks with a third-degree perineal laceration with surgical rupturing of membrane after delivery had begun?

   A. 645.21, 664.21, 73.09
   B. 664.21, V27.0, 73.6
   C. 645.11, V27.0, 73.01
   D. 645.11, 664.21, V27.0, 73.09

ANSWER D: The pregnancy was longer than usual, but not prolonged, so the code must reflect this. The perineal laceration is not considered postpartum as it occurred during delivery. The type of birth must also be coded. The delivery is coded with the rupturing of the membrane after delivery had started.

38. What is the correct code(s) for a patient who was treated with chemotherapy for carcinoma of the muscle of the right leg?

   A. V58.1, 171.3
   B. 195.5, V58.1
   C. 239.2
   D. V58.1, 195.5

ANSWER A: Chemotherapy should be coded first. The carcinoma is of the muscle, which is connective tissue, so this should be coded from that category—connective tissue.

39. What is the correct code(s) for a patient who was suffering from seizures after having ingested her husband's prescription for Benzedrine in addition to a large quantity of alcohol?

   A. 969.7, E939.7
   B. 780.39, 980.9, E939.7, E947.8
   C. 969.7, 980.9, 780.39
   D. 436, E939.7

ANSWER C: This patient ingested her husband's prescription with alcohol, which is against physician advice, so this is coded as a poisoning for the prescription only.

40. What is the correct code(s) for a patient who received a herniorrhaphy for a direct inguinal hernia with graft?

   A. 550.90, 53.01
   B. 550.00, 53.02
   C. 550.90, 53.03
   D. 551, 53.11

ANSWER C: A direct inguinal hernia is repaired, so this is included in the repair code. Herniorrhaphy is coded with codes for hernia repair. Graft is included in the code for the repair.

41. Third-party payer is

   A. The medical office
   B. The patient
   C. The guarantor
   D. The organization responsible for payment

ANSWER D: The organization, such as insurance company, responsible for payment is the third-party payer.

42. Which is not true about a PAR?

   A. Assignment of benefits is not necessary
   B. Payment is sent directly to PAR
   C. Additional percentage of reimbursement
   D. Patients are advised who PARS are

ANSWER A: A PAR is a physician who has chosen to participate in the Medicare program. Assignment of benefits must be signed, which indicates that the physician will accept Medicare payment in full payment and will not charge patients additional amounts.

43. With CHAMPUS, the following is true

   A. Now known as TRICARE
   B. Bills are submitted based on insured's home base
   C. CHAMPUS is always the secondary payer
   D. None of the above

ANSWER A: The bills are submitted by location where services are rendered, not by home base. CHAMPUS is now referred to as **TRICARE**. CHAMPUS is not always the secondary payer as Medicare and Medicaid are more likely to be the secondary payer.

44. Which of the following is not a correct billing procedure?

   A. Writing clearly and legibly
   B. Obtaining proper signatures
   C. Whiting out numbers
   D. Providing patient name and address

ANSWER C: Whiting out or erasing numbers if not allowable. Proper signatures and legible writing are required.

45. Fiscal intermediary is

   A. Guarantor
   B. Company responsible for payment of Medicare bills
   C. Person responsible for payment for medical services
   D. Collection agency

ANSWER B: Medicare contracts with fiscal intermediaries, such as insurance companies, for the provision of billing services, who interact with patients and physicians

46. Lifetime reserve is

   A. Coinsurance
   B. An additional 30 days of hospital coverage
   C. An additional 60 days of hospital coverage
   D. A monetary value that can be applied to medical services

ANSWER B: With Medicare, lifetime reserve is 30 additional days of hospital coverage that can be used only once but can occur over numerous visits.

47. Which of the following is information that does not need to be collected during registration?

   A. Deductible
   B. Insurance company
   C. Copayment
   D. Coinsurance

ANSWER A: Deductible is the amount that the patient must pay in total before insurance begins coverage, such as $200. Medical practices do not keep track of this since it applies to all medical services offered by any facility or practice; instead the insurance company keeps track of this.

48. If a Medicare patient is injured while at work at Microsoft, what would be the proper procedure for reimbursement?

   A. Send bill to employer's insurance company for their employees first
   B. Send bill to Worker's Compensation insurance carrier first
   C. Send bill to Worker's Compensation second
   D. Send bill to Medicare first

ANSWER B: Worker's Compensation would be the first payer for this account.

49. UHDDS is not

   A. Used by long term care facilities
   B. Inclusive of patient diagnoses
   C. Inclusive of E codes
   D. The standard for hospital patient information

ANSWER A: Uniform Hospital Discharge Data Set was developed to ensure that a certain data set of information was collected about patients so that there was consistency in reporting and billing. Long-term care facilities, however, use the MDDS, Minimum Discharge Data Set instead.

50. Which of the following would not be considered an occurrence code on the CMS-1450 form?

   A. Comorbidities
   B. Illness
   C. Auto accident
   D. Last menstrual period

ANSWER A: Occurrence codes indicate significant events that are related to the provision of health care services.

51. Guarantor is

   A. The insurance company responsible for payment
   B. The individual responsible for payment
   C. The insured
   D. The physician providing services

ANSWER B: The guarantor is the individual responsible for payment of services, such as a parent, and is not necessarily the insured.

52. CCI is not

   A. Incorporated within OCEs
   B. Applicable to correct coding combinations
   C. Composed of various edits
   D. A determining factor in the RBRVS

ANSWER D: The CCIs are the Corrective Coding Initiatives and are incorporated within the OCEs,

CPC-H Mock Exam 1   207

Outpatient Code Editors. They provide information about which codes link together properly by providing information about various edits. They are not an element used to determine the Resource-Based Relative Value System for charges.

53. What is the proper APC billing method for outpatient observation services?

   A. Never bill for outpatient observation services

   B. Bill for no more than 24 hours

   C. Bill in increments of one hour

   D. Do not bill with other services that have a status indicator of T

ANSWER D: A status indicator of T indicates that another procedure was done at the time the observation service is listed, therefore, the observation cannot be listed. Observation status can only be billed with certain conditions for outpatient services, such as chest pain.

54. Allowable charge is based on charges that are

   A. Usual

   B. Medically necessary

   C. Customary

   D. Reasonable

ANSWER B: Medically necessary is important in determining if a code is paid at all, but is not the basis for determining the allowable charge. Allowable charge is determined by what is the usual, customary and reasonable fee for service.

55. Medicare does not provide coverage for

   A. Senior citizens

   B. Indigent

   C. Patients with kidney failure

   D. Blind

ANSWER B: Medicaid may provide coverage for indigent people, not Medicare.

56. With APCs, a status indicator of C indicates that

   A. This service is a transitional pass-through

   B. Code has been deleted

   C. This is a multiple procedure

   D. These charges will only be paid as inpatient by Medicare, not outpatient

ANSWER D: Status indicator C with APCS indicates that this is service will only be paid if it is performed as inpatient, not outpatient.

57. If a patient was seen in the Emergency Room in Los Angeles, California, for appendicitis, where would the bill be sent for reimbursement if the patient is enrolled in the U.S. Air Force in Tucson, Arizona, and is presently living in Butte, Montana?

   A. Tucson

   B. Butte

   C. Los Angeles

   D. None of the above

ANSWER C: Los Angeles; for the military, services are paid out of where services were provided.

58. What practices can a medical facility perform in order to collect payment for services rendered?

   A. Call during the day and early evening

   B. Contacting employer about patient's delinquency in payment

   C. Threaten the patient with a lawsuit

   D. Call only during the day

ANSWER A: A medical practice can call during the day, starting at 8 am until 9 pm in the evening as stated in the Fair Debt Collection Practices Act.

59. Form locators are

   A. Blocks to be completed on the CMS-1500 form

   B. Blocks to be completed on the CMS-1450 form

   C. Blocks on billing forms for diagnostic codes

   D. Blocks on billing forms indicating where services were provided

ANSWER B: The blocks on the CMS-1450 (UB-92) form are called form locators.

60. The CMS-1450 form may be referred by some people as

   A. CMS-1500 form

   B. HCFA-1500

   C. UB-92

   D. Physician billing form

ANSWER D: The CMS-1450 form is the new name for the hospital-billing form, which was formerly known as UB-92 and is still sometimes referred to as UB-92.

61. What is the correct code(s) for a patient who was seen again in the Emergency Room for another rigid esophagoscopy for removal of a foreign object, which was performed by the same physician?

   A. 43200

208 Appendix A

B. 43215-76

C. 45332

D. 43215

ANSWER B: The procedure is indicated to be a repeat by the same physician, so modifier 76 should be added.

62. Radioimmunoassay tests are found in what section?

    A. Chemistry

    B. Diagnostic Ultrasound

    C. Nuclear Medicine

    D. Clinical Pathology

ANSWER D: Codes for radioimmunoassay tests are coded from codes 82000–84999 as stated in the Radiology Section.

63. What is the correct code(s) for a patient who is admitted to the Emergency Room for fracture and dislocation of the right bimalleolar ankle with manipulation without anesthesia?

    A. 27810

    B. 27808

    C. 27810, 27830

    D. 27808, 27830

ANSWER A: The dislocation is included in the code for the fracture, which is closed with manipulation.

64. Which of the following coding scenarios for services provided within an ASC is valid?

    A. 99203, 99202-24

    B. 99281, 99202

    C. 99281-27, 99283-27

    D. 99203, 99211

ANSWER C: Modifier 24 cannot be used with ASCs, but modifier 27 can be used, which indicates that there are multiple E/M services on the same day.

65. What is the correct code(s) for a patient who received photodynamic therapy during bronchoscopy for 1 hour for removal of abnormal tissue?

    A. 31641

    B. 31641, 96570, 96571 × 2

    C. 96570, 96571 × 2

    D. 31640

ANSWER B: The bronchoscopy for removal of the tumor using photodynamic therapy must also be coded as stated within the coding notes. The phototherapy must also be coded, which has increments of time that must each be coded separately. Notice that 96570 and 96571 are add-on codes, so this indicates that another code must be coded additionally.

66. Which of the following is not a proper way to code analytes?

    A. Report separately for multiple specimens from different sources

    B. Report separately for specimens obtained at different times

    C. Must be combined into one comprehensive code that includes all analytes

    D. Analytes can be collected from any source

ANSWER C: The description under Chemistry within the Pathology and Laboratory section explains that multiple specimens from different sources or specimens obtained at different times should be coded separately, and that analytes can be collected from any source.

67. For E/M codes, the levels of history for Emergency Room visits do not include

    A. Problem focused

    B. High complexity

    C. Comprehensive

    D. Detailed

ANSWER B: Histories are coded as problem focused, expanded problem focused, detailed, and comprehensive. High complexity relates to medical decision making.

68. What is the correct code(s) for a patient who was treated with hot packs and diathermy for her injured knee?

    A. 97010, 97024

    B. 97024

    C. 97010

    D. 97039

ANSWER A: Both codes must be coded in order to include both types of services because the semicolon excludes the type of modality.

69. What is the correct code(s) for a patient who was tested for presence of cocaine?

    A. 82520

    B. 80101

    C. 80100

    D. 86160

ANSWER B: Since no more information is given, it is understood that the test is qualitative in that it is determining the presence of the drug, not quantity.

70. Administration of contrast material would be coded "with contrast" when
    A. Oral
    B. Injected
    C. Rectal
    D. Infused

ANSWER B: Contrast material is injected, not infused. Rectal and oral contrast materials are not coded as "with contrast" as stated within the coding book.

71. What would not be proper when coding for mapping for intracardiac electrophysiological procedures?
    A. Used to code for tachycardia origination site
    B. Use of multiple catheters
    C. Coded in addition to the procedure
    D. Standard mapping should be reported in addition to 3-D mapping

ANSWER D: As the coding notes indicate, standard mapping should not be coded in addition to 3-D mapping.

72. If a patient is seen in the Emergency Room for complaints of severe chest pain with expanded problem focused history and exam and moderate medical decision making complexity, which level of code would be used for billing with APCs?
    A. 0600
    B. 99283
    C. 99203
    D. 99204

ANSWER B: This is an Emergency Room visit, so it is just coded as an Emergency Room visit, not as a level code for APCs visits.

73. Which of the following conditions would be true if code 90919 is used?
    A. Patient is age 14
    B. Billed for a full month of services
    C. Patient is in the hospital
    D. Non-dialysis services were provided

ANSWER B: Code 90919 is for patients 2 to 11 years old. These codes are used for outpatient dialysis, not inpatient. This code is also used for dialysis services only.

74. What is the correct code(s) for a patient who received a B-scan, which included the mediastinum?

A. 76506
B. 71275
C. 76604
D. 76645

ANSWER C: B-scan codes are coded from the diagnostic ultrasound section. Inclusion of the mediastinum indicates that this scan is of the chest.

75. What are differences between coding for critical care in the hospital for an adult and for a child 2 years old?
    A. Pediatric critical care is coded by days, not hours
    B. Pediatric critical care is coded by age
    C. Pediatric critical care is coded by initial and subsequent
    D. Both A and C.

ANSWER D: Obviously pediatric and adult critical care both are coded by age, so this is not a difference. Pediatric care is coded by initial and subsequent, but adult is not.

76. If surgery for a patient is cancelled prior to administration of anesthesia and surgical preparation, how should this be coded?
    A. With modifier 74
    B. With modifier 73
    C. No code
    D. With modifier 52

ANSWER C: If the anesthesia has not been administered and surgical preparation has not occurred, then nothing is coded.

77. What is the correct code(s) for a patient who received a unilateral duplex scan of the aorta?
    A. 93979
    B. 93880-52
    C. 93978-52
    D. 93882

ANSWER A: The duplex scan is of the aorta, so it is coded as 93978. No modifier for reduced services is necessary, because the code states unilateral, rather than bilateral.

78. The unit of service for surgical pathology is based on
    A. Size
    B. Specimen
    C. Anatomic site
    D. Both B and C

ANSWER B: The specimen is the unit of service for surgical pathology.

79. What is the correct code(s) for a patient who is tested for *H. pylori* by enzyme immunoassay?

   A. 87338
   B. 78267
   C. 83014
   D. 87339

ANSWER D: *H. pylori* can be tested with many methods, including stool, breath, blood, scintillation, and enzyme immunoassay.

80. Which modifier cannot be applied to both physician and ASC procedures and services?

   A. 52
   B. 57
   C. 58
   D. 78

ANSWER B: Modifier 57, decision for surgery, is not available for coding ASC, because this modifier is used for physician services.

81. When coding for cardiac catheterizations, the code should be coded to

   A. Include an additional code for introduction
   B. Include catheters
   C. The highest order of each family
   D. Type of catheter

ANSWER C: With cardiac catheterizations, the correct code is based on the highest order, beginning with the lowest being first and continuing upward to second, third, fourth, etc.

82. What is the correct code(s) for a patient who received a CABG with single venous and single arterial grafts including the saphenous with endarectomy?

   A. 33517, 33533, 33572
   B. 33533
   C. 33517, 33572
   D. 33510, 33533, 33572

ANSWER A: With CABGs that include both venous and arterial grafts, both codes must be coded. The endarectomy is included in the code. Procurement of the grafts is included in the code.

83. What is the correct code(s) for a patient who received wound repair for a 6.2-cm laceration of the abdomen, which required considerable time and enlargement of the wound for adequate treatment including coagulation of various vessels extending into the muscle and fascia and repair of nerve?

   A. 13101, 35221
   B. 20102, 64872
   C. 20102, 35221, 64872
   D. 13101, 35206, 64872

ANSWER C: This service required extensive exploration, so it is coded as 20102 and not a complex repair as outlined in the coding notes in the book. The repair of the blood vessels and nerve must also be coded.

84. What is the correct code(s) for a patient who received a thoracotomy for insertion of permanent epicardial pacemaker?

   A. 33206
   B. 33200
   C. 33210
   D. 33201

ANSWER B: The thoracotomy indicates that this has epicardial electrodes since a thoracotomy is required.

85. Which of the following statements is not true about percutaneous fixation?

   A. X-rays can be used
   B. Implants are included
   C. Is always coded as open
   D. Can be inserted into the bone

ANSWER C: Percutaneous fixation can be performed either as open or closed. Implants consist of materials such as pins or screws.

86. What is the correct code(s) for a patient who received a mastoidectomy and resection with decompression and an infratemporal post-auricular approach for an intradural resection of the neoplasm of the cavernous sinus with graft with transection of the carotid artery?

   A. 61591, 61608
   B. 61591, 61607
   C. 61608
   D. 61591, 61608, 61609

ANSWER D: For the procedure, which is resection of the cavernous sinus with transection of the carotid artery and graft, all of this must be coded with two codes. The approach was infratemporal post-auricular, which must also be coded.

87. What is the correct code(s) for a patient who received care for treatment of hallux valgus that included fusion and transplantation of the metatarsal bone?

A. 28290

B. 28294

C. 28296

D. 28290, 28296

ANSWER B: This is known as a Joplin Procedure with the distinguishing characteristic being the transplantation of the tendon.

88. What is the correct code(s) for a patient who received a nephrectomy and total ureterectomy with the same incision?

A. 50543

B. 50548

C. 50234

D. 50548, 50548

ANSWER C: This is an open procedure since no scope is mentioned with the same incision, so code 50234 must be used.

89. What is the correct code(s) for a patient who had a biopsy of the transverse colon with partial excision with anastomosis through abdominal approach?

A. 44100, 45114

B. 44140

C. 44025, 45114

D. 44140, 45100, 45114

ANSWER B: The excision is the main procedure, so this is coded.

90. What is not true about coding for wound repair?

A. Debridement is not coded

B. Complex repair includes closure

C. Simple repair includes suturing

D. The lengths of the wounds are added together for each anatomic group listed

ANSWER A: Debridement can be coded if it is more extensive than would debridement would normally be, or if no other service is provided.

91. What is the correct code(s) for a patient who received I&D with incisional removal of glass of three fingers at the metacarpophalangeal joint?

A. 26075, 25028

B. 25248, 10180

C. 26075

D. 26070 × 3

ANSWER D: The removal of the glass includes the incision and drainage, so no other code is necessary.

92. What is the correct code(s) for a patient who had repair of a closed unicondylar fracture of the proximal tibia with internal fixation?

A. 27759, 20690

B. 27756

C. 27535

D. 20690

ANSWER C: Internal fixation involves an incision, so it is coded as open. The fact that the injury is closed is irrelevant to the type of care. This not an external fixation, so no additional code is needed for this.

93. What is the correct code(s) for a patient who received a flexible colonoscopy with removal of a tumor by snare including insertion of a stent?

A. 45384

B. 50605

C. 45385, 45387

D. 43256

ANSWER C: The placement of the stent must also be coded as it is not included in the code for the removal of the tumor.

94. What is the correct code(s) for a patient who received curettement of 4 benign lesions?

A. 17000, 17003 × 3

B. 11056

C. 17000, 17003

D. 11056 × 4

ANSWER B: The curettement is not described as surgical, so it should be coded from the 11000 codes. The 11000 codes specify groups of lesions, not each individual, so only one code is necessary.

95. Which of the following is not a HCPCS code?

A. I1020

B. 67316

C. B9000

D. None of the above

ANSWER A: I is not a category for HCPCS.

96. What is the correct code(s) for a patient for whom bone marrow was obtained through fine needle aspiration?

A. 10021

B. 38220

C. 10022

D. 10021, 38220

ANSWER B: Bone marrow aspiration is coded with its own code of 38220.

97. What is the correct code(s) for a patient who received selective catheterization of the right common artery and right and left middle cerebral artery and introduction?

   A. **36217, 36218**
   B. 36215
   C. 36215, 36217, 36218
   D. 36217, 36100

ANSWER A: This is not a right or left heart catheterization, so this is coded from the 36000 codes. The right middle cerebral artery is a third order and the right common artery is its first order, so only the cerebral artery needs to be coded. An additional code is necessary since the left and right cerebral arteries are catheterized. Introduction is included in the code.

98. What is the correct code(s) for a patient who received repair of a strangulated ventral hernia with mesh with repair of the intestinal obstruction?

   A. 49507
   B. 44615
   C. **44615, 49561, 49568**
   D. 44615, 49568

ANSWER C: For the ventral hernia, the mesh is also coded. This hernia is strangulated, so this must be reflected in the code. The repair of the intestines must also be coded as outlined in the coding notes. This is known as enterotomy and enterorrhaphy.

99. What is the correct code(s) for a patient who received a 2-cm Z-plasty with destruction of a malignant lesion of the scalp?

   A. 15120
   B. 14020, 17272
   C. **14020**
   D. 15120, 17262

ANSWER C: The lesion is removed with the procedure for the Z-plasty, so no excision code is needed.

100. Which of the following codes do not include the use of a microscope?

   A. **34001**
   B. 20955
   C. 17304
   D. 44970

ANSWER A: This code, 34001, does not mention the use of a microscope, as indicated by words such as microvascular.

101. What are the correct code(s) for a patient who has occlusive disease of the coronary arteries with a possible MI? The patient has a significant history of ongoing care for COPD and malignant hypertension as well as asthmatic bronchitis. Patient received a retrograde catheterization through the femoral vein of the left main coronary artery with cineangiography of the left coronary arterial system, which was then followed by insertion of a catheter into the right coronary artery with monitoring of arterial pressure, which included a venous bypass graft.

   A. 411.81, 493.00, 496, 93501, 93545, 93556
   B. **411.81, 401.0, 493.90, 93526, 93540, 93545, 93556**
   C. 410.90, 401.9, 493.90, 93510, 93526
   D. 410.90, 401.9, 496.00, 493.00, 93510, 93501, 93540, 93545

ANSWER B: There is an MI associated with the occlusive disease but is described as possible and since this coding is for the physician, the MI is not coded; therefore, code 411.81 is used. The COPD is coded with the asthmatic bronchitis and is not listed separately as described in the coding notes in the coding book. The hypertension is malignant, so this must be included in the code.

102. What are the correct code(s) for a patient who presents with complaints of severe chest pains and was found to have severe coronary artery disease with left ventricular dysfunction and ejection fraction of 30%? CABG was performed times three, which included the left internal mammary artery to distal left anterior descending artery and sequential WYE saphenous vein graft with the main branch from the aorta to the posterolateral artery and the WYE branch from the mid portion of this graft to the posterior descending artery.

   A. 414.01, 33518 × 3
   B. 429.2, 33533 × 3
   C. **414.01, 33533, 33518**
   D. 429.2, 33533, 33511

ANSWER C: The patient has severe coronary artery disease times three, which is coded with 414.01. One of the grafts was arterial, the other two were venous.

103. What are the correct code(s) for a pregnant woman with aseptic meningitis due to adenovirus?

   A. 647.83, 049.1, 322.9

B. 647.83

C. 647.83, 049.1

D. 049.1, V22.2

ANSWER C: There is no pregnancy code for meningitis, but the general code for infection while pregnant can be used. The meningitis must also be coded in addition to the organism, which are included in one code.

104. What are the correct code(s) for a patient who was diagnosed with cellulitis of the left foot with gangrene?

   A. 682.7

   B. 785.4

   C. 682.7, 785.4

   D. 785.4, 682.7

ANSWER B: Gangrene with cellulitis would be coded as 785.4, because cellulitis is included within the gangrene classification.

105. What are the correct code(s) for a patient who received trephination for drainage of brain abscess due to TB?

   A. 324.0, 013.30, 61150

   B. 324.0, 013.30, 61105

   C. 324.0, 61105

   D. 013.30, 61150

ANSWER D: The abscess is caused by tuberculosis. Trepination is used to open the skull to achieve drainage of the abscess.

106. What are the correct code(s) for a patient who presented with symptoms of glomerulonephritis with renal necrosis?

   A. 593.9, 580.4, 584.9

   B. 583.9

   C. 580.9, 584.9

   D. 580.4, 593.9

ANSWER B: The necrosis is included in the code for the glomerulonephritis.

107. What are the correct code(s) for the newborn with a mother with incompetent cervix?

   A. 654.4

   B. 622.5

   C. 622.5, V22.2

   D. 761.0

ANSWER D: Make sure that you code for the baby and not the mother, because her care is affecting the baby for whom we are coding.

108. What are the correct code(s) for a patient who received photocoagulation with an additional cryotherapy for repair of retinal detachment of the left eye with defects?

   A. 361.00, 67101

   B. 361.2, 67105

   C. 361.2, 67105, 67101

   D. 361.02, 67105

ANSWER D: This detachment is with multiple tears as noted by the description of defects. The procedure includes both cryotherapy and photocoagulation, so only the principal modality is coded, not both, as indicated in the coding notes in the coding book.

109. What are the correct code(s) for a patient who was seen in urgent care for pneumonia as a result of measles? Patient is currently being treated for arthritis. Frontal and lateral chest x-rays were taken and patient provided prescription for antibiotics. History and exam were problem focused and medical decision making was low.

   A. 055.1, 99201, 71020

   B. 055.9, 99202, 71020

   C. 055.9, 486, 99201, 71023

   D. 055.1, 486, 99212, 71023

ANSWER A: The pneumonia is due to measles, which are both coded together in one code. The visit was low level because the history and exam were low level. The x-ray was two views without any fluoroscopy or anything else.

110. What are the correct code(s) for a patient who received thoracentesis to remove fluid in the right lung with ultrasound in association with small cell carcinoma of the upper lobe?

   A. 191.1, 511.9, 32002

   B. 162.3, 511.8, 32000

   C. 162.3, 32000

   D. 239.1, 511.9, 32002

ANSWER B: This is cancerous, so it is coded as a primary neoplasm. The fluid in the lung is coded as 511.8, not as pleural effusion NOS, 511.9, as described in the alphabetic index. The procedure does not include insertion of a tube, so it is coded as thoracentesis only.

111. What are the correct code(s) for a patient who had 8 sections of hypertrophied skin removed on both sides of the body?

   A. 701.9, 11200

   B. 733.9, 11200 × 8

C. 701.9, 11200-51

D. 733.9, 11200-50

ANSWER A: The hypertrophied skin is known as skin tags. The removal of skin tags is not coded individually, so the code only needs to be listed once. No modifiers are necessary either.

112. What are the correct code(s) for a patient who received a 50 sq cm porcine graft for scarring on both arms due to third-degree burns sustained two months ago?

A. 709.2, 15000-50, 15100-50

B. 934.30, 15400-50

C. 709.2, 906.7, 15000, 15400

D. 943.30, 15400

ANSWER C: This is a late effect of a burn that has created scar tissue, which is then the main code. It is the scar tissue being grafted that is a late effect of the burn that is coded. The site must be prepared because the burn has healed and the scar tissue must be removed. Porcine indicates that the graft is a xenograft.

113. What are the correct code(s) for a patient who was seen in the Emergency Room for blackouts and vomiting due to intentional overdose with benzodiazepines? History and exam were problem focused, although the patient was not able to provide much information and the medical decision making was low.

A. 780.2, E950.3, 99282

B. 780.02, 969.4, E950.3, 99281

C. 780.02, 99282

D. 969.4, 780.2, E950.3, 99281

ANSWER D: The blackouts are listed under 780.2 with syncope. The ingestion was intentional, so this is coded as suicide attempt. The Emergency Room visit is the lowest level because the history and exam were problem focused.

114. What are the correct code(s) for a patient who received a T11–T12 laminectomy with decompression and T11–T13 posterior spinal fusion with posterior instrumentation using screws for lumbar instability, lumbar degenerative disc disease, and chronic low back pain? This included structural femoral head allografting, and intrathecal Duramorph injection. The patient was found to have left sided hemiparesis from a past CVA.

A. 722.72, 436, 342.90, 63170, 22840, 20930

B. 722.51, 438.20, 63003, 22610, 22614, 22840, 20931

C. 722.72, 63003, 22630, 20931, 22610

D. 722.51, 438.20, 436, 22630, 22840, 29030

ANSWER B: The degeneration of the thoracic disks are coded. The hemiparesis as a residual effect of the CVA is also coded, which is a late effect, but no other code is necessary for late effect since the code is all-inclusive. The dominant side is not known, although the left side is stated, but we do not know if this was the patient's dominant side. The thoracic vertebrae are involved, so the correct code must include the thoracic vertebrae. This laminectomy includes decompression. The fusion is coded as arthrodesis, and there must be two codes, because there are two spaces involved. This code does not include the laminectomy, (although there is a code that does, 22630) because this code is for the lumbar area and does not include the decompression.

The instrumentation must also be coded as stated in the coding notes in the book. The allograft must be coded also as stated in the coding book notes.

115. What are the correct code(s) for a patient who was diagnosed with osteomyelitis due to infection by Salmonella of the first metacarpal with amputation and a flap?

A. 730.2, 003.24, 26910

B. 730.2, 26952

C. 003.24, 26910

D. 003.24, 26952

ANSWER D: The code for the organism includes the code for the osteomyelitis. No other code is necessary. The amputation includes a flap and is of the finger.

116. What are the correct code(s) for a patient who received an arthroplasty with allograft for traumatic left intertrochanteric fracture due to a fall at home? The patient has also been diagnosed with adult onset diabetes mellitus with neuropathy, hypertension, hypercholesterolemia, hyperglyceridemia, and intestinal diverticulosis.

A. 820.21, 250.40, 583.81, 401.9, 272.2, 562.00, 27130

B. 820.8, 250.00, 401.9, 272.0, 272.1, 562.00, 27125

C. 820.21, 250.40, 272.2, 562.01, 27130, 27130, 29031

D. 820.8, 250.40, 583.81, 401.9, 272.0, 272.1, 27130

ANSWER A: The introchanteric fracture is coded as fracture, femur, neck, introchanteric. It is assumed to be closed, because there is no mention of it being open. The diabetes is associated with neuropathy, so both of these codes must be coded. The hypertension should also be coded, as it may influence the patient's care. The hypercholesterolemia and hyperglyceridemia are coded within the same code, so no extra code is required for these two. The arthroplasty includes the allograft so it is not coded separately.

117. What are the correct code(s) for a patient who had a 0.5 cm lesion removed from the upper frenulum of her lip, which was found to be malignant?

    A. 140.0, 11600

    B. 140.3, 40819

    C. 140.3, 11600

    D. 140.0, 40819

ANSWER C: The lesion is malignant so it is coded as primary neoplasm. The location is the upper frenulum of the lip, which is coded as internal lip. The excision of the lip is coded as excision of skin.

118. What are the correct code(s) for a patient who had complicated repair of a 3.5-cm laceration of the right cheek and 4.6-cm complete laceration of the nose, which all required sutures?

    A. 873.51, 873.30, 13132, 13152

    B. 879.9, 13132, 13133

    C. 873.30, 873.51, 12053

    D. 873.51, 873.30, 13132, 13133

ANSWER A: The lacerations, or wounds, are coded by site. The sites use different codes, so they are not added together for repair.

119. What are the correct code(s) for a patient who had a tube inserted into the lungs for drainage of fluid? Patient has been diagnosed with metastatic carcinoma of the right lung from the esophagus and is currently receiving chemotherapy.

    A. V28.1, 162.9, 43267

    B. 150.9, 197.0, 43267

    C. V28.1, 511.9, 162.9, 32020

    D. 150.9, 511.9, 197.0, 32020

ANSWER D: The lung cancer is secondary and the esophageal cancer is primary. There is no mention that the esophageal cancer has been eradicated, so it is coded as current, not as a history. The fluid in the lungs should also be coded, which is known as pleural effusion. The chemotherapy is not coded as the patient is receiving services today for the insertion of the chest tube and not the chemotherapy. Insertion of a chest tube is known as thoracostomy.

120. What are the correct code(s) for a patient who was diagnosed with severe external hemorrhoids, for which a rigid protoscopy was performed for visualization in addition to open complete hemorrhoidectomy with a Harmonic scalpel, which were found to be irreducible?

    A. 455.6, 46230, 46600

    B. 455.3, 46230

    C. 455.5, 46250

    D. 455.3, 46230, 46600

ANSWER C: The hemorrhoids must be coded as external, and they are complicated due to being irreducible. The hemorrhoidectomy was external and complete, which are included in the code. The protoscopy was used to visualize the hemorrhoids and is, therefore, part of the procedure for the hemorrhoid removal.

121. What are the correct code(s) for a patient who had subtemporal craniectomy of ganglion cysts?

    A. 727.43, 61450

    B. 727.41, 61440

    C. 727.43, 20612

    D. 727.41, 61450

ANSWER A: The subtemporal area is of the cranium. This procedure involves a craniectomy to produce section or decompression of the ganglion.

122. What are the correct code(s) for a patient who developed ulcerative proctitis due to radiation for endometrial carcinoma for which she had a vaginal hysterectomy? Today, patient received rigid proctoscopy with topical treatment.

    A. 601.0, 46600

    B. 990, 556.2, 46600

    C. 990, 46604

    D. 556.2, 990, 46600

ANSWER D: The proctitis is coded using the code for effects from radiation for unspecified conditions. The proctoscopy is coded as an anoscopy.

123. What are the correct code(s) for a patient who had a first dorsal compartment decompression fasciotomy, carpal tunnel release, and flexor tenosynovectomy for tendonitis of the left wrist of the first dorsal compartment, carpal tunnel syndrome of the left wrist, and left wrist chronic flexor tenosynovitis?

    A. 354.0, 727.05, 726.90, 25115, 25118, 64721

    B. 354.0, 727.00, 726.00, 64721, 25118

    C. 354.0, 727.05, 25020

    D. 354.0, 727.00, 25115, 25020

ANSWER C: The carpal tunnel syndrome is included in the code 25020 as well as the synovectomy. The tendonitis is included in the code for the tenosynovitis, so it is not coded separately.

124. What are the correct code(s) for a patient who presents with hyperparathyroidism due to renal disease as associated with diabetes? Patient underwent a subtotal parathyroid resection.

   A. 588.81, 250.40, 60240
   B. 252.0, 250.40, 60500
   C. 250.40, 583.81, 588.81, 60500
   D. 250.40, 588.81, 60240

ANSWER C: The hyperparathyroidism is due to the renal dysfunction and is included in the code, but the renal disease and diabetes must also be coded. The excision of the parathyroid is coded also.

125. What are the correct code(s) for a patient who had intradural resection of neoplasm of the parasellar area with infratemporal approach with a cranioplasty with the use of a synthetic graft?

   A. 61607, 61591
   B. 61607, 61590, 61618
   C. 61607, 61591, 61618
   D. 61590, 61607

ANSWER C: The procedure involves the parasellar area as well as the approach. The repair is coded, though it usually is not, because it did involve cranioplasty with a graft.

126. What are the correct code(s) for a patient who presents today with complaints of dyspnea, chest pain, and nauseas? He received an ECG that demonstrated an old MI and enlarged left atrium due to hypertension.

   A. 412, 402.90, 89.52
   B. 410.90, 429.3, 89.51
   C. 410.90, 429.3, 89.52, 401.9
   D. 412, 89.51, 402.90, 401.9

ANSWER A: Code 412 is for an old MI. The hypertrophy should also be coded as involving hypertension, so this is coded from the hypertensive heart disease codes. Nothing else is coded because they are symptoms of the MI. The ECG is not specified as to the number of leads, so this is coded as NOS, which is 89.52.

127. What are the correct code(s) for a patient who has a history of gallstones and presents today again with pancreatitis with cholelithiasis? A cholecystectomy was performed for removal with an intraoperative cholangiogram.

   A. 577.1, 574.20, 47605
   B. 577.0, 575.0, 47563
   C. 577.1, 575.0, 47563
   D. 574.20, 47605

ANSWER A: The pancreatitis has occurred again, so the patient has recurrent pancreatitis. The cholelithiasis is associated with the presence of gallstones, so this is coded. There is no mention of a scope being used, so this is coded as an operation without the scope.

128. What are the correct code(s) for a patient who had abnormal findings of an AIDS test, but the patient is not exhibiting any symptoms at this time?

   A. V01.7
   B. V08
   C. 042
   D. 795.71

ANSWER B: The test is positive but the patient is asymptomatic, so V08 is correct.

129. What are the correct code(s) for a patient who had a graft implanted for repair of an abdominal aneurysm and aorta with endarterectomy?

   A. 442.84, 34800, 33572
   B. 441.2, 35091
   C. 440.4, 34800
   D. 442.84, 35092

ANSWER B: The graft is part of the procedure for repair of a mesenteric abdominal aneurysm, which also includes the endarterectomy. The aneurysm is not reported as ruptured, so this is not coded as such.

130. What are the correct code(s) for a patient who was seen in the urgent care for shortness of breath? She has a 40-year history of smoking two packs a day but quit two months ago. She was diagnosed as having an acute exacerbation of her COPD as complicated by her emphysema with bronchitis. She was given oxygen to continue at home in addition to prescriptions for Flovent and prednisone. The history was expanded problem focused, the exam was detailed, and the decision making was low.

   A. 491.21, V15.82, 99202
   B. 496, 492.8, 305.1, 99282
   C. 496, V15.82, 99282
   D. 496, 491.21, 305.1, 99203

ANSWER A: The COPD is not coded separately but is included in the code for bronchitis and emphysema. The history of tobacco use is important since it is long and since the patient has breathing problems. The visit is at the moderate level due to the lower category for the history.

131. What are the correct code(s) for a diabetic patient who was seen with an infected tib-fib fracture, which was fixated with screws with casting of the right lower extremity in addition to transient hypertension?

   A. 823.80, 823.81, 401.9, 27752, 29405

   B. 823.92, 958.3, 796.2, 27758

   C. 823.82, 958.3, 796.2, 27532, 27759

   D. 823.92, 401.9, 27532, 27780, 29405

ANSWER B: The tibia-fibula fracture is open as indicated by the term "infection." Only one code is necessary to include all of this information. The infection should also be coded. The transient hypertension is coded as high blood pressure and not hypertension, because it is only temporary. The closed reduction is coded as closed manipulation, and the casting is included in that code.

132. What are the correct code(s) for a patient who had ECE of the right eye due to a cataract associated with glaucoma, which included lens implant?

   A. 366.31, 66830

   B. 366.9, 365.9, 66882-RT

   C. 365.9, 366.31, 66984-RT

   D. 36631, 365.9, 66982-RT

ANSWER C: Notice that the glaucoma and cataract are an example of multiple coding, which is indicated by slanted brackets in the alphabetic index, which means the ordering of the codes must be the same as indicated in the index. ECE means extracapsular extraction for cataract removal. Lens implant was also performed, which is included in the code. The right eye should be indicated with a modifier.

133. What are the correct code(s) for a patient who was seen for complaints of distention and progressively severe abdominal pain, which were diagnosed as associated with carcinoma of the descending colon, which had invaded through the bowel wall into the subserosa into the lymphatics in that area for which the patient received a sigmoid colectomy with anastomosis?

   A. 153.2, 196.2, 44140

   B. 198.9, 44140

   C. 154.0, 198.9, 44150

   D. 153.2, 55140

ANSWER A: The carcinoma of the colon is primary and the carcinoma of the lymph nodes is secondary. Notice that there is the selection for colic under lymph nodes, which is applicable in this case, because the invasion is described as moving from the colon.

134. What are the correct code(s) for a patient who delivered a single liveborn at 36 weeks who developed severe hypertension and albuminuria after delivery and who had been under physician care throughout her entire pregnancy?

   A. 650, V27.0, 50400

   B. 642.21, 642.52, V27.0, 59409

   C. 650, 644.21, 401.9, 646.2, 59400

   D. 642.51, 644.21, V27.0, 59400

ANSWER D: The delivery is early, anything before 37 weeks, so early onset of delivery should be coded. There is a fifth digit of 2 because there is delivery, and there is the albuminuria postpartum, which indicates that it began after delivery making it postpartum (within the puerperium period). Code 650 is not used for delivery because of the postpartum condition, which is indicated by notes within the coding book. The outcome of the delivery must also be coded.

135. What are the correct code(s) for a patient who received a right shoulder hemiarthroplasty due to complaints of right shoulder pain within the glenohumeral area, which is complicated with inflammation? She had a cortisone injection 3 weeks ago for the pain but it did not relieve the pain. She is also ESRD with peripheral vascular disease with the shoulder pain attributable to fragmentation with sequestrum of the humeral head due to osteonecrosis with osteomyelitis.

   A. 730.11, 585, 443.9, 23470

   B. 733.40, 443.9, 585, 24400

   C. 730.11, 733.40, 23470

   D. 730.11, 733.40, 585, 23472

ANSWER A: The osteomyelitis is coded instead of the osteonecrosis, because it is associated with the osteomyelitis, which is indicated by the problems of inflammation and sequestrum, which are manifestations of the osteomyelitis. The end stage renal disease should also be coded in addition to the peripheral vascular disease. The hemiarthroplasty should be coded, which includes the glenohumeral joint of the shoulder.

136. What are the correct code(s) for a patient who was diagnosed with ovarian cysts including pelvic adhesions and endometriosis of the tubes, for which her tubes were removed and she was given a TAH in addition to a Burch bladder suspension?

   A. 620.2, 614.6, 615.9, 58210

   B. 620.2, 58150, 617.2, 58140

   C. 620.2, 614.6, 617.2, 58152

   D. 620.2, 568.0, 58150, 58140

ANSWER C: The ovarian cysts should be coded in addition to the pelvic adhesions. The endometriosis is of the tubes, so this should be included in the code. TAH is a total abdominal hysterectomy. Suspension is known as pexy and is included in code 58152 for the TAH.

137. What are the correct code(s) for a patient who was experiencing severe hypoxemia with oxygen saturations of 82% with decreased level of consciousness when seen in the Emergency Room with possible pulmonary embolism due to COPD? The history and exam were detailed and the medical decision making was high complexity. The patient has a history of long time tobacco abuse with COPD, which is contributing to the hypoxia and panic attacks. She is currently being treated for chronic urinary incontinence. Her fingernails are cyanotic.

    A. 496, 788.30, 300.01, V15.82, 99284

    B. 411.89, 496, V15.82, 300.01, 625.6, 99284

    C. 411.89, 625.6, 300.01, 99285

    D. 799.0, 496, 788.30, 99203

ANSWER A: The hypoxia is not coded, as it is a symptom of COPD. The COPD should be coded. It is not associated with any other condition, such as asthma, so it is coded as 496. The history of smoking should be coded as it is significant for the outcome of this patient's care and the current problems. The pulmonary embolism is not coded as it is described as possible and this is an Emergency Room visit. The panic attacks should also be coded. The Emergency Room visit was not the highest level in all three elements, so the next highest code is correct.

138. What are the correct code(s) for a patient who was trimming a callus from his left foot, which has ulcerated as complicated by his diabetes and extends into the muscle fascia, which is excised today in the outpatient clinic at the hospital?

    A. 707.15, 11011

    B. 250.80, 707.15, 11043

    C. 707.15, 250.00, 11011

    D. 250.70, 707.15, 11043

ANSWER B: The code for the diabetes must be listed first and, then the code for the ulcer must also be coded. Diabetes code 250.80 is listed as directed in the notes in the 707 section. No circulatory disorder can be assumed because it is not stated. The debridement does not refer to removal of foreign material, but does extend into the muscle fascia, so these must be included in the code.

139. What are the correct code(s) for a patient who was seen in an outpatient care clinic for an EKG with three leads and two-view chest x-ray due to complaints of rapid pulse and shortness of breath? Patient had angioplasty six months ago and is currently under treatment for hypertension. X-ray demonstrated cardiomegaly. Patient admitted to hospital.

    A. 429.3, 401.9, 71020, 93040

    B. 402.9, 429.3, 93041, 71030

    C. 402.9, 71020, 93040

    D. 429.3, 71020, 93041

ANSWER A: Patient has hypertension but doctor did not describe the cardiomegaly as due to hypertension, so they are coded separately. The EKG was with three leads and interpretation. The X-ray was two-views.

140. What are the correct code(s) for a patient who received an ankle arthroplasty with implant, syndesmosis arthrodesis, harvesting of distal tibial bone graft, removal of external fixator under anesthesia for previous fracture, with which the patient has continued to have problems due to nonunion?

    A. 824.8, 27700, 27870, 20670

    B. 824.8, 27700, 20900, 20670

    C. 733.82, 905.4, 27702, 20900, 27870, 20694

    D. 733.82, 905.4, 27702, 20900

ANSWER C: The fracture is not coded as the effects being treated today are late effects for the fracture that has already been unsuccessfully treated. The failure of the fracture to heal would be considered non-healing, so this should be coded. There is implant, so code 27702 must be used instead of 27700. The harvesting of the autogenous bone graft should also be coded. The syndesmosis arthrodesis is fusion of the bone and ligaments in the ankle. The removal of the fixation device should also be coded as external.

141. What are the correct code(s) for a 4-year-old patient who received repair of an umbilical hernia, which was incarcerated with mesh, and release of the omentum?

    A. 552.1, 49582

    B. 553.1, 49580

    C. 552.1, 49582, 49568

    D. 553.1, 49587, 49568

ANSWER A: Incarcerated indicates that the hernia is obstructed. The umbilical hernia is incarcerated, so this must be included in the code. The patient is 4 years old, so this is a determining factor for the code also. There is no mesh coded because mesh is only coded as extra for incisional and ventral hernias.

142. What are the correct code(s) for a patient who was admitted to the hospital for diaphoresis and purpura with possible thrombocytopenia due to use of

prescription drugs? The patient has been treated for chronic hemolytic anemia in addition.

   A. 287.4, 287.2, 283.0
   B. 287.3, 283.9
   C. 287.4, 283.9
   D. 287.2, 287.3

ANSWER C: The code for thrombocytopenia is coded although it is described as possible, because this is a hospital admission. The thrombocytopenia is also associated with drugs, so this must be included in the code. The diaphoresis and purpura are symptoms of the thrombocytopenia. The hemolytic anemia should also be coded as chronic, which is code 283.9.

143. What are the correct code(s) for a 14-year-old patient who was diagnosed with hypertrophy of the adenoids, which were excised a week ago, but returns today for excision of his tonsils, which were found to be inflamed and enlarged at the patient's follow-up visit?

   A. 474.11, V45.89, 42821
   B. 474.00, V45.89, 42826-78
   C. 47400, 42821-78
   D. 474.11, 42826

ANSWER B: The tonsils are inflamed in addition to the hypertrophy, so this is part of the selection criteria for the correct tonsillitis code. The history of the adenoidectomy should also be coded as it is pertinent to the current care. The adenoids were removed earlier, so they are not coded. The modifier 78 must be attached to the tonsillectomy, as the procedure was performed within the postoperative period of the previous surgery and is related.

144. What are the correct code(s) for a patient who was seen in the Emergency Room for lacerations sustained when he fell through a glass window? The lacerations consisted of a 2.8 cm heavily contaminated wound of the right cheek that required single layer closure, layered closure of the superficial fascia of the chin that is 3 cm, and a layered closure of the superficial fascia of a neck wound that was 3.3 cm and required ligation of muscular blood vessels and removal of glass.

   A. 873.8, 12052, 12013, 13132
   B. 874.8, 12054
   C. 874.9, 873.44, 873.41, 12052, 12013
   D. 874.9, 873.44, 873.41, 12053, 12042

ANSWER D: There is removal of glass and ligating of blood vessels, so the wound is complicated for the chin. The simple repair of the 2.8-cm wound is coded as intermediate repair, because the wound is heavily contaminated, so this must be added to the wound of the chin and is also intermediate repair. The repair of the 3 cm neck wound is intermediate but has its own code, but also needs a code for the ligation.

145. What are the correct code(s) for a patient who had a loose screw removed from her leg for fixation of a fracture?

   A. 996.4, 27500
   B. 996.4, 20680
   C. V52.8, 20680
   D. V52.8, 20650

ANSWER B: The patient had a loose screw, so this was replaced, which is coded from the mechanical complications of surgery and aftercare. For the procedure, the screw indicates that the procedure is deep. The fixation of the fracture is not coded as the treatment is for the removal of the screw.

146. What are the correct code(s) for a patient who received a left heart catheterization due to persistent chest pains, although the stress test was negative with possible MI? The pain is left-sided and not related to exertion or associated with shortness of breath. The angiocath was placed in the right radial artery and into the left main, LAD, and circumflex, with angiographies of the right coronary artery and left ventricle. Findings were normal coronary arteries and ventricular function.

   A. 786.50, 93510, 93545, 93543, 93555, 93556
   B. 414.00, 93510, 93543
   C. 410.9, 93510, 93556
   D. 786.50, 93510, 93555, 93556

ANSWER A: Only the chest pain is coded because nothing else is found and the MI is stated to be possible.

147. What are the correct code(s) for a patient who received debridement with anesthesia for 7 days for burns sustained over 32% of his body with 15% as third-degree burns with infection and the remainder as second degree?

   A. 948.31, 958.3, 16015 × 7
   B. 949.3, 949.2, 16025
   C. 949.2, 949.3, 16015
   D. 948.31, 16015 × 7

ANSWER A: All types of burns constitute 32% of the surface burned, which is the criterion for selecting the fourth digit. However, the fifth digits is based on how much of the area burned is third-degree only. The infection must also be coded. The debridement was for seven times, so the code must be multiplied by 7. It was also done with anesthesia.

148. What are the correct code(s) for a patient who received a bunionectomy with proximal metatarsal osteotomy and internal fixation with wire for metatarsus primus varus, hallux valgus, and bunion?

   A. 28296, 28740
   B. 735.0, 735.1, 28292
   C. 727.1, 735.0, 735.1, 28296
   D. 735.0, 28292

ANSWER B: The hallux valgus and varus should be coded, but the bunion is part of these and is, therefore, not coded separately. This procedure includes an osteotomy of the metatarsal with wire, so this is a Keller-Type procedure.

149. What are the correct code(s) for a patient who had two adjacent malignant skin lesions removed from the chest that were 1.2 cm and 1.4 cm with one excision, which was then covered with a Z-plasty?

   A. 195.111603, 14000
   B. 174.9, 11603
   C. 195.1, 11602 × 2, 14000
   D. 195.1, 14000

ANSWER D: The malignant lesions are coded as primary neoplasms of the chest. There is no need to code for the excision of the lesions as they were cut out as one excision and are included as part of the Z-plasty, so only the Z-plasty is coded.

150. What are the correct code(s) for a patient who was seen in the Emergency Room for urinary tract infection with sepsis due to *Pseudomonas,* for which the patient was provided antibiotics? History and exam for this visit were expanded problem focused and the medical decision making was straightforward.

   A. 995.91, 99282
   B. 599.0, 041.7, 99281
   C. 038.9, 041.7, 99201
   D. 995.91, 99281

ANSWER B: The sepsis of the urinary tract is known as urosepsis. For urosepsis, you can code either 599.0 or 038.43 depending on whether you are coding for a urinary tract infection or sepsis, so this can be confusing. The presence of the *Pseudomonas* indicates that it is sepsis with an organism. The Emergency Room visit is the lowest level because the decision making is straightforward.

Appendix B   CCS MOCK EXAM 1

The following appendix provides a mock example of the Certified Coding Specialist (CCS) exam offered by the American Health Information Management Association (AHIMA). The exam is broken into two parts, one is a 60-question multiple-choice exam, the other is an abstract exam.

MULTIPLE CHOICE MOCK EXAM

No books are allowed on this multiple-choice section of the test.

1. Basic unit of payment for a DRG is based on
   A. Visit
   B. Admission/discharge
   C. Morbidity
   D. Length of stay

2. What is the correct code(s) for a patient who received a fine needle biopsy of a kidney mass with guidance and evaluation in an ASC for possible carcinoma of the kidney?

   593.9 Unspecified disorder of kidney and ueter
   189.0 Kidney malignant neoplasm and unspecified urinary organ
   10022 Fine needle aspiration with imaging guidance
   50200 Renal biopsy, percutaneous by trocar or needle
   88172 Cytopathology, evaluation of fine needle aspirate, immediate cytohistolgic study

   A. 189.0, 50200
   B. 593.9, 88172
   C. 593.9, 10022, 88172
   D. 189.0, 10022

3. What is the correct code(s) for a patient who received a corneal transplant for the outer layer of the cornea due to proliferative insulin-dependent diabetic retinopathy?

   250.50 Diabetes with ophthalmic manifestations, Type II
   250.51 Diabetes with ophthalmic manifestations, Type I
   362.02 Proliferative diabetic retinopathy
   362.29 Other nondiabetic proliferative retinopathy
   65400 Excision of the cornea (keratectomy, lamellar, partial)
   65710 Keratoplasty, lamellar
   65730 Keratoplasty, penetrating
   65771 Radial keratotomy

   A. 362.29, 65771
   B. 250.50, 362.29, 65400
   C. 250.51, 65730
   D. 250.51, 362.02, 65710

4. If a patient is admitted to the Emergency Room with complaints of heartburn and pyrosis and the examination reveals that the patient has esophageal diverticula, but the patient experiences appendicitis for which he is provided emergency surgery, what is the principal diagnosis?
   A. Pyrosis
   B. Heartburn
   C. Diverticula
   D. Appendicitis

5. Which of the following services would not be performed before a patient receives medical services?
   A. Insurance verification
   B. Payment of copay
   C. Payment of deductible
   D. Obtain past medical history

6. What is not true about DRGs?
   A. 25 categories
   B. 115 categories
   C. Divided into MDGs
   D. Derived from ICD-9-CM-CM codes

7. What is the correct code(s) for a patient who had a recurrent ganglion cyst of the wrist removed which was 1.4 cm?

706.2 Sebaceous cyst
727.43 Ganglion unspecified
25111 Excision of ganglion, wrist; primary
25112 Recurrent

A. 706.2, 25112
B. 706.2, 25111, 25112
C. 727.43, 25112
D. 727.43, 25111, 25112

8. What is the correct code(s) for a patient who received wound repair for a 4-cm laceration of the right forearm, which required closure of subcutaneous tissue and extensive debridement and a 3.7-cm laceration of the chest, which required repair of the fascia?

11042 Debridement, skin, partial thickness
12002 Simple repair of superficial wounds of scalp, neck, axillae, external genitalia, trunk and/or extremities, 2.6 cm to 7.5 cm
12004 Simple repair of superficial wounds of scalp, neck, axillae, external genitalia, trunk and/or extremities, 7.6 cm to 12.5 cm
12032 Layer closure of wounds of scalp, axillae, trunk and/or extremities, 2.6 cm to 7.5 cm
12034 Layer closure of wounds of scalp, axillae, trunk and/or extremities, 7.6 cm to 12.5 cm

A. 12034
B. 12032, 12002
C. 11042, 12034
D. 11042, 12002, 12004

9. Advance Beneficiary Notices are important because

A. They insure that payment will be sent to the service provider
B. They inform the patient that they may be responsible for payment
C. They insure that services are covered by insurance
D. They state that the services are covered services under Medicare

10. Which of the following elements would not constitute an element used in determining proper history level for E/Ms?

A. Review of Systems
B. Medical Records
C. Present Illness
D. Past History

11. What is the correct code(s) for a patient who received manipulation of a radial shaft greenstick fracture?

813.81 Fracture, unspecified part, closed, radius
813.21 Fracture, shaft, closed, radius
813.31 Fracture, shaft, open, radius
25390 Osteoplasty, radius or ulna
25505 Closed treatment of radial shaft fracture with manipulation
25515 Open treatment of radial shaft fracture with or without internal or external fixation

A. 813.81, 25390
B. 813.21, 25505
C. 813.31, 25515
D. 813.31, 25505

12. What is the correct code(s) for a patient who was admitted to the hospital with severe chest pain and dyspnea? The patient was diagnosed with third-degree AV block due to digitalis toxicity. A permanent atrial-ventricular pacemaker with transvenous leads was inserted.

426.0 Atrioventricular block complete
426.10 Atrioventricular block, unspecified
428.1 Left heart failure
786.09 Other respiratory abnormalities
796.0 Nonspecific abnormal toxicological findings
972.1 Poisoning by cardiotonic glycosides and drugs of similar action
37.72 Initial insertion of transvenous leads into atrium and ventricle
37.83 Initial insertion of dual-chamber device
39.64 Intraoperative cardiac pacemaker
E942.1 Adverse effect of cardiotonic glycosides and drugs of similar action

A. 426.10, 786.09, 972.1, 39.64
B. 426.0, E942.1, 37.83, 37.72
C. 428.1, 796.0, 37.83
D. 426.0, 786.09, E942.1, 39.64

13. What is the correct code for a baby 4 weeks old who was seen in the Emergency Room for 2½ hours of critical care?

99291 Critical care; first 30 minutes
99292 Each additional 30 minutes
99293 Initial inpatient pediatric critical care

99295 Initial inpatient neonatal critical care
99296 Subsequent inpatient neonatal critical care

A. 99295
B. 99293
C. 99291, 99292 × 3
D. 99291 × 5

14. What is the correct code(s) for a patient who is diagnosed fever and chills due to SARS-related pneumonia?

079.82 SARS-associated coronavirus
480.3 Pneumonia due to SARS-associated coronavirus
486 Pneumonia organism unspecified
V01.82 Exposure to SARS-associated corona virus

A. 486, 079.82
B. 480.3
C. 486, V01.82
D. 079.82

15. Removal of fluid from the abdomen surgically is known as
A. Amniocentesis
B. Thoracocentesis
C. Thoracentesis
D. Paracentesis

16. What is the correct code(s) for a patient who has acute renal failure and malignant hypertension?

401.0 Essential hypertension malignant
403.91 Unspecified hypertensive renal disease
584.9 Acute renal failure, unspecified

A. 403.91
B. 403.91, 401.0
C. 584.9, 401.0
D. 584.9

17. What is the correct code(s) for a patient who was seen today in the Emergency Room for complaints of dyspnea and pain in the left arm with a history of AICD six months ago and possible MI?

410.9 Acute myocardial infarction, unspecified site
729.5 Pain in limb
786.00 Respiratory abnormality unspecified
786.09 Other symptoms involving respiratory system and chest symptoms
V44.3 Colostomy status
V45.02 Postprocedural cardiac device *in situ*, automatic implantable cardiac defibrillator

A. 786.09, 729.5, V45.02
B. 410.9, V44.3
C. 786.09, 729.5
D. 410.9, V45.02

18. Case mix is not based on
A. Patient's past medical history
B. Volume
C. Exacerbating conditions
D. Patient's age

19. What is the correct code(s) for a patient who had a UTI due to *E. coli* and is 6 months pregnant?

008.00 Intestinal infections due to *E. coli* unspecified
041.4 Bacterial infection in conditions classified elsewhere, *E. coli*
599.0 Urinary tract infection, site not specified
646.63 Infections of genitourinary tract in pregnancy

A. 646.63
B. 646.63, 599.0, 041.4
C. 008.00
D. 599.0, 041.4

20. In best practices, professional courtesy
A. Must be applied to all patients
B. Can be applied to friends and family
C. Is illegal
D. Is billed as insurance only

21. If a patient with COPD is admitted to the hospital with hepatomegaly and a severe headache, which was diagnosed as possible hepatitis B, what is the principal diagnosis?
A. Hepatomegaly
B. COPD
C. Headache
D. Hepatitis B

22. What is the correct code(s) for a patient with a perforating ulcer who receives a partial gastrectomy with gastrojejunostomy?

533.10 Acute peptic ulcer site unspecified with perforation without mention of obstruction
533.50 Chronic or unspecified peptic ulcer with perforation without mention of obstruction
43.7 Partial gastrectomy with anastomosis to jejunum
44.32 Percutaneous gastrojejunostomy
44.39 Other gastroenterostomy

   A. 533.10, 44.39
   B. 533.50, 43.7
   C. 533.10, 44.32
   D. 533.50, 44.39

23. Chondromalacia is
   A. Abnormal condition of the muscles
   B. Softening of articular cartilage
   C. Weakening of the joints
   D. Abnormal condition of high cholesterol

24. Which of the following is not an emancipated minor?
   A. Living separately from parents
   B. Married
   C. In the military
   D. College student

25. Which drug would be used to treat PID?
   A. Streptokinase
   B. Lanacort
   C. Doxycycline
   D. Comtrex

26. If the APC for low-level E/M visits is 0600, the middle-level is 0601 and the high-level is 0602, what would the correct APC be for code 99202?
   A. Low-level
   B. 0602
   C. Middle-level
   D. 0600

27. What is the correct code(s) for a patient who had a coronary embolism 8 weeks ago?

410.92 Acute myocardial infarction, unspecified site
414.8 Other specified forms of chronic ischemic heart disease
429.79 Other sequelae of myocardial infarction NOS

   A. 414.8
   B. 410.92
   C. 410.92, 429.79
   D. 429.79

28. Methods of reimbursement do not include
   A. Capitation
   B. Fee for service
   C. Remittance
   D. Fee per visit

29. The slipping of one part of the intestine within another part is known as
   A. Intussusception
   B. Colectomy
   C. Enterolysis
   D. Intubation

30. What is the correct code for a patient who came to the Emergency Room and was then admitted to observation for 10 hours due to MVA with the history and exam comprehensive and the medical decision making low?

0601 Low-level E/M visit APC
0602 Moderate-level E/M visit APC
99203 Office visit new patient, detailed history and exam, low complexity medical decision making
99218 Initial observation care with detailed or comprehensive history and exam and straightforward medical decision making.

   A. 99218
   B. 0602
   C. 0601
   D. 99203

31. Which of the following information would not be available in the Chargemaster?
   A. Inventory control number
   B. Fees
   C. HCPCS codes
   D. Discharge status

32. What is the correct code(s) for a one-year-old female patient who is admitted to the Emergency Room

with severe diarrhea for three days? Cultures reveal *C. jejuni*. Erythromycin is prescribed.

787.91 Diarrhea
008.43 Intestinal infections due to *Campylobacter*
008.49 Intestinal infections due to other organism

A. 787.91
B. 787.91, 008.49
C. 008.43
D. 787.91, 008.43

33. With revenue codes, if pharmacy is indicated by 25X, and 0 is for general drugs, 1 is for generic drugs, 2 is for nongeneric drugs, and 3 is for take-home drugs, what would be the proper revenue code for amitriptyline?

A. 250
B. 25X3
C. 251
D. 252

34. Which of the following is not true about coding for professional services?

A. Can be indicated by modifier 26
B. Not necessary when coding for hospital services
C. Is referred to as supervision and interpretation
D. Must always be coded as separate code from other services when performed by physician

35. Removal of skin tags is based on

A. Each individual tag
B. Type of removal
C. Each additional group of tags
D. Each additional tag

36. What is the correct code(s) for a patient who received a right heart catheterization and arterial coronary angiography?

93501 Right heart catheterization
93508 Catheter placement in coronary artery(s), arterial coronary conduit(s), and/or venous coronary bypass graft(s) for coronary angiography without concomitant left heart catheterization
93541 Injection procedure during cardiac catheterization, for pulmonary angiography

---

A. 93501
B. 93501, 93541
C. 93501, 93508, 93541
D. 93501, 93508

37. What drug would be prescribed for treatment of an ulcer?

A. Citrucel
B. Flagyl
C. Topicort
D. Prevacid

38. What does extrapyramidal mean?

A. Motor pathways to the brain
B. Spinal cord
C. External causes of diseases
D. Synergistic effect

39. What is the correct code(s) for a patient who received a split-thickness graft, 180 sq cm for scarring of the back due to radiation 3 years ago?

709.2 Scar conditions and fibrosis of skin
906.8 Late effect of burn, other site specified
909.2 Late effect of radiation
942.04 Burn of trunk, unspecified degree, back
15000 Surgical preparation for creation of recipient site by excision of open wounds, burn eschar, or scar, first 100 sq cm
15001 Surgical preparation for creation of recipient site by excision of open wounds, burn eschar, or scar, each additional 100 sq cm or each additional percent of body area of infants and children
15100 Split graft, trunk, arms, legs; first 100 sq cm or less, or one percent of body area of infants and children
15120 Split graft, face, scalp, eyelids, mouth, neck, ears, orbits, genitalia, hands, feet and/or multiple digits, first 100 sq cm or one percent of body area of infants and children
15121 Split graft, face, scalp, eyelids, mouth, neck, ears, orbits, genitalia, hands, feet and/or multiple digits, each additional 100 sq cm or each additional one percent of body area of infants and children, or part thereof

---

A. 709.2, 909.2, 15100, 15001
B. 709.2, 909.2, 15100, 15000, 15001
C. 942.04, 15120, 15121
D. 709.2, 906.8, 15000, 15001, 15120, 15100

40. What is the correct code(s) for a patient who received debridement and shaving of the articular cartilage of

the patellofemoral compartment of the knee and a meniscectomy of the lateral compartment?

> 27332 Arthrotomy, with excision of semilunar cartilage, knee; medial or lateral
> 29862 Arthroscopy, hip, surgical, with debridement/shaving of articular cartilage, abrasion arthroplasty and/or resection of labrum
> 29877 Arthroscopy, knee, diagnostic debridement/shaving of articular cartilage

  A. 29877
  B. 27332
  C. 27332, 29877
  D. 29862, 27332

41. ERCP means
  A. Percutaneous erythrocyte count
  B. Endoscopic radioimmunoassay of colon and prostate
  C. Blood test
  D. Endoscopic retrograde cholangiopancreatography

42. What is the correct code(s) for a patient who received an ethmoidectomy with removal of bullae from the anterior sinus and complete examination of the posterior sinus due to inflammation?

> 471.1 Polypoid sinus degeneration
> 478.1 Other disease of the nasal cavity and sinuses
> 31200 Ethmoidectomy, intranasal, anterior
> 31201 Ethmoidectomy, intranasal, total
> 31205 Ethmoidectomy, extranasal, total

  A. 471.1, 31200
  B. 478.1, 31201
  C. 478.1, 31200
  D. 471.1, 31205

43. Decrease in platelets is known as
  A. Achondroplasia
  B. Leukocytosis
  C. Thrombocytopenia
  D. Hemolysis

44. Using CPT codes, the use of oral contrast material for testing is coded as
  A. With contrast material
  B. As two codes
  C. An additional code to the surgery
  D. Without contrast material

45. What is the correct code(s) for a patient who received a TAH due to menorrhagia and pain?

> 58150 Total abdominal hysterectomy with or without removal of tubes, with or without removal of ovaries
> 58200 Total abdominal hysterectomy, including partial vaginectomy with para-aortic and pelvic lymph node sampling with or without removal of tubes with or without removal of ovaries
> 59850 Induced abortion, by one or more intra-amniotic injections including hospital admission and visits
> 59851 Induced abortion, by one or more intra-amniotic injections including hospital admission and visits with dilation and curettage and/or evacuation

  A. 59850
  B. 58150
  C. 59851
  D. 58200

46. What is the correct code(s) for a patient who received an adenoidectomy due to acute and chronic inflammation and was subsequently treated for control of postoperative hemorrhage?

> 463 Acute tonsillitis
> 474.01 Chronic adenoiditis
> 474.8 Other chronic disease of tonsils and adenoids
> 998.11 Hemorrhage or hematoma or seroma complicating a procedure
> 28.6 Adenoidectomy without tonsillectomy
> 28.7 Control of hemorrhage after tonsillectomy and adenoidectomy

  A. 474.8, 28.6, 28.7
  B. 474.01, 998.11, 28.7
  C. 474.8, 463, 28.6
  D. 463, 474.01, 998.11, 28.6, 28.7

47. What is the correct code(s) for a patient who received reduction for a comminuted dislocation of the radial head and fracture of the proximal ulna but the reduction was unsuccessful?

> 813.03 Monteggia's fracture
> 813.82 Fracture, ulna alone

833.00 Closed dislocation of the radius
24620 Closed treatment of Monteggia type of fracture dislocation at elbow with or with manipulation
24635 Open treatment of Monteggia type of fracture dislocation at elbow, with or without internal or external fixation
24640 Closed treatment of radial head subluxation with manipulation
25535 Closed treatment of ulnar shaft fracture, with or without internal or external fixation

A. 813.82, 833.00, 25535, 24640
B. 813.82, 833.00, 25535
C. 813.03, 24620
D. 813.03, 24635

48. What is true about an excisional biopsy?
   A. Incision is made to remove a portion of the lesion
   B. Removal of lesion
   C. Markers are not used
   D. Is coded from the biopsy sections

49. If a patient is admitted to the hospital with edema and hematuria and a history of hypertension and insulin-dependent diabetes with tests indicating glomerulonephritis, but the patient experiences an MI while in the hospital, what is the principal diagnosis?
   A. Glomerulonephritis
   B. MI (myocardial infarction)
   C. Diabetes
   D. Hematuria

50. Which of the following is not a method for cauterization?
   A. Dry ice
   B. Electric current
   C. Exenteration
   D. Chemicals

51. What would the correct code(s) be for a patient admitted to the hospital with bleeding esophageal varices, which were endoscopically injected with sclerosing agent? The patient was also diagnosed with portal hypertension.

456.0 Esophageal varices with bleeding
456.20 Esophageal varices in diseases classified elsewhere with bleeding
572.3 Portal hypertension

38.87 Other surgical occlusion of vessels abdominal veins
42.33 Endoscopic excision or destruction of lesion or tissue of esophagus
99.29 Removal of foreign body without incision from lower limb, except foot

A. 456.20, 38.87, 99.29
B. 456.0, 42.33, 99.29
C. 572.3, 456.20, 42.33
D. 456.20, 572.3, 99.29

52. Drug used for reducing cholesterol is
   A. Lasix
   B. Lipitor
   C. Dilantin
   D. Prevacid

53. Intrathecal is an example of
   A. Endoscopic exam of the gastrointestinal tract
   B. Rectal test
   C. Parenteral administration of drugs
   D. Administration of chemotherapeutic drugs

54. What is the correct code(s) for a patient who was injured in a MVA and was diagnosed with fractured C2–3, which required the removal of fragments and hemilaminectomy with anterior fusion?

805.02 Cervical, closed fracture, second cervical vertebrae
805.08 Cervical, closed fracture, multiple cervical vertebrae
806.02 C1-C4 level cervical, closed, with anterior cord syndrome
806.03 C1-C4 level cervical, closed, with central cord syndrome
806.10 C1-C4 level cervical, open, with unspecified spinal cord injury
80.50 Excision or destruction of intervertebral disc
80.51 Excision of intervertebral disc
81.00 Spinal fusion, NOS
81.02 Other cervical fusion, anterior technique

A. 806.02, 806.03, 80.51
B. 806.10, 80.50, 81.00
C. 805.08, 80.51, 81.02
D. 805.02, 80.50

55. Which is not part of the vascular system?
    A. Axon
    B. Arterioles
    C. Venules
    D. Capillaries

56. If a patient who is asthmatic is seen in the hospital's outpatient clinic with pitting edema and hypertension and is diagnosed with possible renal failure, what is the principal code?
    A. Asthma
    B. Pitting edema
    C. Renal failure
    D. Hypertension

57. What is the correct code(s) for a patient who received a complex fistulectomy with seton guidance of the anus?

    565.1 Anal fistula
    569.81 Fistula of intestine
    46020 Placement of seton
    46262 Hemorrhoidectomy, internal and external, complex or extensive, with fistulectomy with or without fissurectomy
    46280 Surgical treatment of anal fistula, complex or multiple with or without placement of seton

    A. 565.1, 46262, 46020
    B. 569.81, 46262
    C. 565.1, 46280
    D. 569.81, 46280, 46020

58. Which of the following is not a type of diagnostic ultrasound as listed in the CPT book?
    A. Two-dimensional
    B. Real time scan
    C. P-scan
    D. M-mode

59. What is the correct code(s) for a patient who received a needle biopsy of the ovaries?

    54.21 Laparoscopy of abdominal region
    54.24 Closed biopsy of intra-abdominal mass
    65.11 Aspiration biopsy of ovary
    65.13 Laparoscopic biopsy of ovary

    A. 65.11
    B. 54.24
    C. 54.21
    D. 65.13

60. What is the correct code(s) for a patient who received a partial ostectomy with replacement of both femoral head and acetabulum with prosthesis?

    77.89 Other partial ostectomy
    81.51 Total hip replacement
    81.52 Partial hip replacement
    81.80 Total shoulder replacement

    A. 77.89
    B. 81.51, 77.89
    C. 81.51
    D. 81.80

ABSTRACT MOCK EXAM

In the second half of their exam, AHIMA uses an abstract style, which is similar to this exam. For the following scenarios, provide the appropriate ICD-9-CM and CPT/HCPCS codes. There are eleven abstracting cases—six are inpatient and seven are outpatient. The outpatient scenarios require the use of CPT procedural codes, including the coding of modifiers. No HCPCS are to be listed. E codes are only used where applicable to codes for adverse effects of drugs. For these cases, up to four diagnostic codes are allowed and up to four CPT procedural codes. The remainder of Part II consists of six inpatient scenarios. All of these cases are inpatient hospital cases and are, therefore, coded from the ICD-9-CM procedural codes, Volume 3. You are allowed up to nine ICD-9-CM diagnosis codes and six procedural codes, although many scenarios will not require this many codes.

For abstracts requiring an E/M code, the three main components for determining the level of code are listed at the end of the abstract. If necessary, use these components to determine the correct level of code. This means if an E/M code requires you to determine a component level, the levels of the three components are provided for you. Do not try to provide your own interpretation for the E/M levels; accept what is provided as valid.

Modifiers and ordering of codes, e.g., the listing of the principal diagnosis first, must be provided, just as they are required on the CCS exam, which is in contrast to the CCS-P exam in which neither ordering nor modifiers are required on the abstraction part of the exam.

ABSTRACT 1

EMERGENCY ROOM

PATIENT: Ricky Garcia
SEX: Male
DATE: 8/9/04

This 27-year-old male drug user, who has been abusing cocaine for many years, is seen in the Emergency Room for hallucinations associated with continuous dependence. The patient was also diagnosed with cellulitis of the right antecubital area and right elbow. He was started on Unasyn on admission and was taken for incision and drainage of the abscesses. Cultures from surgery are negative. Temperature has come down and his pain is subsiding.

PHYSICAL EXAMINATION: VITAL SIGNS: Temperature 100.2. GENERAL APPEARANCE: The patient does not appear to be well-nourished and is thin. The patient is alert at this time, but orientation is limited. The patient is in no distress. SKIN: No rash. No peripheral embolic signs. EYES: No conjunctival microemboli. CHEST: Clear. HEART: No murmur, rub, click or gallop. ABDOMEN: Soft, nontender, no mass and no organomegaly. EXTREMITIES: Operative dressings were not removed form the right elbow area. Motor strength, sensation, and peripheral pulses are all normal in the right hand and wrist.

IMPRESSION: My suspicion is that conventional flora, group A streptococci, are the causative organisms with sensitivity to Unasyn. Unasyn should be continued. Routine wound care. Depending on progress, we might be able to eventually continue therapy with Augmentin 875 mg q.12h. p.o. The patient is already talking about leaving the hospital but has been advised that this would be unwise for fear of permanent damage to the right elbow and arm.

HISTORY: Expanded problem focused
EXAM: Expanded problem focused
MEDICAL DECISION MAKING: Low complexity

ICD-9-CM:
1.
2.
3.
4.

CPT:
1.
2.
3.
4.

ABSTRACT 2

EMERGENCY ROOM

PATIENT: Sherry Canton
SEX: Female
DATE: 12/2/04

This is a 51-year-old Caucasian female who was admitted to the Emergency Room for complaints of rapid breathing. Patient has a history of chronic obstructive pulmonary disease with possible supraventriuclar tachycardia. The records indicate that the patient had three nonsustained episodes of supraventricular tachycardia that was associated with Bovie use initially, but then she had another episode of 30 to 45 seconds of supraventricular tachycardia, which was not associated with Bovie use but the strips are not available in the chart.

PAST HISTORY: The patient had a permanent pacemaker inserted a year ago. Patient is not sure why the pacemaker was implanted. Patient denies any history of CAD, congestive heart failure, chest pain, shortness of breath, or other symptoms. She also denies any history of hypertension or hyperlipidemia. She denies any episodes of increased heart rate, syncope or dizziness. Patient has past history of chronic obstructive pulmonary disease for which she is on oxygen at home. Patient had a knee replacement 3 years ago.

MEDICATIONS: Pain medications.

ALLERGIES: Penicillin.

SOCIAL HISTORY: Patient previously smoked quite heavily, however, she quit a few years ago. No significant alcohol intake.

FAMILY HISTORY: Noncontributory. Mother and father are alive. Father has diabetes.

REVIEW OF SYSTEMS: CONSTITUTIONAL: Denies any recent fevers, chills, weight gain or weight loss. HEENT: Significant in that she had laser surgery to correct her vision which is good now. Otherwise noncontributory. CARDIOVASCULAR: As per the history of present illness. RESPIRATORY: Significant for chronic obstructive pulmonary disease for which she is oxygen-dependent now.

GASTROINTESTINAL: Negative for any problems with change in habits or blood in stools.

PHYSICAL EXAMINATION: Her vital signs are stable. She is afebrile. GENERAL: She is a well-nourished, well-developed woman in no acute distress. She was alert and oriented times three. HEENT: Pupils equal, round, reactive to light and accommodation bilaterally. Extraocular muscles intact bilaterally. NECK: No jugular venous distention. No bruits. CHEST: Diffusely diminished breath sounds. No wheezes or rhonchi. CARDIOVASCULAR: Regular rate and rhythm without rubs or gallops. She did have a 1-2/6 systolic murmur heard best at the upper left sternal border. ABDOMEN: Soft, nontender, nondistended with bowel sounds present. EXTREMITIES: No clubbing, cyanosis, or edema. Pulses were 2+ and symmetrical.

PERTINENT LABORATORIES AND X-RAY DATA: ECG that showed a normal sinus rhythm. She has multiple telemetry strips that demonstrate ventricular tachycardia with intrinsic ventricular conduction. Chemistries were unremarkable. She is slightly anemic with a hemoglobin of 10.3. Troponin was negative times two and TSH was within normal limits.

HISTORY: Expanded problem focused
EXAM: Detailed.
MEDICAL DECISION MAKING: Moderate complexity

ICD-9-CM:

1.

2.

3.

4.

CPT:

1.

2.

3.

4.

ABSTRACT 3

OUTPATIENT CLINIC
PATIENT: Alice Marple
SEX: Female
DATE: 8/14/04

HISTORY OF PRESENT ILLNESS: This 75-year-old patient is seen in the outpatient clinic for suspicious lump in her right breast, which is causing some discomfort at this time. Patient does not perform regular checks of her breasts, so she does not know when the lump appeared. The patient is positive for hypertension and hypothyroidism. She also has osteoporosis with severe low back pain and arthritis with current significant right knee symptomatology.

PAST HISTORY: The patient is gravida 2, para 2, and is on hormone supplementation. Her daughter, who accompanied her for this visit, is a breast cancer survivor, which developed when she was 38. Her daughter underwent modified radical mastectomy with adjuvant chemotherapy and radiotherapy and had node positive disease. Of note, her daughter also had ovarian carcinoma and is five years from that diagnosis. There were no other family members with breast or ovarian carcinoma. Patient has had lumbar laminectomy 10 years ago with hardware removal a year later. She had another laminectomy three years later.

MEDICATIONS: Hydrocodone, clonazepam, Lotensin, amitriptyline, Premarin, Miacalcin, levothyroxine, baclofen, Lasix, and potassium.

ALLERGIES: Penicillin gives her a rash and she has swelling due to codeine and aspiring with gastritis.

SOCIAL HISTORY: The patient is widowed and does not smoke or drink.

REVIEW OF SYSTEMS: GENERAL: No history of weight loss, fatigue, headache or malaise. HEENT: No history of glaucoma, otitis, pharyngitis, or sinusitis. HEART: No history of chest pain, heart attack, heart murmur, arrhythmias, or heart failure. LUNGS: No history of cough, wheezing, shortness of breath, emphysema, or pneumonia. ABDOMEN: No history of abdominal pain, diarrhea, constipation, or blood in the stools. EXTREMITIES: No history of deep venous thrombosis, leg swelling or trauma to the extremities. NEURO: No history of TIA, strokes, or seizures. MUSCULOSKELETAL: Significant back and knee pain. ENDOCRINE: Hypothyroid and no diabetes. PSYCHIATRIC: No history of depression or anxiety.

PHYSICAL EXAMINATION: This is a pleasant elderly female accompanied by her daughter who is in no apparent distress. Heart has regular rate and rhythm. Lungs are clear to auscultation. Breast examination reveals large pendulous breasts. There is a firm mass in the right breast at the 2 o'clock position, which is nontender. There are no other breast masses bilaterally. Normal axillae bilaterally.

Mammogram with digitization of film radiographic images was performed that demonstrated possible cancerous mass in the right breast, so fine needle aspiration was performed. The aspiration demonstrated that the mass is 1 cm in greatest dimension and is located 6 cm from the nipple at the 12:30 position. After prepping the aspiration site with alcohol preps, two passes were made using 23-gauge 1-inch needles. No local anesthetic was given and the patient tolerated the procedure well with no complications. The biopsy confirmed a diagnosis of right breast lobular carcinoma, Black's nuclear grade I–II. I discussed the options with the patient for surgical staging including breast conservative therapy with lumpectomy and adjuvant radiotherapy versus mastectomy.

ICD-9-CM:

1.

2.

3.

4.

CPT:

1.

2.

3.

4.

ABSTRACT 4

OUTPATIENT CLINIC
PATIENT: Becky Barnes
SEX: Female
DATE: 7/6/04

HISTORY OF PRESENT ILLNESS: This 34-year-old female is seen in the outpatient clinic for complaints of rectal bleeding over the last couple of months. She describes bright red rectal bleeding with bowel movements. This was scant in quantity. There was no pain with defecation. In general, she has been having a more erratic bowel pattern over the last five or six months. She describes a tendency towards constipation. She does not describe to me any associated abdominal pain. She has no upper GI symptoms. She has had a prior sigmoidoscopy five years ago for screening purposes, which was apparently unremarkable. Her weight has been stable.

ALLERGIES: No known drug allergies.

PAST MEDICAL HISTORY: Her additional medical history is fairly minimal. She has had some arthritic symptoms and takes aspirin and glucosamine. She also takes Vioxx on an as needed basis.

PAST SURGICAL HISTORY: She has had no prior abdominal surgeries.

FAMILY HISTORY: There is no known family history of colon cancer or adenomatous polyps.

SOCIAL HISTORY: She does not smoke and denies any significant quantities of alcohol.

PHYSICAL EXAMINATION: GENERAL: Well-nourished and well-developed 2-month pregnant woman in no acute distress. Vital signs are stable. **HEENT:** Unremarkable. PEERLA. No adenopathy or thyromegaly. **CARDIOPULMONARY:** Normal S1 and S2. Lungs clear to P&A. No bruits. No rales or rhonchi. **ABDOMINAL:** Normal bowel sounds, soft, nontender, nondistended and no hepatosplenomegaly.

COURSE OF TREATMENT: The rectal bleeding most likely represents bleeding from a benign anorectal source such as hemorrhoids or fissure. The patient was prepped and draped in the usual manner. She was positioned in a lithotomy position. A colonoscopy was performed to determine the exact cause. The Olympus video colonoscope was introduced into the rectum and advanced under direct vision. Several small hemorrhoids were noted which were ulcerated, so these were ligated. The patient tolerated this procedure well and returned to the recovery area and later discharged to follow up in one week.

ICD-9-CM:

1.

2.

3.

4.

CPT:

1.

2.

3.

4.

ABSTRACT 5

OUTPATIENT CLINIC
PATIENT: Alice Marshall

SEX: Female
DATE: 4/4/04

HISTORY OF PRESENT ILLNESS: This 55-year-old female patient is referred for postmenopausal vaginal bleeding. She is a breast cancer survivor from a year ago. She had undergone a lumpectomy with lymph node dissection and radiation therapy. She has been on Tamoxifen since. Starting 5 months ago, she noticed rare vaginal bleeding q. 3 to 4 months, but it has increased to q. 1 to 2 months, lasting several days with spotting. EMB performed two months ago is reportedly negative by the patient's account. She complains of dyspareunia, primarily secondary to vaginal dryness with only partial relief using a water-based lubricant.

PAST MEDICAL HISTORY: Chronic hypertension. Breast cancer a year ago.

PAST SURGICAL HISTORY: Lumpectomy with axillary node dissection with two subsequent lumpectomies, both benign. Postpartum TL in 1970.

CURRENT MEDICATIONS: Megace 20 mg p.o. b.i.d. for hot flashes. Tamoxifen. Prinivil. Multivitamins.

ALLERGIC TO CODEINE—causes nausea and vomiting.

SOCIAL HISTORY: One-half pack of cigarettes per day. Rare alcohol use. No drug use.

FAMILY HISTORY: Negative for ovarian or uterine cancer. No colon cancer. No osteoporosis. Mother diagnosed with breast cancer at age 51.

PAST OB/GYN HISTORY: Previously regular menses × 5 to 6 days. Menopause at age 47. Started HRT at age 47. Discontinued with diagnosis of breast cancer a year ago. No STDs or abnormal PAP smears. Monogamous and sexually active with her husband. No pain associated with the bleeding. NSVD × 3 without complication. Ultrasound performed several months ago reveals normal sized uterus with 1.2-cm fibroid, which does not generally appear submucosal. EEC has heterogenous consistency with some cystic components consistent with the Tamoxifen, measuring approximately 20 mm.

EXAMINATION: Abdomen soft, nontender. Nondistended. No masses. Pelvic EG/BUS is WNL. Vagina WNL. Vaginal pH 4.5 to 5. Appears moist with normal rugae. Cervix generally normal. Uterus small, mobile, nontender. Ovaries not palpable. No adnexal mass or tenderness.

PROCEDURE: The patient was prepped and draped in the usual manner with Betadine. The uterus was examined and several adhesions were noted, which were removed, but no fibroids. Inspection of the endocervical canal showed was normal. The cervix was dilated and appeared to be normal and not enlarged. The cervix was then further dilated and a curette inserted and rotated. Endometrial biopsies were taken and sent to the lab.

ICD-9-CM:

1.

2.

3.

4.

CPT:

1.

2.

3.

4.

ABSTRACT 6

OUTPATIENT CLINIC
PATIENT: Sandy Kantor
SEX: Female
DATE: 10/14/04

This 44-year-old female was admitted to the hospital's outpatient clinic with complaints of severe dypsnea, which is not relieved by mediation. She says that she is not able to perform many daily functions due to the need to breathe which seems to be compromised. She also states that she has been having chest pain. The patient had a C-section in 1992.

SOCIAL/FAMILY HISTORY: The patient lives with her sister and mother. The mother smokes. The patient did smoke as a teenager but quit due to its aggravation of her asthma. Patient does not drink. Sister and daughter are healthy. Mother has diabetes and COPD.

CURRENT MEDICATIONS: Albuterol nebulizations, q. 2h p.r.n. asthma, Ventolin inhaler two puffs q.i.d., and Vanceril four puffs q.i.d., with spacer.

ALLERGIES: The patient is allergic to codeine, which gives her nausea and vomiting.

PAST HISTORY: The patient has a past history of chronic sinusitis and asthma.

REVIEW OF SYSTEMS: HEENT: She has no major history of problems with her eyes, ears, nose or throat. No thyroid problems. CARDIOVASCULAR: There is no history of heart disease, heart murmur, heart failure, or chest pain. RESPIRATORY: Patient has had asthma since she was a young teenager. She denies any incidents of pneumonia, cough, or emphysema. GASTROINTESTINAL: No history of nausea, vomiting, diarrhea, recent weight loss or gain, or hematemesis. GENITOURINARY: No history of kidney, bladder or other urinary problems. MUSCULOSKELETAL: No problems with musculoskeletal problems with no arthritis. SKIN/LYMPHATICS: No history of skin disease, cancer or lymphadenopathy. NEUROLOGICAL/PSYCHOLOGICAL: The patient does not consider herself overly nervous or stressed. No motor problems.

PHYSICAL EXAMINATION: GENERAL: The patient is a well-developed, well-nourished female. Temperature 97.2, pulse 74, respirations 28, BP 90/60. She is oriented times three. HEENT: PEERLA. Pharynx benign. No thyromegaly. NECK: No jugular venous distention. No bruits or adenopathy. HEART: Regular rate and rhythm without murmur, rub or gallop. LUNGS: The patient does have asthma and chronic sinusitis, but this appears to not be the concern at this time as it is under control with medication. There are no rales or rhonchi at this time; however, there is the suspected mass which may be associated to the patient's difficulty in breathing. ABDOMEN: Nontender, nonpalplable. No pain. EXTREMITIES: No cyanosis, clubbing or edema. LYMPHATICS: Not palpable. SKIN: Benign.

A CT of the abdomen and pelvis was performed with and without contrast with no comparison studies available. 75 cc of ionic contrast was administered intravenously for the enhanced phase of this examination. The dominant finding is a large rounded well-defined mass in the left lower lobe. This mass measures approximately 7.5 cm in greatest dimension and is highly suspicious for possible primary left lung neoplasm. In addition, there are scattered tiny determinate nodules at both lung bases, possibly representing both contralateral and ipsilateral metastases. There is no sign of lower thoracic free fluid. There is no pericardial effusion.

In addition, the CT scan reveals there are multiple small gallstones with no evidence of cholecystitis. The liver, spleen and pancreas appear normal as do both kidneys and the adrenal glands.

The GI tract appears intact. Few scattered small lymph nodes are present in the descending and sigmoid colon. There is no colonic mass or mass effect.

There is no free fluid in the abdomen or pelvis. There is no lymphadenopathy. The bones of the lower chest and abdomen appear intact, displaying no osteolytic or sclerotic metastases. There appears to be a chronic healed fracture of the left transverse process of L1. Patient is scheduled to return for percutaneous biopsy of the left lung with possible ipsilateral and contralateral metastases.

ICD-9-CM:

1.

2.

3.

4.

CPT:

1.

2.

3.

4.

ABSTRACT 7

EMERGENCY ROOM
PATIENT: Frank Padilla
SEX: Male
DATE: 8 /2/04

HISTORY OF PRESENT ILLNESS: The patient is a 49-year-old male who was seen in the Emergency Room complaining of abdominal pain that became acute within the past 24 hours, with possible gastritis.

PAST MEDICAL HISTORY: The patient denies angina or dyspnea on exertion. He denies exercise intolerance, heart murmur, or edema. He had palpitations about 10 years ago when he was going through a divorce, which was very stressful after an 18-year marriage. He has had no episodes of syncope at least over the last 20 years. He does not exercise on a regular basis, but does heavy work including drilling. He previously was a miner and knows that he is exposed to a fair amount of dust with that occupation.

PAST MEDICAL HISTORY: Otherwise unremarkable except for cigarette smoking two packs per day for greater than 20 years. His lipid status is not known.

PREVIOUS OPERATIONS: Previous repair of a left inguinal hernia 12 years ago.

FAMILY HISTORY: Unremarkable other than his father is diabetic.

ALLERGIES: Sulfa which causes asthma.

MEDICATIONS: He is on no medications.

REVIEW OF SYSTEMS: Remarkable for recurrent episodes of chronic bronchitis. EKG a week ago indicates right axis deviation and slow R wave progression across the precordium, but cannot exclude old anteroseptal MI.

PHYSICAL EXAMINATION: GENERAL: The height is 6 feet with weight of 180 pounds. BP 142/82, heart rate 90, and respirations are 18. He is pleasant and well nourished. SKIN: No edema or remarkable lesions. Skin is well perfused. EYES: Pupils are round, equal and reactive to light and accommodation. Normal extraocular movements. CAROTIDS: No carotid bruits. Neck veins are not seen at 30 degree angle. No thyromegaly or adenopathy. LUNGS: Mildly decreased breath sounds with occasional wheezes. CARDIAC: Shows a normal S1 and S2. There is no murmur, gallop or rub. ABDOMEN: No organomegaly or tenderness. NEUROLOGIC: No gross focal findings.

TREATMENT COURSE: The patient was prepped for surgery, but also received a chest x-ray and he was found to have stable left upper lobe pulmonary nodules. Needle biopsy demonstrated small cell carcinoma. After patient was sedated, he was prepped and draped in the usual manner in the Emergency Room. A transverse incision was made above the inguinal ligament. Subcutaneous tissue was dissected and the external oblique fascia was divided laterally. The inguinal nerve was dissected away from surrounding tissue. Both sides of the hernia sac were isolated and dissected. Strangulation was noted. Penrose drain was placed. A 2-0 pursestring suture was placed along the base of the sac with a second suture placed above the other one. Hemostasis was achieved. Prolene mesh was cut to size and tacked. The transversalis fascia was used to complete the repair. The internal oblique fascia was closed as well as subcutaneous tissue. Sterile dressing applied with Steri-Strips.

HISTORY: Expanded problem focused
EXAM: Problem focused
MEDICAL DECISION MAKING: Straightforward

ICD-9-CM:

1.
2.
3.
4.

CPT:

1.
2.
3.
4.

ABSTRACT 8

INPATIENT HOSPITAL

PATIENT: Cathy Edwards
SEX: Female
DATE: 10/22/04

This is a 62-year-old female with history of coronary artery disease with bypass grafting, hypertension, and diabetes mellitus type II. The patient fell at home and was seen for complaints of severe hip pain as well as chest pain. The patient has a history of coronary artery disease with bypass grafting a year ago. The patient had a cardiology consultation prior to surgery for clearance. She was last seen in our office 3 months ago. At that time the patient had been doing very well with no complaints of chest pain, shortness of breath, or dizziness. A 12-lead ECG demonstrated a sinus rhythm with left ventricular hypertrophy with T-wave inversion in the anterolateral leads. The patient indicates she is normally very active at home with no complaints of chest pain, shortness of breath, or dizziness until this week. She indicates she has had no cardiac problems since her last open-heart surgery.

PAST MEDICAL HISTORY: The previous history as mentioned above in addition to hyperlipidemia and hypothyroidism.

MEDICATIONS: Atenolol, Lopid, levothyroxine, lisinopril.

ALLERGIES: NKDA.

PHYSICAL EXAMINATION: The patient is awake and alert with complaints of left hip pain and explains that since the fall she has been experiencing some dyspnea and chest pain. LUNGS: Breath sounds are clear in equal lung fields. CARDIOVASCULAR: Regular rate and rhythm. S1 and S2 with no significant murmurs, rubs or gallops. ABDOMEN: Soft and nontender. No organomegaly is palpated. EXTREMITIES: No peripheral edema, motor, sensory and pulses are intact in left lower extremity. X-rays indicate the patient has sustained a left hip fracture.

A 12-lead ECG shows normal sinus rhythm, left ventricular hypertrophy with T-wave inversion V1 through V6. These changes are unchanged from previous ECGs.

LABORATORY: Urinalysis is positive for urinary tract infection due to *E. coli*. White blood cell cont 5.4, hemoglobin 11.3, hematocrit 33.1, platelets 362, INR 1.17, sodium 144, potassium 3.9, chloride 106k, CO2 23.

PROCEDURE: Patient received an ORIF of the left hip. A longitudinal skin incision was started at the level of the greater trochanter and carried distally parallel to the shaft of the femur. Dissection was carried down through the fascia lata. The fibers of the vastus lateralis were elevated anteriorally. Incision was made in the vastus lateralis 1 cm to the linea aspera. The dissection was carried down through the lateral cortex. A dural hole was placed in the middle cortex and guide wire was inserted into the head and neck. Using the guide wire, it was determined that they were in proper alignment with the fluoroscopic unit. The head and neck were then reamed over the guide wire. A Richards hip compression screw was inserted. This was 85 mm in length and over the guide wire was inserted a side plate at an angle of 135 degrees. This was a four-hole side plate. This was held in place with Bell clamp. Four drill holes were then drilled and the appropriate length cortical screws were then utilized in stabilizing the plate to the shaft of the femur.

Postoperative portable pelvis, one view, in PACU radiology report found postoperative changes of a hip replacement on the left. The femoral head has been removed. Impressions were satisfactory hip replacement on the left with resection of the femoral head and placement of a ball and intermedullary type of prosthesis.

PLAN: The patient was discharged home to return for follow-up in one week. Patient will be provided rehabilitation therapy, as her movements are limited at this time. Home health care will be provided several hours a day to assist patient in household tasks.

ICD-9-CM:

1.
2.
3.
4.
5.
6.
7.
8.
9.

CPT:

1.
2.
3.
4.
5.
6.

ABSTRACT 9

INPATIENT HOSPITAL

PATIENT: Anna May
SEX: Female
DATE: 6/18/04

PRESENT ILLNESS: This patient presents today with complaints of paresthesia, edema, and fatigue, which patient suspects is related to her COPD. She states that she has not been diagnosed in the past with CAD, although this is possible. She claims that the paresthesia began several days ago but it has not worsened. Her fatigue has continued even after sleeping long hours, she states.

PAST HISTORY: No known drug allergies. She has an ongoing history with hypertension, peptic ulcer disease and osteoarthritis. The patient has smoked for 25 years, which aggravates her asthma.

SOCIAL/FAMILY HISTORY: Drinks only occasionally. Father is deceased but mother is alive with insulin-dependent diabetes.

REVIEW OF SYSTEMS: HEENT: No history of problems. No thyromegaly. RESPIRATORY: Patient does have history of asthma. She has had pneumonia twice in the past. She has no cough at this time. CARDIOVASCULAR: Patient states that she has not had cardiac problems in the past and no accompanying signs or symptoms. No heart murmur or chest pains in the past. GASTROINTESTINAL: Patient does have history of peptic ulcer disease for which she has prescriptions. No history of kidney, bladder or liver problems. MUSCULOSKELETAL: Has been diagnosed with osteoarthritis 2 years ago. She states that she does not experience severe pain unless the weather changes. SKIN: No history of skin disease although she states she had problems with acne as a teenager and into early adulthood. NEUROLOGICAL: Patient is experiencing paresthesia today, but states that in the past has had no problems with numbness or tingling.

PHYSICAL EXAMINATION: Patient is alert and oriented times three. She is in mild distress, but states she is feeling better, but that the symptoms have persisted for several days now. Neck is supple. No thyromegaly. PERRLA. Heart rate is regular, but 12-lead ECG demonstrated blockage in the anterior descending artery and obtuse marginal and right coronary arteries. Lungs are clear to P&A. Abdomen shows no palpable masses. Patient does have history of peptic ulcer disease for the past four years. Distended. Nontender. No tenderness or rebound. Extremities show edema and paresthesia.

OPERATIVE FINDINGS: The heart was of normal size, but there is severe CAD with hypertension of the left anterior descending artery, obtuse marginal and right coronary artery. The circumflex minor coronary artery was about a 2-mm vessel. The LAD was a 1.5-mm vessel and the right was about a 1.0 to 1.5-mm vessel.

OPERATIVE PROCEDURE: The patient was placed supine on the operating table. The chest, abdomen, and both lower extremities were prepped with Betadine and draped in the usual fashion. The chest was opened with median sternotomy and the left internal mammary artery was taken off the chest wall. Vein was harvested from the left lower extremity. After giving 27,000 units of heparin, cannulation of the ascending aorta and right atrium was performed and the patient was placed on cardiopulmonary bypass. She was cooled with inflow temperature of 30 degrees centigrade. Topical cooling of the heart was achieved by continuous irrigation of the pericardial cavity with cold saline. The aorta was cross-clamped and high K/low K crystalloid cardioplegia was given through the root of the aorta. Individual vein grafts were placed to the obtuse marginal and to the right coronary artery and finally the left internal mammary artery was anastomosed to the left anterior descending artery with 7-0 Prolene suture. After all the distal anastomoses had been completed, the aortic cross-clamp was unclamped. A partial occluding clamp was applied to the ascending aorta. Two aortotomies were made. They were enlarged with a 4.8-cm aortic punch. The proximal anastomosis between the vein graft and ascending aorta was performed with a 6-0 Prolene suture. The patient was warmed to 36 degrees centigrade and was taken of cardiopulmonary bypass without any difficulty. Two chest tubes were placed in the mediastinum and brought out through a separate stab wound. Temporary transvenous pacemaker was implanted. Heparin was neutralized with 300 mg of Protamine and decannulation was performed. Once we were satisfied with hemostasis, the chest was closed with interrupted sutures.

The wound was then thoroughly irrigated with antibiotic solution and closed in layers.

ICD-9-CM:

1.

2.

3.

4.

5.

6.

7.

8.

9.

CPT:

1.

2.

3.

4.

5.

6.

ABSTRACT 10

INPATIENT HOSPITAL

PATIENT: Rose Andrews
SEX: Female
DATE: 7/18/04

HISTORY OF PRESENT ILLNESS: This 77-year-old female patient had earlier presented to the hospital for an arthroplasty due to traumatic hip fracture.

PROCEDURE: Eight hours after surgery, she required intubation for pulmonary insufficiency. She was monitored overnight, kept n.p.o. after midnight, and seen this morning for medical clearance. She apparently has continuing history of hypertension, aortic insufficiency with diastolic dysfunction, without any known congestive heart failure, coronary artery disease or cerebrovascular disease. She also has a history of URIs.

Her intraoperative course was reportedly unremarkable, with uneventful placement of an LMA, and intraoperative oxygen saturation initially of 99% with end-tidal CO_2 of 30%. The intraoperative oxygen saturation ranged from 92 to 99%. The estimated blood loss was approximately 150 ml. Initial fluid intake was 1500 cc with urine output as 150 cc. Today the patient received an additional 500 cc of lactated ringers with only about 100 ml fluid out. She was noted to have some apparent gastric juices sustaining her intraoperatively. The patient was provided with Reglan and Zofran.

After extubation, an arterial blood gas was 735/29/43 while on 15-liter (100%) nonrebreather mask. The patient was subsequently endotracheally intubated by anesthesia. She was placed on assist control 10, tidal volume 500, FI02 1, and PEEP of 5, with saturations of only 81% upon my arrival. Blood pressure was 110/60, pulse 105 with sinus tachycardia on the monitor. Respiratory rate was 10 with a ventilator.

The ventilator was subsequently adjusted by me, increasing respiratory rate, tidal volume and PEEP, ultimately arriving at assist control of 16, tidal volume 700, FI02 1 and PEEP of 12, giving oxygen saturation that steadily rose into the 90s, now 97%. The patient's blood pressure did fall to 90/50, responding to initial fluid bolus of 250 ml of lactated ringers. She remains sedated, now with blood pressure of 140/70, sinus tachycardia 105, breathing with the ventilator at 16 breaths per minute.

Of note, the patient reportedly has no history of gastroesophageal reflux disease or reflux symptoms. She is currently not able to provide medical history.

PAST MEDICAL HISTORY: Hypertension.

PAST SURGICAL HISTORY: Otherwise reportedly negative.

ALLERGIES: None known.

MEDICATIONS: Norvasc and Diovan usually. Current medications ordered include Compazine prn, Xofran prn, laxative of choice, Tylenol prn, Maalox prn, Benadryl prn, iron sulfate, Fragmin, prophylaxis, Percocet prn.

MULTISYSTEM REVIEW: Unobtainable.

EXAM: The patient is barely responsive, but starting to waken. Temperature 98.8, blood pressure now 140/70, pulse 105, sinus tachycardia, respiratory rate 16. HEENT: Arcus senilis. PEERLA. Oropharynx moist. No missing teeth or oropharyngeal trauma evident. NECK: Supple without tracheal deviation, thyromegaly or masses. NODES: No supraclavicular, cervical, axillary or inguinal adenopathy. BACK: No CVA or spine tenderness. LUNGS: Notable for right coarse breath sounds with inspiratory squeaks and scattered crackles. CARDIAC: Regular rhythm. ABDOMEN: Soft, quiet, not overtly tender. Good bowel sounds. EXTREMITIES: Slightly cool. No cyanosis, clubbing, asymmetry, cods or obvious tenderness. Left hip surgical site is dry and intact. GU: Foley intact.

LABS: Preop WBC 13.7, hemoglobin 15.2, hematocrit 44.3, platelet count 199. Urinalysis showed leukocyte esterase 3+, nitrate negative, 229 white cells, 19 red cells. Urine culture pending due to probable UTI. Basic metabolic panel with sodium 134, potassium 3.6, chloride 94, total CO2 26, BUN 14, creatinine 0.8, glucose 163, INR 0.94, PTT 32.

Preoperative ECG with normal sinus rhythm, 85 beats per minute, single premature ventricular contraction, first degree AV block, normal axis, low voltage in the frontal leads and poor R wave progression.

Chest x-ray: AP portable, pre-intubation, extensive perihilar and right lower lob infiltrate. Postintubation film reveals the same with endotracheal tube just above the carina.

PLAN:

1. Ventilator adjusted as above. Will check arterial blood gases and adjust further as necessary.

2. Will watch fluids, avoiding congestive heart failure with known history of hypertension, diastolic dysfunction.

3. IV metoprolol in lieu of usual antihypertensives for now.

4. Unasyn IV as above.

5. Will withdraw endotracheal tube 2 cm.

6. The patient does not require any traction or weight bearing limitation. Bed to chair as soon as appropriate. For now, will keep comfortable on the ventilator with propofol.

7. Pepcid for stress ulcer prophylaxis.

8. The patient will begin Fragmin for hip surgery deep venous thrombosis prophylaxis.

ICD-9-CM:

1.

2.

3.

4.

5.

6.

7.

8.

9.

CPT:

1.

2.

3.

4.

5.

6.

ABSTRACT 11

INPATIENT HOSPITAL

PATIENT: Joanne Garcia
SEX: Female
DATE: 8/8//04

HISTORY: The patient is a 43-year-old female who was admitted to the hospital today for a hysterectomy. She had been experiencing lower abdominal pain and menorrhagia for the past several months, which was found to be due to a prolapse, which had not responded to nonmedical therapy.

PAST HISTORY: The patient is G2, P2, Ab 0. She did have an appendectomy when she was younger. She has no problems since then.

FAMILY HISTORY: Both parents are alive and well.

SOCIAL HISTORY: The patient has never smoked. She drinks occasionally on the weekend.

ALLERGIES: Motrin.

MEDICATIONS: None at this time.

REVIEW OF SYSTEMS: Unremarkable other than as previously stated.

PHYSICAL EXAMINATION: General appearance: The patient is a well-developed, well-nourished female with some distress. Temperature is 99.6 degrees. HEENT: PEERLA. Normocephalic. NECK: Supple with no adenopathy. CHEST: Lungs equal expansion. Decreased breath sounds. No rales or rhonchi. HEART: Normal S1 and S2. No murmurs, gallops, or rubs. ABDOMEN: Nontender, nonpalpable masses. NEUROLOGICAL: Nonfocal.

HOSPITAL COURSE: The patient was admitted due to menorrhagia with prolapse of the uterus and vagina. Her vagina was prepped and draped in the usual manner for surgery. Incision was made around the patient's cervix and cul-de-sac. The uterosacral ligaments were ligated. There was difficulty in ligating the bladder from the uterus. Suprapubic catheter was placed for drainage. A cystocele was visualized and repaired. The uterine vessels and cardinal ligaments were also cut, with the uterus then being removed. The patient was then sent to Recovery where she did well with no complications. The patient was discharged two days later for follow up.

ICD-9-CM:

1.

2.

3.

4.

5.

6.

7.

8.

9.

CPT:

1.

2.

3.

4.

5.

6.

ABSTRACT 12

INPATIENT HOSPITAL

PATIENT: Duane Kennedy
SEX: Male
DATE: 9/22/04

SPECIMEN: None

ESTIMATED BLOOD LOSS: 250 cc

FLUIDS: Approximately 3 liters of crystalloid

DRAINS: Medium Hemovac placed superficially

COMPLICATIONS: None.

SPONGE AND INSTRUMENT COUNT: Correct

CONDITION: The patient was taken awake to the recovery room.

MEDICAL HISTORY: This is a 63-year-old male one-year status postlumbar fusion for severe spondylolisthesis L5 on S1. He had done very well postoperatively, but began having pain in his back some two months ago due to nonunion. He also had some complaints of right lower extremity pain versus tingling. Plain x-rays demonstrated what appeared to be a loose screw at the L4 level, and the decision was made to proceed with further operative intervention.

OPERATIVE PROCEDURE: After obtaining appropriate informed consent, the patient was taken to the operating room. After adequate general anesthesia was obtained, the patient was positioned prone on the Wilson frame. Great care was taken to pad all bony prominences, and the shoulders were placed no more than 70 degrees of abduction. The C-arm image intensifier was placed into the field, and the lateral position of the back was prepped and draped in normal sterile fashion. The large midline lumbar scar was excised, and 10 cm in the mid portion was ellipsed out thoroughly down to the lumbar dorsal fascia. The fascia was then incised using the Bovie electrocautery, exposing the hardware and posterior elements. A meticulous exposure was made until the transverse process of L4 on the left was thoroughly exposed. The exposure continued until all the hardware was exposed. The spanning rods bilaterally were removed. The right-sided L4 pedicle screw was grossly loose and it was removed. The L4 transverse process was then thoroughly exposed. The L5-S1 segment was solidly fused. There was some very minimal evidence of L5 pedicle screw loosening, and both L5 pedicle screws were removed. The S1 pedicle screws were then solidly fixed. A thorough exploration of the effusion was then undertaken. The L5-S1 was again demonstrated as thoroughly fused. The L4-5 segment was grossly unstable and not united. Attention was then turned towards replacement of pedicle screws at the L4 level. A 5.5 × 50.0 mm screw was placed on the right at L4 and 6.5 × 55 mm screw was placed on the left. Pedicle screw placement was conducted using the technique of decortication posteriorly, blunt and fine probing tapping with the 5.5 mm tap and then placing it above the described screws. The C-arm image intensifier was employed in the lateral plane during the placement process. Attention was then turned towards thorough decortication. The L4 transverse processes as well as the L5 transverse processes and proximal aspect of the fusion mass were thoroughly decorticated.

The small amount of locally harvested autograft was taken off the field. The bone morphogenic protein was prepared appropriately onto its collagen matrix. The locally harvested autograft was placed within the collagen bone morphogenic protein construct. A femoral head allograft was then thoroughly morselized into very small fragments and prepared for placement in the intertransverse space. The locally harvested autograft was then placed in the intertransverse space L4 to L5 bilaterally followed by the bone morphogenic collagen carrier and its autograft and then the femoral head allograft. A spanning construct was made from S1 to L4 bilaterally. A single cross-length was paced and then the remainder of the femoral head allograft was placed below the rods bilaterally. Satisfied with the position of the pedicle screws and the construct built, attention was turned to closure.

The wound had been copiously irrigated with normal saline Bacitracin throughout the procedure. The wound was inspected for hemostasis. The fascia was closed using 0 Vicryl employed in a figure-of-eight interrupted fashion. The superficial layers were closed over a medium Hemovac drain using a 2-0 Vicryl in a similar fashion. The skin was closed using a running Monocryl subcuticular type suture. 30 cc of 1/2 % Marcaine with epinephrine was infiltrated into the wound site for immediate postop analgesia. A sterile dressing was applied and the procedure was terminated.

PLAN: The patient was discharged in good condition and will follow up in one week.

ICD-9-CM:

1.
2.
3.
4.
5.
6.
7.
8.
9.

CPT:

1.
2.
3.
4.
5.
6.

ABSTRACT 13

INPATIENT HOSPITAL

PATIENT: Terese Langdon
SEX: Female
DATE: 12/20/04

INDICATIONS: Preoperative evaluation showing

HISTORY OF PRESENT ILLNESS: This patient became light-headed with associated dyspnea and had a temporary loss of consciousness at work. She was taken to a local hospital and admitted for an abnormal heart rhythm, which was diagnosed as a Mobitz Type 2 atrioventricular (AV) block. While under observation, she had further deterioration of her heart's conduction system, going into a complete third-degree AV block with an escape ventricular heart rate in the 20's. Her past medical history was positive for a prior ECG that showed a right bundle branch block with frequent PVCs.

PAST HISTORY: Patient does not smoke. Drinks occasionally socially. Mother is alive and well, but father is deceased due to a car accident. Patient is currently being treated for hypertension.

HOSPITAL COURSE: During this hospitalization, her echocardiogram was essentially normal, and risk factors for heart disease were not high. Current medications included one aspirin daily. The procedure was performed with mild sedation and a local anesthetic. Patient was not put to sleep. A 2 inch incision was made parallel to and just below the collar bone. Transvenous leads were then inserted just under the collarbone and advanced through that vein under fluoroscopic guidance into the heart. The other ends of the wires were connected to the generator that was implanted under the skin beneath the collarbone. The atrial-ventricular generator was about half an inch deep and one and a half inches wide. The skin was then sutured closed. Incisional pain is mild and transient and usually responds to Tylenol. It is possible to feel the pacer generator under the skin and a slight deformity of the skin can be visually noticed. The patient was discharged home with follow-up in one week.

ICD-9-CM:

1.
2.
3.
4.
5.
6.
7.
8.
9.

ICD-9-CM (procedures):

1.
2.
3.
4.
5.
6.

ANSWERS

CCS MOCK EXAM 1

1. Basic unit of payment for a DRG is based on
   A. visit
   B. admission/discharge
   C. morbidity
   D. length of stay

ANSWER B: DRGs are used for billing inpatient stays, which are, therefore, based on admission and discharge.

2. What is the correct code(s) for a patient who received a fine needle biopsy of a kidney mass with guidance and evaluation in an ASC for possible carcinoma of the kidney?

593.9 Unspecified disorder of kidney and ueter
189.0 Kidney malignant neoplasm and unspecified urinary organ
10022 Fine needle aspiration with imaging guidance
50200 Renal biopsy, percutaneous by trocar or needle
88172 Cytopathology, evaluation of fine needle aspirate, immediate cytohistolgic study.

   A. 189.0, 50200
   B. 593.9, 88172
   C. 593.9, 10022, 88172
   D. 189.0, 10022

ANSWER C: The mass is coded, but not the neoplasm since this is an ASC visit in which possible conditions are not coded. The biopsy is obtained with fine needle aspiration, which is coded differently from the regular biopsy code. The specimen was also examined, so this should be coded also.

3. What is the correct code(s) for a patient who received a corneal transplant for the outer layer of the cornea due to proliferative insulin-dependent diabetic retinopathy?

> 250.50 Diabetes with ophthalmic manifestations, Type II
> 250.51 Diabetes with ophthalmic manifestations, Type I
> 362.02 Proliferative diabetic retinopathy
> 362.29 Other nondiabetic proliferative retinopathy
> 65400 Excision of the cornea (keratectomy, lamellar, partial)
> 65710 Keratoplasty, lamellar
> 65730 Keratoplasty, penetrating
> 65771 Radial keratotomy

    A. 362.29, 65771
    B. 250.50, 362.29, 65400
    C. 250.51, 65730
    D. 250.51, 362.02, 65710

ANSWER D: The diabetes is insulin-dependent but controlled. There is involvement of the eye, so this must be included in the diabetes code as well. As is normal for diabetes codes, both the diabetes and the manifestation are coded. The outer layer of the cornea is involved, so this is coded as lamellar.

4. If a patient is admitted to the Emergency Room with complaints of heartburn and pyrosis and the examination reveals that the patient has esophageal diverticula, but the patient experiences appendicitis for which he is provided emergency surgery, what is the principal diagnosis?

    A. Pyrosis
    B. Heartburn
    C. Diverticula
    D. Appendicitis

ANSWER D: With outpatient visits, the primary diagnosis is related to the main reason service is provided, which in this case was appendicitis.

5. Which of the following services would not be performed before a patient receives medical services?

    A. Insurance verification
    B. Payment of copay
    C. Payment of deductible
    D. Obtain past medical history

ANSWER C: The deductible is not determined by the medical facility, but is tracked by the insurance company.

6. What is not true about DRGs?

    A. 25 categories
    B. 115 categories
    C. Divided into MDGs
    D. Derived from ICD-9-CM-CM codes

ANSWER B: MDCs are major diagnostic categories of which there are 25. They are derived from the ICD-9-CM-CM codes.

7. What is the correct code(s) for a patient who had a recurrent ganglion cyst of the wrist removed which was 1.4 cm?

> 706.2 Sebaceous cyst
> 727.43 Ganglion unspecified
> 25111 Excision of ganglion, wrist; primary
> 25112 Recurrent

    A. 706.2, 25112
    B. 706.2, 25111, 25112
    C. 727.43, 25112
    D. 727.43, 25111, 25112

ANSWER C: The recurrent code, 25112, includes the section of 25111 up until the semicolon, so primary is replaced with recurrent.

8. What is the correct code(s) for a patient who received wound repair for a 4-cm laceration of the right forearm, which required closure of subcutaneous tissue and extensive debridement and a 3.7-cm laceration of the chest, which required repair of the fascia?

> 11042 Debridement, skin, partial thickness
> 12002 Simple repair of superficial wounds of scalp, neck, axillae, external genitalia, trunk and/or extremities, 2.6 cm to 7.5 cm
> 12004 Simple repair of superficial wounds of scalp, neck, axillae, external genitalia, trunk and/or extremities, 7.6 cm to 12.5 cm
> 12032 Layer closure of wounds of scalp, axillae, trunk and/or extremities, 2.6 cm to 7.5 cm
> 12034 Layer closure of wounds of scalp, axillae, trunk and/or extremities, 7.6 cm to 12.5 cm

    A. **12034**
    B. 12032, 12002
    C. 11042, 12034
    D. 11042, 12002, 12004

ANSWER A: Extensive debridement with a laceration that requires simple repair is consequently coded as intermediate repair without an additional code for the debridement. Therefore, the two codes are added together as they are both intermediate repair and of the same anatomic sites listed in the one code.

9. Advance Beneficiary Notices are important because

   A. They insure that payment will be sent to the service provider

   B. They inform the patient that they may be responsible for payment

   C. They insure that services are covered by insurance

   D. They state that the services are covered services under Medicare

ANSWER B: Advance Beneficiary Notices are statements signed by the patient stating that they are responsible for coverage if the insurance carrier, such as Medicare, will not pay for services.

10. Which of the following elements would not constitute an element used in determining proper history level for E/Ms?

   A. Review of Systems

   B. Medical Records

   C. Present Illness

   D. Past History

ANSWER B: Medical records are used as an element in the Medical Decision Making, not part of the History.

11. What is the correct code(s) for a patient who received manipulation of a radial shaft greenstick fracture?

   813.81 Fracture, unspecified part, closed, radius
   813.21 Fracture, shaft, closed, radius
   813.31 Fracture, shaft, open, radius
   25390 Osteoplasty, radius or ulna
   25505 Closed treatment of radial shaft fracture with manipulation
   25515 Open treatment of radial shaft fracture with or without internal or external fixation

   A. 813.81, 25390

   B. 813.21, 25505

   C. 813.31, 25515

   D. 813.31, 25505

ANSWER B: Greenstick fracture indicates that the fracture is closed. In addition, manipulation must be included in the code, and there is no indication that the treatment was open, so it is coded as closed.

12. What is the correct code(s) for a patient who was admitted to the hospital with severe chest pain and dyspnea? The patient was diagnosed with third-degree AV block due to digitalis toxicity. A permanent atrial-ventricular pacemaker with transvenous leads was inserted.

   426.0 Atrioventricular block complete
   426.10 Atrioventricular block, unspecified
   428.1 Left heart failure
   786.09 Other respiratory abnormalities
   796.0 Nonspecific abnormal toxicological findings
   972.1 Poisoning by cardiotonic glycosides and drugs of similar action
   37.72 Initial insertion of transvenous leads into atrium and ventricle
   37.83 Initial insertion of dual-chamber device
   39.64 Intraoperative cardiac pacemaker
   E942.1 Adverse effect of cardiotonic glycosides and drugs of similar action

   A. 426.10, 786.09, 972.1, 39.64

   B. 426.0, E942.1, 37.83, 37.72

   C. 428.1, 796.0, 37.83

   D. 426.0, 786.09, E942.1, 39.64

ANSWER B: The AV block is a third-degree block due to digitalis toxicity, so both of these must be coded. For toxicity, the alphabetic index indicates that the Table for Drugs and Chemicals should be used when there are symptoms, which there are here. There is no mention that the digitalis was taken against medical advice; toxicity does not indicate this, so this is coded as an E code for therapeutic use. The pacemaker is permanent and is dual chamber with tranvenous leads, which all must be coded. Notice that the insertion of the leads must also be coded.

13. What is the correct code for a baby 4 weeks old who was seen in the Emergency Room for 2 1/2 hours of critical care?

   99291 Critical care; first 30 minutes
   99292 Each additional 30 minutes
   99293 Initial inpatient pediatric critical care
   99295 Initial inpatient neonatal critical care
   99296 Subsequent inpatient neonatal critical care

   A. 99295

   B. 99293

C. **99291, 99292 × 3**

D   99291 × 5

ANSWER C: When a neonate/child under 24 months is seen for critical care in the Emergency Room or physician's office, it is coded from the regular critical care codes, which are based on half hour segments.

14. What is the correct code(s) for a patient who is diagnosed fever and chills due to SARS-related pneumonia?

079.82 SARS-associated coronavirus
480.3 Pneumonia due to SARS-associated coronavirus
486 Pneumonia organism unspecified
V01.82 Exposure to SARS-associated corona virus

A. 486, 079.82

B. 480.3

C. 486, V01.82

D. 079.82

ANSWER B: Only the one code is necessary as it contains all of the information, and there are no notes requiring any additional coding of the organism.

15. Removal of fluid from the abdomen surgically is known as

A. Amniocentesis

B. Thoracocentesis

C. Thoracentesis

D. Paracentesis

ANSWER D: Parecentesis is removal of surgical removal of fluid from the abdomen. Thoracentesis and thoracocentesis are the same procedure, just different spelling, and are removal of fluid from the chest. Amniocentesis is used for testing the amnion fluid when a mother is pregnant.

16. What is the correct code(s) for a patient who has acute renal failure and malignant hypertension?

401.0 Essential hypertension malignant
403.91 Unspecified hypertensive renal disease
584.9 Acute renal failure, unspecified

A. 403.91

B. 403.91, 401.0

C. 584.9, 401.0

D. 584.9

ANSWER C: This is acute renal failure and there is no indication that it is related to the hypertension, so both are coded separately. Although hypertension and renal failure commonly are assumed to be related, particularly with chronic renal failure/disease, acute renal failure can occur for other reasons.

17. What is the correct code(s) for a patient who was seen today in the Emergency Room for complaints of dyspnea and pain in the left arm with a history of AICD six months ago and possible MI?

410.9 Acute myocardial infarction, unspecified site
729.5 Pain in limb
786.00 Respiratory abnormality unspecified
786.09 Other symptoms involving respiratory system and chest symptoms
V44.3 Colostomy status
V45.02 Postprocedural cardiac device *in situ*, automatic implantable cardiac defibrillator

---

A. 786.09, 729.5, V45.02

B. 410.9, V44.3

C. 786.09, 729.5

D. 410.9, V45.02

ANSWER A: The myocardial infarction is not coded as it is described as possible, which is not, therefore, coded for outpatient purposes. The postprocedural status of the automatic implantable cardiac defibrillator is coded as it is significant to the current complaints. The dypsnea and arm pain would be coded.

18. Case mix is not based on

A. Patient's past medical history

B. Volume

C. Exacerbating conditions

D. Patient's age

ANSWER A: Case mix is based on current hospital and admission factors that influence the level of care.

19. What is the correct code(s) for a patient who had a UTI due to *E. coli* and is 6 months pregnant?

008.00 Intestinal infections due to *E. coli* unspecified
041.4 Bacterial infection in conditions classified elsewhere, *E. coli*
599.0 Urinary tract infection, site not specified
646.63 Infections of genitourinary tract in pregnancy

A. 646.63
B. 646.63, 599.0, 041.4
C. 008.00
D. 599.0, 041.4

ANSWER B: The notes in the coding book indicate that the urinary tract infection should also be coded and the organism too.

20. In best practices, professional courtesy
    A. Must be applied to all patients
    B. Can be applied to friends and family
    C. Is illegal
    D. Is billed as insurance only

ANSWER A: In the past, professional courtesy was the reduction or elimination of fees only for certain people, such as other physicians and their families; however, since these benefits were not offered to all patients, particularly Medicare patients, it could be construed as fraud since the government is charged more for some patients rather than the required standard amounts.

21. If a patient with COPD is admitted to the hospital with hepatomegaly and a severe headache, which was diagnosed as possible hepatitis B, what is the principal diagnosis?
    A. Hepatomegaly
    B. COPD
    C. Headache
    D. Hepatitis B

ANSWER D: The hepatitis B is related to the chief complaints on admission, so this is the principal diagnosis even though it is described as possible. With hospital coding, remember that diagnoses described as possible are coded.

22. What is the correct code(s) for a patient with a perforating ulcer who receives a partial gastrectomy with gastrojejunostomy?

    533.10 Acute peptic ulcer site unspecified with perforation without mention of obstruction
    533.50 Chronic or unspecified peptic ulcer with perforation without mention of obstruction
    43.7 Partial gastrectomy with anastomosis to jejunum
    44.32 Percutaneous gastrojejunostomy
    44.39 Other gastroenterostomy

    A. 533.10, 44.39
    B. 533.50, 43.7
    C. 533.10, 44.32
    D. 533.50, 44.39

ANSWER B: The ulcer is perforating and not stated to be chronic or acute, so it is coded as unspecified which is included with the chronic code. There is no mention of obstruction, so a fifth digit of 0 is used. The procedure includes a partial gastrectomy, so this must be included in the code, as some codes exclude the gastrectomy.

23. Chondromalacia is
    A. Abnormal condition of the muscles
    B. Softening of articular cartilage
    C. Weakening of the joints
    D. Abnormal condition of high cholesterol

ANSWER B: Chondro means cartilage and malacia means softening.

24. Which of the following is not an emancipated minor?
    A. Living separately from parents
    B. Married
    C. In the military
    D. College student

ANSWER D: College students are not necessarily emancipated as the age may differ from state to state (18 to 21), and some college students may, therefore, be under the age of emancipation.

25. Which drug would be used to treat PID?
    A. Streptokinase
    B. Lanacort
    C. Doxycycline
    D. Comtrex

ANSWER C: Doxycycline is used to treat pelvic inflammatory disease. It is a derivative of tetracycline.

26. If the APC for low-level E/M visits is 0600, the middle-level is 0601 and the high-level is 0602, what would the correct APC be for code 99202?
    A. Low-level
    B. 0602
    C. Middle-level
    D. 0600

ANSWER D: The question asked for the APC, which would be 0600, low-level.

27. What is the correct code(s) for a patient who had a coronary embolism 8 weeks ago?

410.92 Acute myocardial infarction, unspecified site
414.8 Other specified forms of chronic ischemic heart disease
429.79 Other sequelae of myocardial infarction NOS

A. 414.8
B. 410.92
C. 410.92, 429.79
D. 429.79

ANSWER B: The embolism occurred 8 weeks ago, so it is still coded as acute myocardial infarction.

28. Methods of reimbursement do not include
    A. Capitation
    B. Fee-for-service
    C. Remittance
    D. Fee per visit

ANSWER C: Capitation is a set amount that is paid each month to the physician. Fee-for-service is payment at the time the service is provided. Fee per visit is a charge for each time the patient sees the physician. Remittance means payment and is not a method of reimbursement.

29. The slipping of one part of the intestine within another part is known as
    A. Intussusception
    B. Colectomy
    C. Enterolysis
    D. Intubation

ANSWER A: Intussusception occurs when one part of an intestine slips into another part, often seen in the ileocecal area.

30. What is the correct code for a patient who came to the Emergency Room and was then admitted to observation for 10 hours due to MVA with the history and exam comprehensive and the medical decision making low?

0601 Low-level E/M visit APC
0602 Moderate-level E/M visit APC
99203 Office visit new patient, detailed history and exam, low complexity medical decision making
99218 Initial observation care with detailed or comprehensive history and exam and straightforward medical decision making.

A. 99218
B. 0602
C. 0601
D. 99203

ANSWER C: The patient was in observation status within a hospital, so this is billed as observation, not as level APC codes.

31. Which of the following information would not be available in the Chargemaster?
    A. Inventory control number
    B. Fees
    C. HCPCS codes
    D. Discharge status

ANSWER D: The question is asking what type of information is available from the Chargemaster, not what personal patient data may be entered during hospitalization.

32. What is the correct code(s) for a one-year-old female patient who is admitted to the Emergency Room with severe diarrhea for three days? Cultures reveals *C. jejuni*. Erythromycin is prescribed.

787.91 Diarrhea
008.43 Intestinal infections due to *Campylobacter*
008.49 Intestinal infections due to other organism

A. 787.91
B. 787.91, 008.49
C. 008.43
D. 787.91, 008.43

ANSWER C: The diarrhea is coded with the organism, which is *Campylobacter jejuni*.

33. With revenue codes, if pharmacy is indicated by 25X, and 0 is for general drugs, 1 is for generic drugs, 2 is for nongeneric drugs, and 3 is for take-home drugs, what would be the proper revenue code for amitriptyline?
    A. 250
    B. 25X3
    C. 251
    D. 252

ANSWER C: Amitriptyline is a generic drug, so a third digit of 1 would replace the X.

34. Which of the following is not true about coding for professional services?

A. Can be indicated by modifier 26
B. Not necessary when coding for hospital services
C. Is referred to as supervision and interpretation
D. **Must always be coded as separate code from other services when performed by physician**

ANSWER D: It is not always a separate code but may be included within another code, so be careful to read the code's description.

35. Removal of skin tags is based on
    A. Each individual tag
    B. Type of removal
    C. **Each additional group of tags**
    D. Each additional tag

ANSWER C: Removal of skin tags is based first on the first 15 lesions, and the second code is on the next ten.

36. What is the correct code(s) for a patient who received a right heart catheterization and arterial coronary angiography?

> 93501 Right heart catheterization
> 93508 Catheter placement in coronary artery(s), arterial coronary conduit(s), and/or venous coronary bypass graft(s) for coronary angiography without concomitant left heart catheterization
> 93541 Injection procedure during cardiac catheterization, for pulmonary angiography

    A. 93501
    B. 93501, 93541
    C. 93501, 93508, 93541
    D. **93501, 93508**

ANSWER D: The right heart catheterization must be coded. In addition, the coronary angiography can be coded because there is no left heart catheterization with which it cannot be coded additionally as stated in the CPT notes.

37. What drug would be prescribed for treatment of an ulcer?
    A. Citrucel
    B. Flagyl
    C. Topicort
    D. **Prevacid**

ANSWER D: Prevacid is known as lansoprazole and is used short term to treat ulcers.

38. What does extrapyramidal mean?
    A. **Motor pathways to the brain**
    B. Spinal cord
    C. External causes of diseases
    D. Synergistic effect

ANSWER A: Extrapyramidal refers to the nervous system that controls rhythmic motor behavior, such as walking and balance. It includes the nerve cells, nerve tracts and pathways that connect to the brain and spinal cord.

39. What is the correct code(s) for a patient who received a split-thickness graft, 180 sq cm for scarring of the back due to radiation 3 years ago?

> 709.2 Scar conditions and fibrosis of skin
> 906.8 Late effect of burn, other site specified
> 909.2 Late effect of radiation
> 942.04 Burn of trunk, unspecified degree, back
> 15000 Surgical preparation for creation of recipient site by excision of open wounds, burn eschar, or scar, first 100 sq cm
> 15001 Surgical preparation for creation of recipient site by excision of open wounds, burn eschar, or scar, each additional 100 sq cm or each additional percent of body area of infants and children
> 15100 Split graft, trunk, arms, legs; first 100 sq cm or less, or one percent of body area of infants and children
> 15120 Split graft, face, scalp, eyelids, mouth, neck, ears, orbits, genitalia, hands, feet and/or multiple digits, first 100 sq cm or one percent of body area of infants and children
> 15121 Split graft, face, scalp, eyelids, mouth, neck, ears, orbits, genitalia, hands, feet and/or multiple digits, each additional 100 sq cm or each additional one percent of body area of infants and children, or part thereof

    A. 709.2, 909.2, 15100, 15001
    B. **709.2, 909.2, 15100, 15101, 15000, 15001**
    C. 942.04, 15120, 15121
    D. 709.2, 906.8, 15000, 15001, 15120, 15100

ANSWER B: The site is scarred so it must be prepared as indicated by the two codes 15000 and 15001. The first code covers the first 100 sq cm and the second code covers the additional 80 sq cm. The late effect of radiation should also be coded.

40. What is the correct code(s) for a patient who received debridement and shaving of the articular

cartilage of the patellofemoral compartment of the knee and a meniscectomy of the lateral compartment?

> 27332 Arthrotomy, with excision of semilunar cartilage, knee; medial or lateral
> 29862 Arthroscopy, hip, surgical, with debridement/ shaving of articular cartilage, abrasion arthroplasty and/or resection of labrum
> 29877 Arthroscopy, knee, diagnostic debridement/shaving of articular cartilage

A. 29877
B. 27332
C. **27332, 29877**
D. 29862, 27332

ANSWER C: The meniscectomy is of the lateral compartment of the knee and the chondroplasty (debridement and shaving) are of the patellofemoral compartment, so they are coded separately. If they were of the same compartment, they would be coded as a meniscectomy only.

41. ERCP means
    A. Percutaneous erythrocyte count
    B. Endoscopic radioimmunoassay of colon and prostate
    C. Blood test
    D. **Endoscopic retrograde cholangiopancreatography**

ANSWER D: ERCP is an endoscopic retrograde cholangiopancreatography in which contrast material is introduced through a catheter into the mouth and down through parts of the digestive system including the stomach and duodenum.

42. What is the correct code(s) for a patient who received an ethmoidectomy with removal of bullae from the anterior sinus and complete examination of the posterior sinus due to inflammation?

> 471.1 Polypoid sinus degeneration
> 478.1 Other disease of the nasal cavity and sinuses
> 31200 Ethmoidectomy, intranasal, anterior
> 31201 Ethmoidectomy, intranasal, total
> 31205 Ethmoidectomy, extranasal, total

A. **471.1, 31200**
B. 478.1, 31201
C. 478.1, 31200
D. 471.1, 31205

ANSWER A: Polypoid sinus degeneration includes ethmoiditis. This is only a partial ethmoidectomy since the bullae were only removed from the anterior.

43. Decrease in platelets is known as
    A. Achondroplasia
    B. Leukocytosis
    C. **Thrombocytopenia**
    D. Hemolysis

ANSWER C: Thrombo means clot, cyt means cell and penia means deficiency, so thrombocytopenia is a deficiency in the cells that are responsible for clotting, which are the platelets.

44. Using CPT codes, the use of oral contrast material for testing is coded as
    A. With contrast material
    B. As two codes
    C. An additional code to the surgery
    D. **Without contrast material**

ANSWER D: Rectal and oral contrast material is coded as without contrast material as noted in the CPT coding book.

45. What is the correct code(s) for a patient who received a TAH due to menorrhagia and pain?

> 58150 Total abdominal hysterectomy with or without removal of tubes, with or without removal of ovaries
> 58200 Total abdominal hysterectomy, including partial vaginectomy with para-aortic and pelvic lymph node sampling with or without removal of tubes with or without removal of ovaries
> 59850 Induced abortion, by one or more intra-amniotic injections including hospital admission and visits
> 59851 Induced abortion, by one or more intra-amniotic injections including hospital admission and visits with dilation and curettage and/or evacuation

A. 59850
B. **58150**
C. 59851
D. 58200

ANSWER B: TAH is a total abdominal hysterectomy, which includes the corpus and cervix.

46. What is the correct code(s) for a patient who received an adenoidectomy due to acute and chronic inflammation and was subsequently treated for control of postoperative hemorrhage?

463 Acute tonsillitis
474.01 Chronic adenoiditis
474.8 Other chronic disease of tonsils and adenoids
998.11 Hemorrhage or hematoma or seroma complicating a procedure
28.6 Adenoidectomy without tonsillectomy
28.7 Control of hemorrhage after tonsillectomy and adenoidectomy

A. 474.8, 28.6, 28.7
B. 474.01, 998.11, 28.7
C. 474.8, 463, 28.6
D. **463, 474.01, 998.11, 28.6, 28.7**

ANSWER D: The hemorrhage is postoperative, so be sure to select the code from the proper codes that are the postoperative codes. Both chronic and acute adenoiditis must be coded. Be careful because the inflammation of the adenoids is termed adenoiditis and not adenitis. The control of the postoperative hemorrhage must also be coded, both surgically and diagnostically, as it required additional services and was not part of the surgical package for the adenoidectomy.

47. What is the correct code(s) for a patient who received reduction for a comminuted dislocation of the radial head and fracture of the proximal ulna but the reduction was unsuccessful?

813.03 Monteggia's fracture
813.82 Fracture, ulna alone
833.00 Closed dislocation of the radius
24620 Closed treatment of Monteggia type of fracture dislocation at elbow with or with manipulation
24635 Open treatment of Monteggia type of fracture dislocation at elbow, with or without internal or external fixation
24640 Closed treatment of radial head subluxation with manipulation
25535 Closed treatment of ulnar shaft fracture, with or without internal or external fixation

A. 813.82, 833.00, 25535, 24640
B. 813.82, 833.00, 25535
C. **813.03, 24620**
D. 813.03, 24635

ANSWER C: Comminuted means closed fracture. There is no mention if the procedure was open or closed, so it is coded as closed. Reduction indicates there is manipulation. This fracture/dislocation is coded together in one code and is known as a Monteggia fracture. It is located at the elbow.

48. What is true about an excisional biopsy?
   A. Incision is made to remove a portion of the lesion
   B. Removal of lesion
   C. Markers are not used
   D. Is coded from the biopsy sections

ANSWER B: In an excisional biopsy, the entire lesion is removed and biopsied. Markers, such as wires, can be used to indicate the lesion's location. There is no biopsy section. The biopsy codes are included in various parts throughout the codes.

49. If a patient is admitted to the hospital with edema and hematuria and a history of hypertension and insulin-dependent diabetes with tests indicating glomerulonephritis, but the patient experiences an MI while in the hospital, what is the principal diagnosis?
   A. Glomerulonephritis
   B. MI (myocardial infarction)
   C. Diabetes
   D. Hematuria

ANSWER A: The patient is admitted for symptoms that are diagnosed to be associated with the glomerulonephritis, so this is the principal diagnosis, not the MI or hematuria. With hospital coding, the principal diagnosis is the primary reason for the admission after tests have determined the cause.

50. Which of the following is not a method for cauterization?
   A. Dry ice
   B. Electric current
   C. Exenteration
   D. Chemicals

ANSWER C: Dry ice, electric current and chemicals all produce the means to burn, so they are forms of cauterization. Exenteration involves excision and is, therefore, not a method for cauterization.

51. What would the correct code(s) be for a patient admitted to the hospital with bleeding esophageal varices, which were endoscopically injected with sclerosing agent? The patient was also diagnosed with portal hypertension.

456.0 Esophageal varices with bleeding
456.20 Esophageal varices in diseases classified elsewhere with bleeding
572.3 Portal hypertension
38.87 Other surgical occlusion of vessels abdominal veins
42.33 Endoscopic excision or destruction of lesion or tissue of esophagus
99.29 Removal of foreign body without incision from lower limb, except foot

   A. 456.20, 38.87, 99.29
   B. 456.0, 42.33, 99.29
   C. 572.3, 456.20, 42.33
   D. 456.20, 572.3, 99.29

ANSWER C: With the portal hypertension and bleeding, the alphabetic index indicate that two diagnostic codes are required and the order is specified. Notice that the injection of therapeutic substance, code 99.29, excludes the sclerosing substance for esophageal varices.

52. Drug used for reducing cholesterol is
   A. Lasix
   B. Lipitor
   C. Dilantin
   D. Prevacid

53. Intrathecal is an example of
   A. Endoscopic exam of the gastrointestinal tract
   B. Rectal test
   C. Parenteral administration of drugs
   D. Administration of chemotherapeutic drugs

ANSWER C: Intrathecal is the administration of drugs into the spinal cord, so it is a type of parenteral administration, which is achieved through injections.

54. What is the correct code(s) for a patient who was injured in a MVA and was diagnosed with fractured C2-3, which required the removal of fragments and hemilaminectomy with anterior fusion?

805.02 Cervical, closed fracture, second cervical vertebrae
805.08 Cervical, closed fracture, multiple cervical vertebrae
806.02 C1-C4 level cervical, closed, with anterior cord syndrome
806.03 C1-C4 level cervical, closed, with central cord syndrome
806.10 C1-C4 level cervical, open, with unspecified spinal cord injury
80.50 Excision or destruction of intervertebral disc
80.51 Excision of intervertebral disc
81.00 Spinal fusion, NOS
81.02 Other cervical fusion, anterior technique

   A. 806.02, 806.03, 80.51
   B. 806.10, 80.50, 81.00
   C. 805.08, 80.51, 81.02
   D. 805.02, 80.50

ANSWER C: The spinal cord fracture is without injury, so it is coded as cervical fracture. There are more than one vertebrae fractured, so the fifth digit of 8 must be used, not the separate listing of the two vertebrae with separate codes. It is assumed that it is closed since there no mention of it being open injury. The fusion must also be coded.

55. Which is not part of the vascular system?
   A. Axon
   B. Arterioles
   C. Venules
   D. Capillaries

ANSWER A: Axons are part of the neuron in the nervous system.

56. If a patient who is asthmatic is seen in the hospital's outpatient clinic with pitting edema and hypertension and is diagnosed with possible renal failure, what is the principal code?
   A. Asthma
   B. Pitting edema
   C. Renal failure
   D. Hypertension

ANSWER B: The edema is the primary diagnosis as this is outpatient and so the possible diagnosis is not coded.

57. What is the correct code(s) for a patient who received a complex fistulectomy with seton guidance of the anus?

565.1 Anal fistula
569.81 Fistula of intestine
46020 Placement of seton
46262 Hemorrhoidectomy, internal and external, complex or extensive, with fistulectomy with or without fissurectomy
46280 Surgical treatment of anal fistula, complex or multiple with or without placement of seton

A. 565.1, 46262, 46020
B. 569.81, 46262
C. 565.1, 46280
D. 569.81, 46280, 46020

ANSWER C: The code 46280 includes the seton placement, so this is not coded additionally. The fistulectomy is excision of a fistula, so this code is correct.

58. Which of the following is not a type of diagnostic ultrasound as listed in the CPT book?
    A. Two-dimensional
    B. Real time scan
    C. P-scan
    D. M-mode

ANSWER C: Ultrasounds can be A-mode, M-mode, B-scan or real time scan which are one or two dimensional.

59. What is the correct code(s) for a patient who received a needle biopsy of the ovaries?

54.21 Laparoscopy of abdominal region
54.24 Closed biopsy of intra-abdominal mass
65.11 Aspiration biopsy of ovary
65.13 Laparoscopic biopsy of ovary

A. 65.11
B. 54.24
C. 54.21
D. 65.13

ANSWER A: Needle indicates aspiration, so this is coded as aspiration biopsy. The biopsy is of the ovaries, which has its own code and is not coded from the code for biopsy of intra-abdominal area.

60. What is the correct code(s) for a patient who received a partial ostectomy with replacement of both femoral head and acetabulum with prosthesis?

77.89	Other partial ostectomy
81.51	Total hip replacement
81.52	Partial hip replacement
81.80	Total shoulder replacement

A. 77.89
B. 81.51, 77.89
C. 81.51

ANSWERS

ABSTRACT 1

ICD-9-CM:
1. 682.3
2. 304.21
3. 292.12
4. 970.8

CPT:
1. 99282-25
2. 23930
3.
4.

ANSWER: The cellulitis code includes the abscesses, so only the cellulitis is coded. The organism is not coded as it is probable, but this is an Emergency Room visit so probables are not coded. The drug using is affecting the condition and care of the patient, so it should also be coded. Notice that drug dependence and addiction use the same code. The hallucinations should be coded. This is an Emergency Room visit for the hallucinations, so it should be coded as such, and not a regular visit. A modifier 25 attached to this indicates to indicate that a separate service is provided.

ABSTRACT 2

ICD-9-CM
1. 427.1
2. V45.01
3. 496
4. V46.2

CPT:
1. 99283
2.
3.
4.

ANSWER: Although the ventricular tachycardia is described as possible and would not be coded, later it is diagnosed in the report as ventricular tachycardia.

The implantation of the pacemaker a year ago is also listed as it may be pertinent to the present condition. The patient also has COPD. The history element is not high enough to select a higher level of code, so the lower level of 99283 is selected.

ABSTRACT 3

ICD-9-CM

   1. 174.9

   2. 401.9

   3. 244.9

   4. V16.3

CPT:

   1. 10021

   2. 76090-RT

   3. 76082

   4.

ANSWER: Cancer of the right breast was confirmed, so the significant cancerous history of her daughter should be coded. The hypertension and hypothyroidism should also be coded. The fine needle aspiration needs to be coded in addition to the mammogram. The mammogram included the use of a computer for detection, so this must also be coded.

ABSTRACT 4

ICD-9-CM

   1. 671.83

   2.

   3.

   4.

CPT:

   1. 46221

   2.

   3.

   4.

ANSWER: The patient is pregnant, so the code must come from the 600 codes for complications of the puerperium. Note that the puerperium is for postpartum, but there is a note that these conditions can occur during childbirth or pregnancy.

There is no note to code the hemorrhoids additionally, so no other code is required. The hemorrhoidectomy is by ligation, which must be included in the code.

ABSTRACT 5

ICD-9-CM:

   1. 623.8

   2. 621.5

   3. 625.0

   4. V10.3

CPT:

   1. 58558

   2.

   3.

   4.

ANSWER: There is no definite diagnosis, so the vaginal bleeding is coded in addition to the adhesions and dyspareunia. The adhesions were found in the uterus, not the vagina, so they are coded with the uterus. The history of cancer should also be coded as it may be relevant to the diagnosis. The adhesions were removed by dilation and curettage, so these must be included in the code for the procedure in addition to the biopsies.

ABSTRACT 6

ICD-9-CM:

   1. 786.6

   2. 574.20

   3. 493.90

   4. 473.9

CPT:

   1. 99203

   2. 74170

   3. 72194

   4.

ANSWER: The patient does have the lung mass although it is not determined at this time if it is neoplastic, so it is coded only as a mass since this is not a hospital admission, therefore probable conditions are not coded. There are gallstones, so cholelithiasis is coded. The patient also has asthma and chronic sinusitis, which must be coded. The visit must be coded but does not meet a higher level because the exam is not extensive enough. The CT of the abdomen should also be coded as being with and without contrast material.

ABSTRACT 7

ICD-9-CM:

1. 550.10

2. 162.3

3. 491.9

4.

CPT:

1. 49507

2. 99281

3. 71010

4. 32405

ANSWER: The hernia is strangulated, so it is coded as obstructed. The patient was also diagnosed with small cell carcinoma of the lung, but this is not the primary reason for admission, so the hernia is coded as the primary code. Chronic bronchitis should also be coded. The mesh is not coded additionally for inguinal hernias. The x-ray is coded in addition to the Emergency Room visit.

ABSTRACT 8

ICD-9-CM (diagnoses):

1. 820.8

2. 786.50

3. 244.9

4. 272.4

5. 599.0

6. 041.4

7. V45.81

8.

9.

ICD-9-CM (procedures)

1. 79.35

2. 88.26

3.

4.

5.

6.

ANSWER: The left hip fracture is coded as the principal diagnosis although the patient was admitted with complaints of chest pain as well as the hip pain. Chest pain is coded because there is no more definitive diagnosis for this. The hip pain is not coded, however, since it is determined there is a fracture. The history of the bypass graft is listed as it was done only a year ago and the patient did have the complaints of chest pain with ECG provided, but no diagnosis was described. The urinary tract infection is mentioned in the laboratory report, so it should be coded also as well as the organism. The ECG is not coded for the CCS exam.

ABSTRACT 9

ICD-9-CM (diagnoses):

1. 414.01

2. 493.20

3. 715.90

4. 533.70

5.

6.

7.

8.

9.

ICD-9-CM (procedures):

1. 36.13

2. 39.61

3. 39.64

4.

5.

6.

ANSWER: The principal diagnosis is coronary artery disease, which is linked to hypertension, so this must be included in the code. The patient does have COPD, but is associated with asthma so it is coded as asthma with COPD, not just COPD (496). The patient also has peptic ulcer disease, which has been ongoing for four years, so it is coded as chronic. The cardiopulmonary bypass must also be coded in addition to the temporary pacemaker.

ABSTRACT 10

ICD-9-CM (diagnoses):

1. 518.5
2. 820.8
3. 424.1
4. 401.9
5.
6.
7.
8.
9.

ICD-9-CM (procedures):

1. 96.70
2. 96.04
3.
4.
5.
6.

ANSWER: The patient has pulmonary insufficiency after surgery, so this is coded. Although described as a history, it is a continuing history, so the aortic insufficiency and hypertension should be coded in addition to the hip fracture. The history of upper respiratory infections is not coded as it is stated to be a history. The patient was intubated and provided ventilation, so these must be coded.

ABSTRACT 11

ICD-9-CM (diagnoses):

1. 618.4
2. 626.2
3.
4.
5.
6.
7.
8.
9.

ICD-9-CM (procedures):

1. 68.59
2. 70.51
3.
4.
5.
6.

ANSWER: Both the vagina and uterus are prolapsed, so both of these need to be coded together in one code. This code also includes the cystocele. The menorrhagia must also be coded. The repair of the cystocele must also be coded.

ABSTRACT 12

ICD-9-CM (diagnoses):

1. 733.82
2. 724.4
3. 724.2
4. 701.4
5. 996.75
6.
7.
8.
9.

ICD-9-CM (procedures):

1. 81.07
2. 84.51
3. 78.69
4.
5.
6.

ANSWER: The nonunion is coded because this is an inpatient service and it is described as suspected, which is similar to probable or possible. The radiculitis of the lower extremity should also be coded, as well as of the lower back. Notice that a lumbar scar was removed, which should be coded. The loose screw has resulted in pain, which must be coded as a complication of surgery. When you check under late effects in the alphabetic index, you will find that for surgery complications it directs you to use the 990 codes and not other late effect codes. The spinal fusion, known as arthrodesis, should be coded. This includes the autograft, which is referenced in the alphabetic index. The use of instrumentation for the arthrodesis should also be coded, which included the removal of the fixation device from the spine, i.e., the screw.

ABSTRACT 13

ICD-9-CM (diagnoses):

1. 426.0
2. 401.9
3.
4.
5.
6.
7.
8.
9.

ICD-9-CM (procedures):

1. 37.83
2. 37.72
3.
4.
5.
6.

ANSWER: This patient has a third-degree Mobitz type 2 atrioventricular block in addition to hypertension. The patient had a dual-chamber pacemaker implanted with transvenous leads inserted.

INDEX

A

Abbreviations, 16
  HCPCS, 144
Abnormal, ICD-9-CM, 39
  lab findings, ICD-9-CM, 39
Abortion, ICD-9-CM, 58
  complications, 58
Abstracting, 22, 155–169. *See also* Principal diagnosis
  application rules, 158–159
  code also, 157
  code instructions, 158
  defined, 155
  histories, 156–158
  importance, 155
  late effects, 158
  mock exam, 230–242, 252–256
  procedure, 158
  relevant patient care and outcomes, 157
  report formats, 156
  required coding, 157
  secondary condition, 155–156
  severity of injuries, 157
  V codes, 156
Abuse
  billing error, 178
  defined, 6, 10
  documentation, 8
  profile, 10
  types, 10
Accepting assignment, defined, 171
Acts of omission, 10
Acute condition, ICD-9-CM, 40
Acute renal disease, 44
Additional, Current Procedural Terminology, 97
Adenoiditis, ICD-9-CM, 53
Admission, CMS-1450 form, 180
Advance Beneficiary Notice, 172–173
Alphabetic index, 16, 24, 145
  Current Procedural Terminology, 95
    defined, 92
  details, 25
  E codes, 2
  ICD-9-CM, 31, 34
    defined, 30
    diagnosis, 34
    precise details, 33
    procedural coding, 148

Ambulatory payment classification, 174, 175–176
  Current Procedural Terminology, 3
  defined, 155
  discounting, 175
  global packaging, 175
  HCPCS, payment status indicators, 175–176
Ambulatory surgery center, 176–177
  geographic adjustment, 176
  global payment, 176
  new technology, 176, 177
  observation, 177
  partial hospitalization, 177
  temporary codes, 176–177
  transitional pass-throughs, 177
American Association of Professional Coders (AAPC), 4–5
  Certified Professional Coder, 1, 4
  Certified Professional Coder–Hospital, 4
    application information, 4–5
    books, 14–15
    coding scenarios, 23
    Current Procedural Terminology, 23
    exam, 22–23
    HCPCS, 23
    ICD-9-CM, 23
    mock exam, 189–221
    multiple-choice questions, 23
  continuing education, 4
  continuing education unit, 27
  website, 4
American Health Information Management Association (AHIMA), 3–4
  certification, 3
  Certified Coding Associate (CCA), 3
  Certified Coding Specialist (CCS), 3
    abstracting, 22
    data management, 22
    data quality, 22
    diagnostic coding guidelines, 21
    health information documentation, 20–21
    ICD-9-CM, 22
    procedural coding guidelines, 21
    regulatory guidelines for hospital-based outpatient services, 21
    regulatory guidelines for inpatient hospitalizations, 21

reporting requirements for hospital-based outpatient services, 21
reporting requirements for inpatient hospitalizations, 21
Certified Coding Specialist–Physician (CCS-P), 1, 3
coding exam, 4
application information, 4
books, 14–15
competencies, 20–22
multiple choice 22, 20–22
withdrawal, 4
continuing education unit, 27
services, 3–4
website, 3, 4
Anastomosis, Current Procedural Terminology, 109
Anatomic site
Current Procedural Terminology, 97
digit, 38
ICM-9-CM, 37–38
neoplasm, 45
And (coding term), ICD-9-CM, 42–43
Anemia, ICD-9-CM, 51
Anesthesia, Current Procedural Terminology, 98
additional codes and modifiers, 98
Angina, ICD-9-CM, 49
Angioplasty, Current Procedural Terminology, 108
Ankylosing spondylitis, ICD-9-CM, 47
Anti-kickback laws, 11
Appeal, 26–27
defined, 26
Arteriosclerotic heart disease, ICD-9-CM, 50
Arteriovenous fistula, Current Procedural Terminology, 108
Arthritis, ICD-9-CM, 47
Arthropathy, ICD-9-CM, 47
Asthma, ICD-9-CM, 53
Atrioventricular heart block, ICD-9-CM, 50
Attitude, coding exam, 15
Auditing, 7–8
Auditory system, Current Procedural Terminology, 112
Authorization to Test letter, 4

B

Beneficiary, defined, 170
Benign hypertension, 44
Billing error
abuse, 178
fraud, 178
penalties and fines, 10–11
Billing form, 177–182
clean claim, 177
Biopsy
Current Procedural Terminology, 100
procedural coding, ICD-9-CM, 149
Birthday Rule, defined, 170
Blood pressure, elevated, 44
Bone graft, Current Procedural Terminology, 105
Bracket, ICD-9-CM, 33
Bundle branch block, ICD-9-CM, 50
Bundling
Current Procedural Terminology, 95–96
defined, 92
evaluation and management code, defined, 136
Burn
Current Procedural Terminology, 103
ICD-9-CM, 46–47, 59
Bypass graft, Current Procedural Terminology, 106

C

Capitation, defined, 171
Cardiac catheterization, Current Procedural Terminology, 107–108
Cardiac dysrhythmia, ICD-9-CM, 50
Cardiovascular system
Current Procedural Terminology, 106–108
procedural coding, ICD-9-CM, 150
Cardioverter-defibrillator, Current Procedural Terminology, 107
Case mix, 175
Causal relationship, ICD-9-CM, 43
Centers for Medicare and Medicaid Services, 3, 170
Cerebrovascular accident
ICD-9-CM, 50
late effect, 41
Certification, vii, 26
American Health Information Management Association, 3
coder, legal protection, 7
defined, 26
hospital coding, vii
national organizations, 3–4
national recognition, 27–28
renewing, 27
Certification exam. *See also specific exam*
American Association of Professional Coders
application information, 4–5
books, 14–15
American Health Information Management Association, 4
application information, 4
books, 14–15
withdrawal, 4
appeal process, 26–27
application packets, 14
areas of study, 16
assumptions, 18

attitude, 15
basics vs. changes, 15
checking before and after code, 24
details, 32–33
disability, 4
documentation, 23
E codes, 15
educated guess, 24
failure, 18
knowledge base, 1–2
materials, 19–20
medical specialty area, 15
mental preparation, 17–18
mock exams, vii, 189–221, 223–256
modifier, 15
multiple-choice section, 20, 22, 23
notes, 23
online, 20
practice, 16
preparing for, 14
prior experience, 1
reading directions, 20
rechecking, 20
results, 26
retesting, 27
skimming, 18
skipping questions, 23
snacks, 20
specificity, 23
study group, 17
studying for, 15–16
studying techniques, 17
study schedule, 17
test-taking technique, 18
Thomson Prometric, 4
thoroughness, 32–33
V codes, 15
Certified Coding Associate (CCA), American Health Information Management Association, 3
Certified Coding Specialist (CCS), American Health Information Management Association, 3
abstracting, 22
competencies, 20–22
data management, 22
data quality, 22
diagnostic coding guidelines, 21
exam, 15
health information documentation, 20–21
ICD-9-CM, 22
mock exam, 223–256
procedural coding guidelines, 21
regulatory guidelines for hospital-based outpatient services, 21
reporting requirements for hospital-based outpatient services, 21
Certified Coding Specialist–Physician (CCS-P), American Health Information Management Association, 1, 3
Certified in Health Care Privacy, 29
Certified in Health Care Privacy and Security, 29
Certified in Health Care Security, 29
Certified Professional Coder. *See* American Association of Professional Coders
Certified Professional Coder–Hospital. *See* American Association of Professional Coders
Civilian Health and Medical Program for Uniformed Services (CHAMPUS). *See* TRICARE
Civilian Health and Medical Program of the Veterans Administration (CHAMPVA), 172
Change, vii
Chargemaster, CMS-1450 form, 179
Chemical, ICD-9-CM, 60
    Appendix C, 60
Chemotherapy
    Current Procedural Terminology, 114
    ICD-9-CM, 40
Chronic disease, 156, 157
    ICD-9-CM, 40
Chronic obstructive pulmonary disease, ICD-9-CM, 53, 157
Chronic renal failure, 44
    ICD-9-CM, 55
Circulatory system, ICD-9-CM, 49–50
    symptoms vs. diagnosis, 49
Civil penalties, 11
Clarification, upcoding, differentiated, 39
Clean claim, defined, 170, 179
CMS-1450 form, 178
    admission, 180
    bill type, 180
    Chargemaster, 179
    coding, 179–183
    condition codes, 180–181
    Current Procedural Terminology codes, 183
    dates/times, 180
    defined, 170
    diagnosis code, 182–183
    discharge, 180
    form locators, 180–183
    HCPCS, 183
    importance, 170–171
    indicators, 180
    ICD-9-CM, 183
    names, 180
    occurrence code, 181
    payer information, 182

physician information, 183
procedure code, 182–183
revenue code, 181–182
scrubber, 179
value codes, 181
CMS-1500 form, 93
coding, 178
defined, 170
importance, 170–171
ICD-9-CM, 179
Coagulation problem, ICD-9-CM, 51
Code also, 2
abstracting, 157
ICD-9-CM, 42–43
procedural coding, 148
Code first, ICD-9-CM, 42–43
Code first underlying disease, ICD-9-CM, 34
Code linkage. *See* Linkage
Coder
background, 1
career opportunities, vii
certification, legal protection, 7
compliance program, 7
contributions, 7–11
educating physicians and staff, 6
employment opportunities, 28–29
employment settings, 28
ethical/legal responsibilities, 6–8
importance, 6–11
medical practice, 28
reimbursement, 6
salary average, 28
specialization, 28
Coding
economic impact, 6, 8
impact of poor, 8
reimbursement within, 170–171
what to code, 36
Coding book, 2, 14–15
Coding Clinic, 15
Coding Edge, defined, 26
Coding guidelines, 7
Coinsurance, defined, 171
Combination codes
ICD-9-CM, 42
vs. multiple coding, 42–43
Compliance, 10–11
defined, 6
Compliance program
coder, 7
components, 7–8
Office of Inspector General, 7
Complicated by, ICD-9-CM, 42

Complications, ICD-9-CM, 59
Condition codes, CMS-1450 form, 180–181
Congestive heart failure, ICD-9-CM, 49
Connecting terms, ICD-9-CM, 34
Continuing education, 29
American Association of Professional Coders, 4
Continuing education unit, 27
activities that qualify, 27
American Association of Professional Coders, 27
American Health Information Management Association, 27
defined, 26
Coordination of benefits, 173
Copayment, defined, 171
Coronary artery bypass graft, Current Procedural Terminology, 106
CCI. *See* National Correct Coding Initiative
CPT Assistant, 15
Criminal indictment, 11
Critical care, evaluation and management code, 140
Cross references
Current Procedural Terminology, 97
ICD-9-CM, 34
Current Procedural Terminology, 3, 114
additional, 97
alphabetic index, 95
defined, 92
Ambulatory Payment Classification, 3
anastomosis, 109
anatomical site, 97
anesthesia, 98
additional codes and modifiers, 98
angioplasty, 108
appendices, 115–116
arteriovenous fistula, 108
auditory system, 112
biopsy, 100
bone graft, 105
bundling, 95–96
burn, 103
bypass graft, 106
cardiac catheterization, 107–108
cardiovascular system, 106–108
cardioverter-defibrillator, 107
chemotherapy, 114
CMS-1450 form, 179–183
codes categorized, 92
codes classified by, 92
conventions, 95
coronary artery bypass graft, 106
cross references, 97
debridement, 102
details, 94

diagnosis-related groupings, 155
dialysis, 114
digestive system, 109
dislocation, 104–105
documentation, 93
downcoding, 93, 155
drainage, 101
each, 97
embolectomy, 108
endoscopy, 109
eponym, 147
essential modifiers, 147
excision, 101
fistula, 108
five-digit codes, 3
flap, 103
foot surgery, 105
form, 92
fracture, 104–105
global procedure, 95–96
graft, 102–103
guidelines, 93
   defined, 92
HCPCS, linkage, 93
hernia repair, 109
hysterectomy, 111
ICD-9, linkage, 93
ICD-9-CM, 147
immunization, 114
incision, 101
indices, 95
infusion, 114
injection, 105, 114
integumentary system, 101–103
interpretation, 97
introduction procedure, 105
laboratory, 113–114
   lab panels, 113–114
laceration, 102
lesion, 101–102
Level I HCPCS codes, 93
medicine, 114
modifiers, 115–116, 144
   bilateral, 116
   defined, 92, 136
Moh's surgery, 101–102
musculoskeletal system, 104–105
nervous system, 110
obstetrics/gynecology, 111
ocular system, 112
pacemaker, 107
pathology, 113–114
   lab panels, 113–114
pregnancy, 111
professional component, 97
radiology, 112–113
   contrast material, 113
   positioning, 113
   professional services, 112–113
   radiation oncology, 113
   types of tests, 112
removal procedure, 105
Resource Based Relative Value Scale, 175
sections, 92
semicolon, 96
separate, 97
spine, 104–105
stents, 100
supervision, 97
surgery, 99–103
   classification, 99
   scopes, 99
   separate procedures, 99
   significant procedures, 99
surgical destruction, 101
tabular code, 95
tabular index, defined, 92
technical component, 97
terminology, 97
thrombectomy, 108
two-digit modifiers, 3
unbundling, 95–96
unlisted procedure, 97
upcoding, 93
uses, 3, 92
vascular access device, 108
wound repair, 102, 104
yearly changes, 3
Current Procedural Terminology-4, neighboring codes, 16

D

Debridement, Current Procedural Terminology, 102
Deductible, defined, 171
Department of Health and Human Services. *See* Medicare
Delivery, ICD-9-CM, 57
Diabetes insipidus, ICD-9-CM, 54
Diabetes mellitus, ICD-9-CM, 54
Diagnosis codes, 33, 34
   CMS-1450 form, 182–183
   coding guidelines, 21
   guidelines, 21
   ICD-9-CM, 2, 38
      primary, 36
      principal, 36

qualified, 38
secondary, 36
principal
abstracting, 155–157
diagnosis-related groups, 174
up to five digits, 2
Diagnosis-related groups, 174–175
case mix, 175
ICD-9-CM, 174
groupers, 174
major diagnostic categories, 174–175
outlier, 175
Dialysis, Current Procedural Terminology, 114
Digestive system, Current Procedural Terminology, 109
Digit, 32, 33, 34
ICD-9-CM, 31, 32
defined, 30
Disability, testing, 4
Discharge, CMS-1450 form, 180
Discounting, ambulatory payment classification, 175
Dislocation, Current Procedural Terminology, 104–105
Diverticulum, ICD-9-CM, 55
Documentation, 16–17, 20–21, 23
abuse, 6, 7, 8
coding exam, 23
Current Procedural Terminology, 93
evaluation and management code, defined, 136
fraud, 8
illegible writing, 23
not documented, not done, 23
Downcoding
Current Procedural Terminology, 93
defined, 155, 170
Drainage, Current Procedural Terminology, 101
Drugs. *See also* Medication
ICD-9-CM, 60
Appendix C, 60
Table of Drugs and Chemicals, 60
Drug abuse
E codes, 2
ICD-9-CM, 56
Drug dependence, ICD-9-CM, 56
Drug withdrawal, ICD-9-CM, 56
Due to (coding term), ICD-9-CM, 42
Durable medical equipment, defined, 144
Dysrhythmia, ICD-9-CM, 50

E

Each, Current Procedural Terminology, 97
E codes, 2
alphabetic index, 2
coding exam, 15

drug use/abuse, 2
ICD-9-CM, 62
defined, 30
Table of Drugs and Chemicals, 60
Educated guess, coding exam, 24
Either/or coding, ICD-9-CM, 38
Elevated, ICD-9-CM, 39
Embolectomy, Current Procedural Terminology, 108
Emergency room visit, evaluation and management code, 140
Endoscopy
Current Procedural Terminology, 109
procedural coding, ICD-9-CM, 150
End-stage renal disease, ICD-9-CM, 55, 173
Epilepsy, ICD-9-CM, 56
Etiology of condition, ICD-9-CM, 34
Evaluation and management codes
bundling, defined, 136
code selection, 139–140
criteria, 137–139
critical care, 140
defined, 136
documentation, defined, 136
emergency room visit, 140
examination, 138–139
history, 137–138
level selection, 138
past, family, and social, 138
of present illness, 137
hospital visit, 140
level, 136
medical decision making, 139
method of categorizing, 136
modifier 25, 136
observation, 140
per day, 136–137
preventive medicine services, 140
prolonged services, 140
review of systems, 138
risk, 139
time, 137
uses, 136
Excision, Current Procedural Terminology, 101
Excludes, ICD-9-CM, 2

F

Failure to properly bill, penalties and fines, 10–11
Fear of failure, 18
Fibrillation, ICD-9-CM, 50
Fifth digit, 12, 32, 33, 34
Fistula, Current Procedural Terminology, 108
Flap, Current Procedural Terminology, 103
Follow-up, ICD-9-CM, 39

Foot surgery, Current Procedural Terminology, 105
Form locators, CMS-1450 form, 180–183
Fourth digit, 32, 33, 34
Fracture
   Current Procedural Terminology, 104–105
   ICD-9-CM, 47–48
Fraud
   billing error, 179
   defined, 6, 10
   documentation, 8
   profile, 10
   types, 10

G

Gastrointestinal tract, ICD-9-CM, 54–55
   procedural coding, 149–150
General insurance, 171–172
Generalized, ICD-9-CM, 39
Geographic adjustment, ambulatory surgery center, 176
Gestational diabetes, ICD-9-CM, 54
Global, defined, 144
Global packaging, ambulatory payment classification, 175
Global payment, ambulatory surgery center, 176
Global procedure, Current Procedural Terminology, 95–96
Graft, Current Procedural Terminology, 102–103
Guarantor, defined, 171
Guidelines, Current Procedural Terminology, 93
   defined, 92

H

Health Care Common Procedural Coding System (HCPCS), 3
   abbreviations, 144
   alphabetic letter followed by four numeric digits, 144
   ambulatory payment classification, payment status indicators, 175–176
   appendices, 145
   CMS-1450 form, 181–182
   codes, 93
   Current Procedural Terminology, linkage, 93
   defined, 144
   ICD-9, linkage, 93
   ICD-9-CM, 179
   Level I, 3
   Level II, 3
   Level III, 3
   mandated for Medicare claims, 144
   medication
      measurements, 145
      method of administration, 145
      table of drugs, 145
   modifiers, 145
   uses, 144
HCPCS. *See* Health Care Common Procedural Coding System
Health Insurance Portability and Accountability Act (HIPAA), 7, 10
Health maintenance organization, 171
Heart attack. *See* Myocardial infarction
Hemiparesis, ICD-9-CM, 56
Hemiplegia, ICD-9-CM, 56
Hemorrhoid, ICD-9-CM, 50
Hepatitis, ICD-9-CM, 52
Hernia, ICD-9-CM, 55
Hernia repair, Current Procedural Terminology, 109
Histories, abstracting, 156–158
History codes, ICD-9-CM, 36
HIV/AIDS, ICD-9-CM, 52
Hospital-based outpatient services
   regulatory guidelines, 21
   reporting requirements, 21
Hospital coding, vii
   certification, vii
   physician coding, compared, 4
   requirements, 15
Hospitalization
   coding, vii
   partial hospitalization, ambulatory surgery center, 177
   regulatory guidelines, 21
   reporting requirements, 21
   vs. outpatient hospital services, 16
Hospital visit, evaluation and management code, 140
Hypertension
   ICD-9-CM, 49
   renal disease, 55
   table, ICD-9-CM, 43–44
      defined, 30
Hypoglycemia, ICD-9-CM, 54
Hysterectomy, Current Procedural Terminology, 111

I

ICD-9. *See* International Classification of Diseases, Ninth Revision
ICD-9-CM. *See* International Classification of Diseases, Ninth Revision, Clinical Modification
Ill-defined findings, ICD-9-CM, 39
Illegible writing, 8
   documentation, 23
Immunization, Current Procedural Terminology, 114
Impending, ICD-9-CM, 39

Incidental condition, ICD-9-CM, 36
Incidental pregnancy, ICD-9-CM, 57
Incision, Current Procedural Terminology, 101
Infection, ICD-9-CM, 46
Infusion, Current Procedural Terminology, 114
Injection, Current Procedural Terminology, 105, 114
Inpatient. *See* Hospitalization
In situ, 45
Insurance
   general, 171–172
   terminology, 171
   types, 171–173
Integumentary system
   Current Procedural Terminology, 101–103
   ICD-9-CM, 46–47
   procedural coding, 149
International Classification of Diseases, Ninth Revision (ICD-9), 2
   Current Procedural Terminology, linkage, 93
   HCPCS, linkage, 93
International Classification of Diseases, Ninth Revision, Clinical Modification (ICD-9-CM), 2–3
   abnormal, 39
   abnormal lab findings, 39
   abortion, 58
      complications, 58
   acute and chronic condition, 53
   acute condition, 40
   adenoiditis, 53
   alphabetic index, 31, 34
      defined, 30
      diagnosis, 34
      precise details, 33
   anatomic sites, 37–38
      digit, 38
   and (coding term), 42, 43
   anemia, 51
   angina, 49
   ankylosing spondylitis, 47
   arteriosclerotic heart disease, 50
   arthritis, 47
   arthropathy, 47
   asthma, 53
   atrioventricular heart block, 50
   bracket, 33
   bundle branch block, 50
   burn, 46–47, 59
   cardiac dysrhythmia, 50
   causal relationship, 43
   cerebrovascular accident, 50
   chemical, 60
      Appendix C, 60
   chemotherapy, 40
   chronic condition, 40
   chronic obstructive pulmonary disease, 53
   chronic renal failure, 55
   circulatory system, 49–50
      symptoms vs. diagnosis, 49
   CMS-1450 form, 183
   CMS-1500 form, 179
   coagulation problem, 51
   code also, 42–43
   code first, 42–43
   code first underlying disease, 34
   code linkage, 36
   combination code, 42
   complicated by, 42
   complications, 59
   congestive heart failure, 49
   connecting terms, 34
   contents, 31
   conventions, 33
   cross reference, 34
   Current Procedural Terminology, 179
   defined, 30
   delivery, 57
   diabetes insipidus, 54
   diabetes mellitus, 54
   diagnosis, 38
   diagnosis-related groups, 174
      groupers, 174
   diagnostic codes, 2
   digit, 2, 16, 31, 32
      defined, 30
   diverticulum, 55
   drugs, 60
      Appendix C, 60
   drug abuse, 56
   drug dependence, 56
   drug withdrawal, 56
   due to (coding term), 42
   dysrhythmia, 50
   E codes, 2, 62
      defined, 30
   either/or coding, 38
   elevated, 39
   end-stage renal disease, 55
   epilepsy, 56
   etiology of condition, 34
   extra digits, 16
   fibrillation, 50
   five-digit codes, 2
   follow-up, 39
   four-digit codes, 2
   fracture, 47–48
   gastrointestinal tract, 54–55

generalized, 39
gestational diabetes, 54
HCPCS, 179
hemiparesis, 56
hemiplegia, 56
hemorrhoid, 50
hepatitis, 52
hernia, 55
history codes, 36
HIV/AIDS, 52
hypertension, 49
hypertension table, 43–44
   defined, 30
hypoglycemia, 54
ill-defined findings, 39
impending, 39
incidental condition, 36
incidental pregnancy, 57
indentations, 31–32
infection, 46
integumentary system, 46–47
labor, 57
late effect, 41
miscarriage, 58
modifiers, 34
morphology codes, 62
   defined, 30
multiple coding, 42
multiple injuries, 42
musculoskeletal system, 47–48
myocardial infarction, 49
neoplasm table, 45
   defined, 30
neurological system, 56
nevus, 46
nondescript term, 39
nonspecific conditions, 61
nonspecific term, 35
not elsewhere classified, 34
notes, 16
not otherwise specified, 34
organism, 39, 50, 52
osteoarthritis, 47
otitis, 56
pneumonia, 53
pneumothorax, 53
poisoning, 60
positive, 39
possible, 38
pre-eclampsia, 57
pregnancy, 54
   antepartum, delivery, or postpartum visits, 57
   baby vs. mother, 57

complications, 57–58
primary diagnosis, 36
principal diagnosis, 36
probable (coding term), 38
procedural codes, 2
pulmonary edema, 53
qualified diagnosis, 38
radiotherapy, 40
rehabilitation, 40
renal failure, 55
renal system, 55
residual effect, 41
respiratory system, 53
rule out, 38
secondary condition, 36
secondary diagnosis, 36
seizure, 56
sepsis, 51
sequelae, 41
surgical complication, 59
   defined, 30
suspected, 38
symptoms, 38
systemic inflammatory response syndrome, 51
Table of Drugs and Chemicals, 60
   Appendix C, 60
   defined, 31
tabular index, defined, 31
tabular list, 31
   categories, 31
   sections, 31
   subcategories, 31
   subclassifications, 31
tachycardia, 50
therapy codes, 40
threatened, 39
three-digit codes, 2
tonsillitis, 53
transient ischemic attack, 50
tuberculosis, 53
ulcer, 46
unspecified, 39, 40
upper respiratory infection, 53
urinary tract infection, 55
use additional code, 42–43
   if desired, 34
uses, 31
varices, 55
V codes, 2, 36, 39, 40, 61
   defined, 31
   follow-up, 61
   history codes, 61
   post-surgical status, 61

screenings, 61
status, 61
vice-versa coding, 38
Volume 1, ICD-9-CM, 2, 31
Volume 2, ICD-9-CM, 2, 31
Volume 3, ICD-9-CM. *See* Procedural coding, ICD-9-CM
what to code, 36
wound, 46
with (coding term), 42, 43
Interpretation, Current Procedural Terminology, 97
Introduction procedure, Current Procedural Terminology, 105

J
Jail time, 11
Joint Commission on Accreditation of Health Care Organizations (JCAHO), 10

L
Labor, ICD-9-CM, 57
Laboratory, Current Procedural Terminology, 113–114
lab panels, 113–114
Laceration, Current Procedural Terminology, 102
Late effect
abstracting, 158
cerebrovascular accident, 41
ICD-9-CM, 41
parenthesis, 41
Lesion, Current Procedural Terminology, 101–102
Letter of recommendation, 4–5
Lifelong education, defined, 26
Linkage, 179
Current Procedural Terminology, 93
defined, 170
ICD-9, 93
ICM-9-CM, 36

M
Malignant hypertension, 44
Materials, 3
Mechanical ventilation. *See* Procedural coding, ICD-9-CM
Medicaid, 173
eligibility, 173
services covered, 173
Medical dictionary, 14, 15
Medical specialty area, coding exam, 15
Medical terminology, 16
Medicare, 172–173
Level III local codes, 3
medical necessity, 172
Part A, 172
Part B, 172
preauthorization, 172–173
preventive health services, 172
as secondary payer, 173
Medication. *See also* Drugs
HCPCS
measurements, 145
method of administration, 145
table of drugs, 145
Table of Drugs and Chemicals, 60
Medigap insurance, 173
Medi-Medi, 173
Miscarriage, ICD-9-CM, 58
Mistakes, 10
Modifiers
coding exam, 15
Current Procedural Terminology, 115–116, 144
bilateral, 116
defined, 92, 136
HCPCS, 145
ICD-9-CM, 34
procedural coding, 148
Moh's surgery, Current Procedural Terminology, 101–102
Morphology codes
ICD-9-CM, 62
defined, 30
neoplasm, 45, 62
Multiple-choice section, 22
coding exam, 20, 22, 23
Multiple coding
combination code, differentiated, 42–43
ICD-9-CM, 42
Multiple injuries, ICD-9-CM, 42
Multiple procedures. *See* Procedural coding, ICD-9-CM
Musculoskeletal system
Current Procedural Terminology, 104–105
ICD-9-CM, 47–48
procedural coding, 149
Myocardial infarction, ICD-9-CM, 49

N
National codes, HCPCS Level II, 3
National Correct Coding Initiative, 171
Neoplasm
anatomic site, 45
morphology code, 62
Neoplasm table, ICD-9-CM, 45
defined, 30
Nervous system, Current Procedural Terminology, 110

Neurological system, ICD-9-CM, 56
Nevus, ICD-9-CM, 46
New technology, ambulatory surgery center, 177
Nondescript term, ICD-9-CM, 39
Nonessential modifiers, 148
Nonspecific term, ICD-9-CM, 35
Not elsewhere classified, ICD-9-CM, 34
Not otherwise specified, ICD-9-CM, 34

O

Observation
    ambulatory surgery center, 177
    evaluation and management code, 140
Obstetrics/gynecology
    Current Procedural Terminology, 111
    procedural coding, ICD-9-CM, 150–151
Occurrence code, CMS-1450 form, 181
Ocular system, Current Procedural Terminology, 112
Office of Inspector General
    compliance program, 7
    website, 11
Omit code. *See* Procedural coding, ICD-9-CM
Organism, ICD-9-CM, 39, 50, 52
Osteoarthritis, ICD-9-CM, 47
Otitis, ICD-9-CM, 56
Outlier, 175
Outpatient code editor, 176
Outpatient coding, vii
Outpatient hospital services, inpatient hospital services, differentiated, 16
Outpatient Prospective Payment System, 174

P

Pacemaker, Current Procedural Terminology, 107
Participating provider agreement, defined, 171
Pathology, Current Procedural Terminology, 113–114
    lab panels, 113–114
Payment status indicator, 175–176
Personal health information, 10
Personal patient information, 7
Physician coding, vii, 1
    hospital coding
        compared, 4
        rules diffferences, 1
Pneumonia, ICD-9-CM, 53
Pneumothorax, ICD-9-CM, 53
Poisoning, ICD-9-CM, 60
Positive, ICD-9-CM, 39
Positive thinking, 17
Possible, ICD-9-CM, 38
Pre-eclampsia, ICD-9-CM, 57
Pregnancy
    Current Procedural Terminology, 111
    ICD-9-CM, 54
        antepartum, delivery, or postpartum visits, 57
        baby vs. mother, 57
        complications, 57–58
Premium, defined, 171
Presenting problem/condition, 24
Preventive health services
    evaluation and management codes, 140
    Medicare, 172
Primary diagnosis, ICD-9-CM, 36, 155
Principal diagnosis
    abstracting, 156–159
        chronic disease, 156, 157
        histories, 156, 157
        inpatient hospital coding, 156–157
        outpatient hospital coding, 156–157
        symptoms, 157
    diagnosis-related groups, 174–175
    ICD-9-CM, 36
Prison sentence, 11
Privacy, 7, 10
    types of communication, 7
Probable (coding term), ICD-9-CM, 38
Procedural coding, ICD-9-CM, 9
    alphabetic index, 147
    biopsy, 149
    cardiovascular system, 150
    CMS-1450 form, 179–183
    code also, 148
    coding guidelines, 21
    conventions, 148
    endoscopic procedure, 150
    gastrointestinal tract, 149–150
    integumentary system, 149
    mechanical ventilation, 150
    modifiers, 148
    multiple procedures, 149
    musculoskeletal system, 149
    obstetrics/gynecology, 150–151
    omit code, 148
    see (coding term), 148
    see also (coding term), 148
    staged repair, 149
    tabular index, 148
    terminology, 148
    Uniform Hospital Discharge Data Set, 148
Professional component, Current Procedural Terminology, 97
Professional courtesy, 11
Profile, abuse and fraud, 10
Prolonged services, evaluation and management codes, 140
Prospective Payment System, 174–177

Provider, defined, 171
Pulmonary edema, ICD-9-CM, 53

Q
Qualified diagnosis, ICD-9-CM, 38

R
Radiation oncology, 113
Radiology, Current Procedural Terminology, 112–113
   contrast material, 113
   positioning, 113
   professional services, 112–113
   types of tests, 112
Radiotherapy, ICD-9-CM, 40
Recommendation letter, 4–5
Rehabilitation, ICD-9-CM, 40
Reimbursement
   coder, 6
   coding, 169–170
   errors, 183
Release of information, 10
Removal procedure, Current Procedural Terminology, 105
Renal disease, hypertension, 55
Renal failure, ICD-9-CM, 55
Renal system, ICD-9-CM, 55
Residual effect, ICD-9-CM, 41
Resource Based Relative Value Scale (RBRVS), Current Procedural Terminology, 175
Respiratory system, ICD-9-CM, 53
Revenue code, CMS-1450 form, 181–182
Review of systems, evaluation and management code, 138
Rule out, ICD-9-CM, 38

S
Scrubber, CMS-1450 form, 179
Secondary condition
   abstracting, 155–156
   ICD-9-CM, 36
Secondary diagnosis, ICD-9-CM, 36
Secondary payer. See Medicare
See (coding term). See Procedural coding, ICD-9-CM
See also (coding term). See Procedural coding, ICD-9-CM
Seizure, ICD-9-CM, 56
Semicolon, Current Procedural Terminology, 96
Separate, Current Procedural Terminology, 97
Sepsis, ICD-9-CM, 51
Sequelae, ICD-9-CM, 41
Sequencing, 148
Signature, lack of, 8
Specialty area, coding exam, 15

Specificity, coding exam, 23
Spine, Current Procedural Terminology, 104–105
Staged repair. See Procedural coding, ICD-9-CM
Standards, vii
Stark laws, 11
Stents, Current Procedural Terminology, 100
Stroke. See Cerebrovascular accident
Study group, 17
Superbill, 178
   defined, 170
   problems, 178
Supervision, 8
   Current Procedural Terminology, 97
Supplies, 3
Surgery, Current Procedural Terminology, 99–103
   classification, 99
   scopes, 99
   separate procedures, 99
   significant procedures, 99
Surgical complication, ICD-9-CM, 59
   defined, 30
Surgical destruction, Current Procedural Terminology, 101
Suspected, ICD-9-CM, 38
Symptoms, 33, 157
   ICD-9-CM, 38
Systemic inflammatory response syndrome, ICD-9-CM, 51

T
Table of Drugs and Chemicals, ICD-9-CM, 60
   Appendix C, 60
   defined, 31
Tabular code, Current Procedural Terminology, 95
Tabular index
   Current Procedural Terminology, defined, 92
   ICD-9-CM, defined, 31
   procedural coding, 148
Tabular list, ICD-9-CM, 31
   categories, 31
   sections, 31
   subcategories, 31
   subclassifications, 31
Tachycardia, ICD-9-CM, 50
Technical component, Current Procedural Terminology, 97
Terminology, 16
Therapy codes, ICD-9-CM, 40
Third-party payer, defined, 171
Thomson Prometric, testing, 4
Threatened, ICD-9-CM, 39
Thrombectomy, Current Procedural Terminology, 108
Tonsillitis, ICD-9-CM, 53

Transient hypertension, 44
Transient ischemic attack, ICD-9-CM, 50
TRICARE, 172
Tuberculosis, ICD-9-CM, 53

U

UB-92. *See* CMS-1450 form
Ulcer, ICD-9-CM, 46
Unbundling, Current Procedural Terminology, 95–96
Uniform Hospital Discharge Data Set (UHDDS), 171. *See also* Procedural coding, ICD-9-CM
Unlisted procedure, Current Procedural Terminology, 97
Unsigned report, 8
Unspecified, ICD-9-CM, 39, 40
Upcoding
   vs. clarification, 39
   Current Procedural Terminology, 93
   defined, 155, 170
Upper respiratory infection, ICD-9-CM, 53
Urinary tract infection, ICD-9-CM, 55
Use additional code, ICD-9-CM, 42–43
   if desired, 34
Utilization Review Accreditation Commission (URAC), 10

V

Value codes, CMS-1450 form, 181
Varices, ICD-9-CM, 55
Vascular access device, Current Procedural Terminology, 108
V codes, 2
   abstracting, 156
   coding exam, 15
   ICD-9-CM, 36, 39, 40, 61
      defined, 31
      follow-up, 61
      history codes, 61
      post-surgical status, 61
      screenings, 61
      status, 61
   worried well (coding term), 156
Vice-versa coding, ICD-9-CM, 38

W

With (coding term), ICD-9-CM, 42, 43
Wound, ICD-9-CM, 46
Wound repair, Current Procedural Terminology, 102, 104

IMPORTANT! READ CAREFULLY: This End User License Agreement ("Agreement") sets forth the conditions by which Thomson Delmar Learning, a division of Thomson Learning Inc. ("Thomson") will make electronic access to the Thomson Delmar Learning–owned licensed content and associated media, software, documentation, printed materials, and electronic documentation contained in this package and/or made available to you via this product (the "Licensed Content"), available to you (the "End User"). BY CLICKING THE "I ACCEPT" BUTTON AND/OR OPENING THIS PACKAGE, YOU ACKNOWLEDGE THAT YOU HAVE READ ALL OF THE TERMS AND CONDITIONS, AND THAT YOU AGREE TO BE BOUND BY ITS TERMS, CONDITIONS, AND ALL APPLICABLE LAWS AND REGULATIONS GOVERNING THE USE OF THE LICENSED CONTENT.

1.0 SCOPE OF LICENSE

1.1 <u>Licensed Content</u>. The Licensed Content may contain portions of modifiable content ("Modifiable Content") and content that may not be modified or otherwise altered by the End User ("Non-Modifiable Content"). For purposes of this Agreement, Modifiable Content and Non-Modifiable Content may be collectively referred to herein as the "Licensed Content." All Licensed Content shall be considered Non-Modifiable Content, unless such Licensed Content is presented to the End User in a modifiable format and it is clearly indicated that modification of the Licensed Content is permitted.

1.2 Subject to the End User's compliance with the terms and conditions of this Agreement, Thomson Delmar Learning hereby grants the End User a nontransferable, nonexclusive, limited right to access and view a single copy of the Licensed Content on a single personal computer system for noncommercial, internal, personal use only. The End User shall not (i) reproduce, copy, modify (except in the case of Modifiable Content), distribute, display, transfer, sublicense, prepare derivative work(s) based on, sell, exchange, barter or transfer, rent, lease, loan, resell, or in any other manner exploit the Licensed Content; (ii) remove, obscure, or alter any notice of Thomson Delmar Learning's intellectual property rights present on or in the Licensed Content, including, but not limited to, copyright, trademark, and/or patent notices; or (iii) disassemble, decompile, translate, reverse engineer, or otherwise reduce the Licensed Content.

2.0 TERMINATION

2.1 Thomson Delmar Learning may at any time (without prejudice to its other rights or remedies) immediately terminate this Agreement and/or suspend access to some or all of the Licensed Content, in the event that the End User does not comply with any of the terms and conditions of this Agreement. In the event of such termination by Thomson Delmar Learning, the End User shall immediately return any and all copies of the Licensed Content to Thomson Delmar Learning.

3.0 PROPRIETARY RIGHTS

3.1 The End User acknowledges that Thomson Delmar Learning owns all rights, title, and interest, including, but not limited to, all copyright rights therein, in and to the Licensed Content, and that the End User shall not take any action inconsistent with such ownership. The Licensed Content is protected by U.S., Canadian, and other applicable copyright laws and by international treaties, including the Berne Convention and the Universal Copyright Convention. Nothing contained in this Agreement shall be construed as granting the End User any ownership rights in or to the Licensed Content.

3.2 Thomson Delmar Learning reserves the right at any time to withdraw from the Licensed Content any item or part of an item for which it no longer retains the right to publish, or which it has reasonable grounds to believe infringes copyright or is defamatory, unlawful, or otherwise objectionable.

4.0 PROTECTION AND SECURITY

4.1 The End User shall use its best efforts and take all reasonable steps to safeguard its copy of the Licensed Content to ensure that no unauthorized reproduction, publication, disclosure, modification, or distribution of the Licensed Content, in whole or in part, is made. To the extent that the End User becomes aware of any such unauthorized use of the Licensed Content, the End User shall immediately notify Thomson Delmar Learning. Notification of such violations may be made by sending an e-mail to delmarhelp@thomson.com.

5.0 MISUSE OF THE LICENSED PRODUCT

5.1 In the event that the End User uses the Licensed Content in violation of this Agreement, Thomson Delmar Learning shall have the option of electing liquidated damages, which shall include all profits generated by the End User's use of the Licensed Content plus interest computed at the maximum rate permitted by law and all legal fees and other expenses incurred by Thomson Delmar Learning in enforcing its rights, plus penalties.

6.0 FEDERAL GOVERNMENT CLIENTS

6.1 Except as expressly authorized by Thomson Delmar Learning, Federal Government clients obtain only the rights specified in this Agreement and no other rights. The Government acknowledges that (i) all software and related documentation incorporated in the Licensed Content is existing commercial computer software within the meaning of FAR 27.405(b)(2); and (ii) all other data delivered, in whatever form, is limited rights data within the meaning of FAR 27.401. The restrictions in this section are acceptable as consistent with the Government's need for software and other data under this Agreement.

7.0 DISCLAIMER OF WARRANTIES AND LIABILITIES

7.1 Although Thomson Delmar Learning believes the Licensed Content to be reliable, Thomson Delmar Learning does not guarantee or warrant (i) any information or materials contained in or produced by the Licensed Content, (ii) the accuracy, completeness or reliability of the Licensed Content, or (iii) that the Licensed Content is free from errors or other material defects. THE LICENSED PRODUCT IS PROVIDED "AS IS," WITHOUT ANY WARRANTY OF ANY KIND AND THOMSON DELMAR LEARNING DISCLAIMS ANY AND ALL WARRANTIES, EXPRESSED OR IMPLIED, INCLUDING, WITHOUT LIMITATION, WARRANTIES OF MERCHANTABILITY OR FITNESS OR A PARTICULAR PURPOSE. IN NO EVENT SHALL THOMSON DELMAR LEARNING BE LIABLE FOR: INDIRECT, SPECIAL, PUNITIVE, OR CONSEQUENTIAL DAMAGES INCLUDING FOR LOST PROFITS, LOST DATA, OR OTHERWISE. IN NO EVENT SHALL THOMSON DELMAR LEARNING'S AGGREGATE LIABILITY HEREUNDER, WHETHER ARISING IN CONTRACT, TORT, STRICT LIABILITY, OR OTHERWISE, EXCEED THE AMOUNT OF FEES PAID BY THE END USER HEREUNDER FOR THE LICENSE OF THE LICENSED CONTENT.

8.0 GENERAL

8.1 <u>Entire Agreement</u>. This Agreement shall constitute the entire Agreement between the Parties and supercedes all prior Agreements and understandings oral or written relating to the subject matter hereof.

8.2 <u>Enhancements/Modifications of Licensed Content</u>. From time to time, and in Thomson Delmar Learning's sole discretion, Thomson Delmar Learning may advise the End User of updates, upgrades, enhancements, and/or improvements to the Licensed Content and may permit the End User to access and use, subject to the terms and conditions of this Agreement, such modifications, upon payment of prices as may be established by Thomson Delmar Learning.

8.3 <u>No Export</u>. The End User shall use the Licensed Content solely in the United States and shall not transfer or export, directly or indirectly, the Licensed Content outside the United States.

8.4 <u>Severability</u>. If any provision of this Agreement is invalid, illegal, or unenforceable under any applicable statute or rule of law, the provision shall be deemed omitted to the extent that it is invalid, illegal, or unenforceable. In such a case, the remainder of the Agreement shall be construed in a manner as to give greatest effect to the original intention of the parties hereto.

8.5 <u>Waiver</u>. The waiver of any right or failure of either party to exercise in any respect any right provided in this Agreement in any instance shall not be deemed to be a waiver of such right in the future or a waiver of any other right under this Agreement.

8.6 <u>Choice of Law/Venue</u>. This Agreement shall be interpreted, construed, and governed by and in accordance with the laws of the State of New York, applicable to contracts executed and to be wholly performed therein, without regard to its principles governing conflicts of law. Each party agrees that any proceeding arising out of or relating to this Agreement or the breach or threatened breach of this Agreement may be commenced and prosecuted in a court in the State and County of New York. Each party consents and submits to the nonexclusive personal jurisdiction of any court in the State and County of New York in respect of any such proceeding.

8.7 <u>Acknowledgment</u>. By opening this package and/or by accessing the Licensed Content on this Web site, THE END USER ACKNOWLEDGES THAT IT HAS READ THIS AGREEMENT, UNDERSTANDS IT, AND AGREES TO BE BOUND BY ITS TERMS AND CONDITIONS. IF YOU DO NOT ACCEPT THESE TERMS AND CONDITIONS, YOU MUST NOT ACCESS THE LICENSED CONTENT AND RETURN THE LICENSED PRODUCT TO THOMSON DELMAR LEARNING (WITHIN 30 CALENDAR DAYS OF THE END USER'S PURCHASE) WITH PROOF OF PAYMENT ACCEPTABLE TO THOMSON DELMAR LEARNING, FOR A CREDIT OR A REFUND. Should the End User have any questions/comments regarding this Agreement, please contact Thomson Delmar Learning at delmarhelp@thomson.com.

SYSTEM REQUIREMENTS

Operating System: Microsoft® Windows® 98 or XP
Processor: Pentium or faster
Memory: 54 MB of RAM
Hard Disk Space: 20 MB or more
CD-ROM drive: 2x or faster

SETUP INSTRUCTIONS

Medical Coding Specialist's Exam Review should automatically open to a welcome and instructions for "Setup" when disk is inserted into the CD-ROM drive. If the program does not open to a "Welcome" screen, please:
1. From the Start Menu, choose **RUN.**
2. In the ***Open*** text box, enter **d:setup.exe**, then click ***OK*** (substitute the letter of your CD-ROM drive for d:).
3. A "Welcome" screen and installation instructions will prompt you for setup.